EMAS 2013 Workshop: 13th European Workshop on Modern Developments and Applications in Microbeam Analysis

AA009071

IOP Conference Series: Materials Science and Engineering Volume 55

Porto, Portugal
12 – 16 May 2013

ISBN: 978-1-63439-866-4
ISSN: 1757-8981

Printed from e-media with permission by:

Curran Associates, Inc.
57 Morehouse Lane
Red Hook, NY 12571

Some format issues inherent in the e-media version may also appear in this print version.

Copyright© by the Institute of Physics
All rights reserved.

Printed by Curran Associates, Inc. (2015)

For permission requests, please contact the Institute of Physics
at the address below.

Institute of Physics
Dirac House, Temple Back
Bristol BS1 6BE UK

Phone: 44 1 17 929 7481
Fax: 44 1 17 920 0979

techtracking@iop.org

Additional copies of this publication are available from:

Curran Associates, Inc.
57 Morehouse Lane
Red Hook, NY 12571 USA
Phone: 845-758-0400
Fax: 845-758-2634
Email: curran@proceedings.com
Web: www.proceedings.com

TABLE OF CONTENTS

IOP Conf. Series: Materials Science and Engineering **55** (2014) 011001 doi:10.1088/1757-899X/55/1/011001

13th European Workshop of the European Microbeam Analysis Society on Modern Developments and Applications in Microbeam Analysis (EMAS 2013), Porto, Portugal, 12-16 May 2013

Xavier Llovet[1], **Michael B. Matthews**[2], **François Brisset**[3], **Fernanda Guimarães**[4] **and Joaquim M. Vieira**[5]

[1] Universitat de Barcelona, Centres Cientifics i Tecnologics (CCiT), C/ Lluis Sole i Sabaris, 1-3, ES-08028 Barcelona, Spain

[2] AWE Plc., Aldermaston, Reading RG7 4PR, Great Britain

[3] CNRS - Université Paris-Sud, ICMMO - Institut de Chimie Moléculaire et des Matériaux d'Orsay, 15 rue Georges Clémenceau, bât. 410, FR-91405 Orsay, France

[4] Laboratório Nacional de Energia e Geologia, I.P., Laboratorio Ciencia e Tecnologia Mineral, Rua da Ameira, P.O. Box 1089, PT-4466-956 S. Mamede de Infesta, Portugal

[5] Universidade de Aveiro, Departamento de Engenharia Cerâmica e do Vidro, Campus Universitario de Santiago, PT-3810-193 Aveiro, Portugal

Preface

This volume of the IOP Conference Series: Materials Science and Engineering contains papers from the 13th Workshop of the European Microbeam Analysis Society (EMAS) on Modern Developments and Applications in Microbeam Analysis which took place from the 12th to the 16th of May 2013 in the Centro de Congressos do Alfândega, Porto, Portugal.

The primary aim of this series of workshops is to assess the state-of-the-art and reliability of microbeam analysis techniques. The workshops also provide a forum where students and young scientists starting out on a career in microbeam analysis can meet and discuss with the established experts. The workshops have a very specific format comprising invited plenary lectures by internationally recognized experts, poster presentations by the participants and round table discussions on the key topics led by specialists in the field. This workshop was organized in collaboration with LNEG - Laboratório Nacional de Energia e Geologia and SPMICROS - Sociedade Portuguesa de Microscopia. The technical programme included the following topics: electron probe microanalysis, future technologies, electron backscatter diffraction (EBSD), particle analysis, and applications.

As at previous workshops there was also a special oral session for young scientists. The best presentation by a young scientist was awarded with an invitation to attend the 2014 Microscopy and Microanalysis meeting at Hartford, Connecticut. The prize went to Shirin Kaboli, of the Department of Metals and Materials Engineering of McGill University (Montréal, Canada), for her talk entitled "*Plastic deformation studies with electron channelling contrast imaging and electron backscattered diffraction*".

The continuing relevance of the EMAS workshops and the high regard in which they are held internationally can be seen from the fact that 74 posters from 21 countries were on display at the

Content from this work may be used under the terms of the Creative Commons Attribution 3.0 licence. Any further distribution of this work must maintain attribution to the author(s) and the title of the work, journal citation and DOI.
Published under licence by IOP Publishing Ltd

meeting and that the participants came from as far away as Japan, Canada and the USA. A selection of participants with posters was invited to give a short oral presentation of their work in three dedicated sessions. The prize for the best poster was an invitation to participate in the 22nd Australian Conference on Microscopy and Microanalysis (ACMM 23) at Adelaide, South Australia. The prize was awarded to Pierre Burdet of the EM Group of the Department of Materials Science and Metallurgy of the University of Cambridge (UK), for the poster entitled: "*3D EDS microanalysis by FIB-SEM: advantages of a low take-off angle*".

This proceedings volume contains the full texts of 8 of the invited plenary lectures and of 13 papers on related topics originating from the posters presented at the workshop. All the papers have been subjected to peer review by a least two referees.

January 2014

Acknowledgements
On behalf of the European Microbeam Analysis Society I would like to thank all the invited speakers, session chairs and members of the discussion panels for making the meeting such a great success. Special thanks go to Fernanda Guimarães and Luc Van't dack who directed the organisation of the workshop giving freely of their time and talents. As was the case for previous workshops, the EMAS board *in corpore* was responsible for the scientific programme. The Workshop also included a commercial exhibition where many leading instrument suppliers were represented. Several companies that exhibited provided financial support, either by sponsoring an event or by advertising.

Below, in alphabetical order, is a list of exhibiting companies and sponsors of the workshop.

- Ametek GmbH, Edax Business Unit
- Bruker Nano GmbH
- Cameca SA
- Câmara Municipal do Porto
- European Institute for Transuranium Elements (Germany)
- FEI Company
- IZASA Group Werfen
- Jeol (Europe) SAS
- Porto Gran Cruz
- Oxford Instruments NanoAnalysis Ltd.
- Probe Software, Inc.
- Tescan, a.s.

Michael B. Matthews
EMAS President

Peer review statement

All papers published in this volume of *IOP Conference Series: Materials Science and Engineering* have been peer reviewed through processes administered by the proceedings Editors. Reviews were conducted by expert referees to the professional and scientific standards expected of a proceedings journal published by IOP Publishing.

Content from this work may be used under the terms of the Creative Commons Attribution 3.0 licence. Any further distribution of this work must maintain attribution to the author(s) and the title of the work, journal citation and DOI.
Published under licence by IOP Publishing Ltd

Study of multi-carbide B_4C-SiC/(Al, Si) reaction infiltrated composites by SEM with EBSD

B A Almeida[1], M C Ferro[1], A Ravanan[1], P M F Grave[1], H-Y Wu[2], M-X Gao[2], Y Pan[2], F J Oliveira[1], A B Lopes[1] and J M Vieira[1]

[1] University of Aveiro, Department of Materials Engineering and Ceramics, Centre for Research in Ceramics and Composite Materials (CICECO), Campus Universitário de Santiago, PT-3810-193 Aveiro, Portugal
[2] Zhejiang University, Department of Materials Science and Engineering & State Key Laboratory of Silicon Materials, CN-310027 Hangzhou, P.R. China

E-mail: jvieira@ua.pt

Abstract. In the definition of conceptual developments and design of new materials with singular or unique properties, characterisation takes a key role in clarifying the relationships of composition, properties and processing that define the new material. B_4C has a rare combination of properties that makes it suitable for a wide range of applications in engineering: high refractoriness, thermal stability, high hardness and abrasion resistance coupled to low density. However, the low self-diffusion coefficient of B_4C limits full densification by sintering. A way to overturn this constraint is by using an alloy, for example Al-Si, forming composites with B_4C. Multi-carbide B_4C-SiC/(Al, Si) composites were produced by the reactive melt infiltration technique at 1200 - 1350 °C with up to 1 hour of isothermal temperature holds. Pressed preforms made from C-containing B_4C were spontaneously infiltrated with Al-Si alloys of composition varying from 25 to 50 wt% Si. The present study involves the characterisation of the microstructure and crystalline phases in the alloys and in the composites by X-ray diffraction and SEM/EDS with EBSD. Electron backscatter diffraction is used in detail to look for segregation and spatial distribution of Si and Al containing phases during solidification of the metallic infiltrate inside the channels of the ceramic matrix when the composite cools down to the eutectic temperature (577 °C). It complements elemental maps of the SEM/EDS. The production of a flat surface by polishing is intrinsically difficult and the problems inherent to the preparation of EBSD qualified finishing in polished samples of such type of composites are further discussed.

1. Introduction

Materials science can be represented by a tetrahedron of relationships with the fundamental vertices of composition, properties, processing and applications. In the definition of conceptual developments and design of new materials characterisation often takes a key role in clarifying what the new material is. Efforts to combine the great benefits obtained by the arrangement of metals and ceramics lead to development of composites. This takes a significant role in the modern industry where the permanent need to identify more economic and effective processing methods increases the research on composites.

Boron carbide (B_4C) is extremely hard, surpassed only by diamond and cubic boron nitride, and has unique properties such as low density (2.52 $g.cm^{-3}$), refractoriness, chemical inertness, thermal

Content from this work may be used under the terms of the Creative Commons Attribution 3.0 licence. Any further distribution of this work must maintain attribution to the author(s) and the title of the work, journal citation and DOI.
Published under licence by IOP Publishing Ltd

stability and high hardness with abrasion resistance, making it the first choice for a wide range of applications [1]. Boron carbide has an atomic structure that can be described as composed of two atom arrangements: a primary structure formed by 12-atom icosahedra with B as the corners, linked by 3-atom C-B-C linear chains along the (111) rhombohedral axis, corresponding to the model stoichiometry of ($B_{13}C_2$), the icosahedra sitting on the vertices of a lattice with trigonal symmetry – space group $R\bar{3}m$; or in terms of an hexagonal lattice in which the [0001] direction axis matches the [111] rhombohedral direction [2]. B_4C structure has been extensively described in the literature [3-7] and although the crystal symmetry is obtained by diffraction methods, the resemblance in terms of nuclear scattering of boron and carbon atoms [8-9] makes almost indistinguishable the two atoms. Zoning of the B_4C solid solution displaying fluctuations of B/C ratio often yields $B_{13}C_2$-domains formed between B_4C zones of composition close to the $B_{12}C_3$ stoichiometry, the average size of the $B_{13}C_2$-domains being smaller than those of B_4C [10]. The zoned domains in the crystalline grains of the B_4C solid solution increase the difficulties of identification of diffraction pattern by automatic procedures as in electron backscatter diffraction (EBSD) results of B_4C which in the presence of Al as in the AA1100 - 16 vol% B_4C composite tend to be misidentified with the Al phase itself [11].

Disadvantages of the B_4C ceramics are the low self-diffusion coefficient, limiting densification by sintering and also brittleness, both being due to the covalent bonding of B_4C. Several methods were developed to prepare B_4C ceramics such as hot pressing [12-13] and pressureless sintering [14-15], however the extremely high temperatures needed of 2000 - 2200 °C increase the global cost and make the materials less attractive. A way to overturn such inherent problems to produce useful materials based on B_4C is by using a different approach such as the infiltration method with a metal as infiltrate forming composites [1, 16-17].

Aluminium is an example of a metal that can be used to fill in the porosity of B_4C preforms. The low density of Al combined with high specific strength, easy machinability, good resistance to corrosion and high thermal and electrical conductivity, makes Al a coherent choice to use as infiltrate. In addition, Al-B_4C composites have low-cost casting for their production. However, the poor wettability of Al below 1000 °C constitutes an obstacle for adequate mixing of the ceramic particles with the liquid phase [18-19]. It is possible to heat treat or coat the B_4C particles to improve the wettability factor [20] or control the formation of ternary phases of the Al-B-C system, of which nine ternary phases are known at least [21]. The alternative is alloying Al with other metallic elements such as Si making it more stable during infiltration. Si alloying also reduces the thermal expansion and improves mechanical properties of the alloy, mainly hardness [1].

The present study aims to characterize the microstructure and crystalline phases of the composites by X-ray diffraction and SEM/EDS with EBSD. It is part of the research work intending to evaluate the infiltration process, mostly on understanding the newly formed phases, specially the potential growth of ternary phases of Al-B-C rims on B_4C cores. It intends to look in detail for the segregation of Si in Al during solidification of the Al-Si metallic infiltrate constrained by the channels of the ceramic matrix when the composite cools down to the eutectic point (577 °C, 12.5 wt% Si) [22].

2. Experimental procedure

As-supplied B_4C (Electron Microscopy Sciences EMS#50510-10, lot #BC80A22#36) powders with average particle size of 10 μm, and 99.9 % pure Al (ABCR GmbH & Co., CAS# 7429-90-5, lot# 143946-11) and 99.99 % pure Si (Emerk, Germany - 12497.0250, 201N628697), were used as raw materials. Fusion for production of smelted Al-Si alloys with Si content of 25 and 35 wt% was done at 1000 °C - batches (1) and (2). Both processes, melting of the alloy as well as infiltration, were carried out on a graphite furnace under 50 kPa pressure of Ar gas above atmospheric pressure, with heating and cooling rates of 20 and 50 °C/min, respectively. Batch (3) of the Al-Si alloy containing 25 wt% Si was smelted twice. Cylindrical pellets, 7.5 mm height, of the B_4C powders with polyvinyl alcohol (PVA) as binder were made by uniaxial pressing followed by isostatic pressing at 60 MPa and 196 MPa respectively, yielding green relative densities close to 50 %. Cut pieces of the Al-Si alloy in small chunks were pressed into cylindrical shapes and placed on top of B_4C preforms in amounts as needed to fill the ceramic pore volume by reactive infiltration. Pressureless infiltration of the as-pressed preforms was carried on at 1200 °C, 1300 °C or 1350 °C after a hold of 5 min to 1 h at the maximum infiltration temperature.

Density of the composites was evaluated by the Archimedes method in water. Crystalline phases were determined by X-ray diffraction (XRD, Rigaku) on polished surfaces of the Al-Si alloys and of the longitudinal cross-sections of the cylindrical pellets of the composites also used for SEM. Characterisation of the microstructure was performed by scanning electron microscopy (SEM, SU-70, Hitachi), combining elemental chemical analysis with energy-dispersive X-ray spectroscopy (high yield EDS, B-U Bruker QUANTAX 400) with analysis of crystalline phase distribution by electron backscatter diffraction (EBSD, Bruker CrystAlign QC 400). EBSD analysis is particularly sensitive to relief of surface finishing. For multiphase materials, in particular those containing phases of large differences of hardness, a final step of polishing with colloidal silica improves the finishing of the (soft) metallic phases, but it may also intensify the surface relief with negative effects on the hit rates of the EBSD analysis [11].

Mechanical polishing of the Al-Si alloy and of the composite was carried out by using diamond lapping papers of 30, 15, 9, 6 3 and 1 μm particle sizes. Surface finishing was done by a final step of polishing with colloidal silica suspension of 0.06 μm particle size for 60 minutes with light load of hand holding. In order to remove the plastically strained layer on the surface of the Al phase, the polished surfaces of the samples were submitted to a subsequent step of chemical etching at room temperature from 30 s to 1 min with a 25 % HNO_3 and 75 % methanol (vol%) solution, effective in the electrochemical polishing of aluminium [23]. By courtesy of Hitachi High-Technologies Europe GmbH (Krefeld, Germany), for comparison of finishing procedures one trial sample of the multi-carbide B_4C-SiC/(Al, Si) composite was mechanically polished and submitted to ion beam polishing for 70 minutes at 10° incident angle, 135 μA ion current in two steps: first 35 minutes with 4 kV accelerating voltage and a final step of 35 minutes at 6 kV. SEM observation, EDS and EBSD analysis of the samples of the Al-Si alloy and of ion beam polished surface of the composite were done without C coating. For the remaining surface finishing conditions of the composite samples light C coating was applied to eliminate electrical charging.

3. Results and discussion

3.1 Al-Si alloy

Figure 1 shows the X-ray diffraction results for the three alloys prepared as infiltrate for the composites. Batches (1) and (2) were prepared in same conditions, with just one melting step, and their difference is only in terms of composition, 25 and 35 wt% of Si, respectively. The third batch was prepared using a mixture containing 25 wt% Si, but in this case in two step melting, with double fusion. The main difference between the XRD spectrum of batches (1) and (3) is the absence of residues of Al_2O_3 oxide in the sample of the twice smelted Al-Si alloy.

Figure 2 gives the results of EDS/SEM and EBSD mapping for the Al-Si alloy of the twice smelted batch. The SEM image, figure 2a, and elemental mapping by EDS, figure 2b, indicate the formation of thin Si primary dendrites aside of the Al-rich phase. EBSD maps of the analysed area in figures 2c and d corroborate the EDS results. Figures 2c and d show many Si phase (~20 - 35 vol%) being detected in the Al areas, and this is due to the software that is unable to select clearly the right Al phase. There is also number of misindexing points in the Si phase. From the automated image analysis by the EBSD software there is 6 and 11 % zero solutions (black dots) in the results of figures 2c and d, respectively. This means that the system could not index any of the available phases on such points. The map in figure 2d, obtained after chemical etching, shows the increase of zero solutions. Although short, the time of chemical etching, 30 seconds, could have been too long. The elemental composition of the Al phase was determined by quantitative standardless EDS analysis after energy calibration of the detector, with the fixed gauge area as given in red in figure 2a in red. Up to five EDS spectra were collected on different areas of the Al matrix, yielding the average composition of

98 ± 1 at% Al (the remaining being Si). This value overlaps with the solid solution limit of α-Al (1.65 wt% Si) [22].

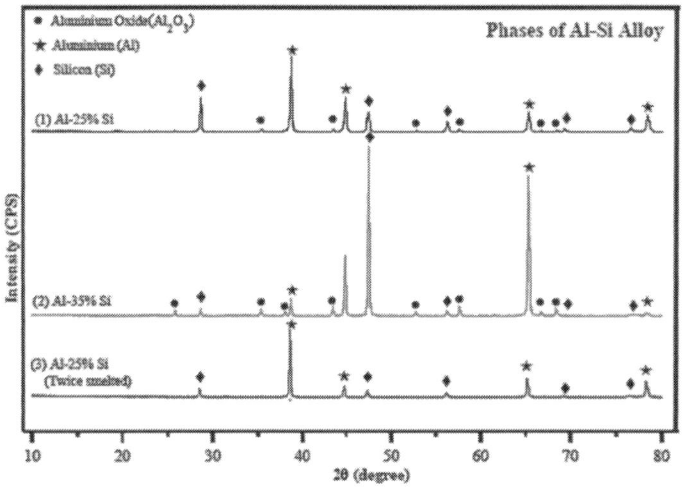

Figure 1. XRD results for the three Al-Si alloys prepared by varying the composition and/or the melting process.

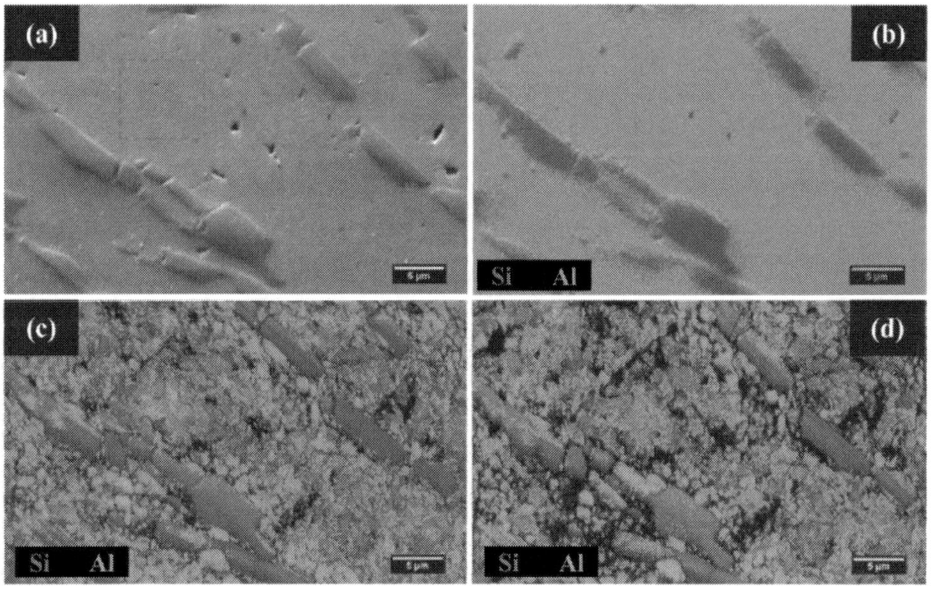

Figure 2. Analysis of microstructure and composition of the twice smelted Al-25 wt% Si alloy: (a) SEM microstructure with region selected for EDS quantification in red; (b) EDS elemental map; EBSD phase map for the alloy (same colour code as in b); (c) before chemical etching; and (d) after chemical etching (map area of 83.7 x 62.78 μm^2 with 0.065 μm step size).

From the EBSD report, the composition of the Al-Si alloy for the total area as shown in figure 2 ranges from 35 vol% of Si phase for the as polished surface in figure 2c to 40 vol% of Si phase after the chemical etching, figure 2d; these values being not far away from the elemental composition of 32 ± 2 at% Si from EDS analysis. As the solid solubility of Al in the Si crystalline lattice is very low [22], these results are significantly above the average Si content of the alloy of 22 vol% Si phase. The composition of the small area sampled in figure 2 neatly deviates from the average composition of the alloy. The results of XRD analysis should probably be more accurate than those of SEM with EDS or EBSD analysis as a much wider volume and more representative of the composition of the sample is analysed by XRD.

The discrepancies that persist between the Al phase content of the alloy obtained from the EBSD analysis compared to the EDS results can be attributed to disturbing effects of low temperature passivation of the Al surface forming a 2 - 4 nm thick amorphous oxide layer [24] that cannot be resolved by EBSD, to smearing of the surface layer from plastic deformation during polishing that blurs the Kikuchi line patterns [11] and to the relief of the surface. The thickness of the passivation layer of Al being a minor fraction of the backscattered electron interaction depth [25], the resulting perturbation of EBSD identification is seemingly minute. Protrusion of the hard Si particles of the alloy and of Si and B_4C particles of the composites can shadow the Al phase once the electron beam intersects the high angle tilted sample. Both the Al and Si phases have face-centred cubic crystalline lattices, the Al belonging to the $Fm\bar{3}m$ space group with a_0 = 4.05 Å and the Si to the $Fd\bar{3}m$ space group with a_0 = 5.43 Å. The likeness of lattice constants and space groups of Al and Si may also explain the difficulty of the EBSD technique to distinguish between these phases. Such resemblance can mislead the EBSD software to select Si instead of Al as the correct phase in significant proportion of the hits. Hence changes in the EBSD resolution associated to smearing or surface relief would necessarily conduct the EBSD software to confuse the identification of the phases in increasing rates. But, the failure of EBSD to correctly index the Al phase may also come from the presence of sub-grains in the Al phase originated from the rapid solidification [26]. Figure 3a displays the grain orientation in X direction with inverse pole figures for both Si and Al phases, the size histogram of the Al phase being given in figure 3b, a fine sub-grain structure being observed inside the Al-rich matrix.

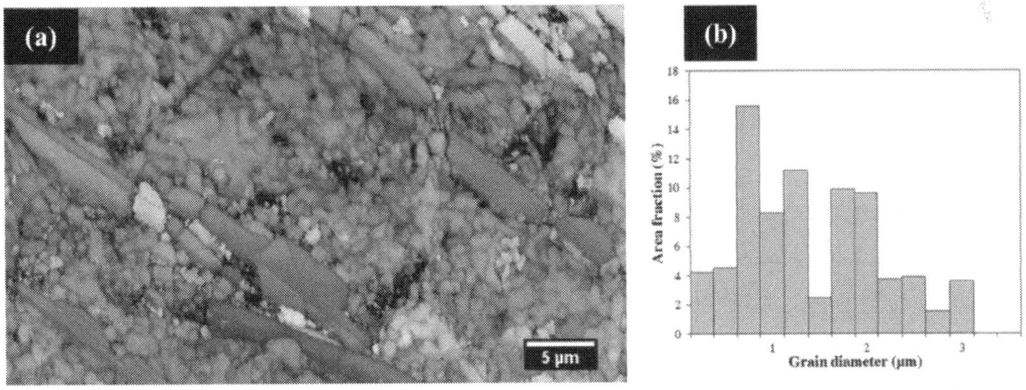

Figure 3. (a) Inverse pole figure in X direction (IPF-X) of figure 2; and (b) the size histogram of the Al phase on an area of approximately 83.7 x 62.78 μm^2, 0.065 µm step size.

The histogram in figure 3b shows that the metallic matrix of Al is largely constituted by fine domains with sizes ranging from 0.25 to 2.0 µm. The particles with sizes above this threshold seen in figure 3a are mainly Si primary dendrites. The fine sub-grain size of the α-Al solid solution is an additional difficulty for crystalline phase identification by EBSD in SEM. This becomes even more relevant on composite structures constituted by many phases. The need for more studies to understand the effect of the sub-micrometre grain sizes on EBSD phase mapping is acknowledged in the bibliography [11, 26].

3.2 Multi-carbide B_4C-SiC/(Al, Si) composites
In order to evaluate the distribution of crystalline phases through the sample, the composite was cut in 3 parts along the axis of the cylindrical pellet for the XRD analysis: top of infiltration (entry of metal), middle and bottom (end). Figure 4 presents the XRD results for composite "C" in the three analysed regions. Analysis of XRD diffractograms using standards shows that the four main phases of the composite have volume fractions close to the following: Al (44 vol%), Si (19 vol%), B_4C (33 vol%) and SiC (4 vol%). The volume fractions were converted to the molar fractions given in the first row of Table 1 with the values of molecular weight and density of each phase.

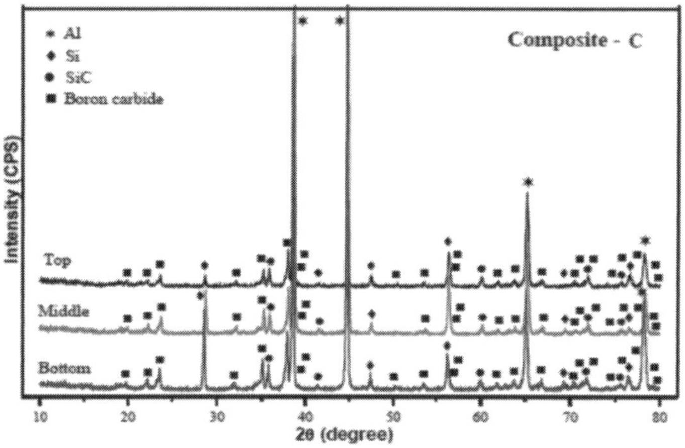

Figure 4. XRD results of composite "C" in three regions: top of infiltration, middle and bottom (end).

Table 1. Molar fraction of main crystalline phases of the B_4C-SiC/(Al, Si) composite determined from XRD analysis and calculated by EBSD image analysis.

	Al	Si	B_4C	SiC	Zero solutions
XRD analysis (mol %)	56.4	20.2	19.3	4.1	---
Figure 5 phase map (d) (vol%)	31.3	27.8	38.8	2.1	24.9 %
Figure 5 phase map (f) (vol%)	39.4	18.6	38.4	3.6	33.0 %

The microstructure of the same multi-carbide B_4C-SiC/(Al, Si) composite "C", with surface finishing obtained by two methods, conventional polishing with chemical etching and ion beam polishing, are presented in figure 5. Elemental maps for Al, Si and B of the composite are displayed in figures 5b and c. EBSD phase maps for conventional polished and ion beam polished surfaces are exhibited in figures 5d and e, respectively. The EBSD analysis of the composite was performed with the intent of determining the segregation of the Si phase from the Al during solidification of the metallic infiltrate inside the channels of the ceramic matrix when the composite cools to the eutectic point (577 °C). The results of the corresponding reports on phase composition based on the EBSD image analysis are given in Table 1. The volume fraction of the phases from EDSD results of the chemically etched sample and ion beam polished sample rank the phase content by the same order, but the Al is detected in relatively lower amounts than by the XRD whereas the opposite is observed for the B_4C phase.

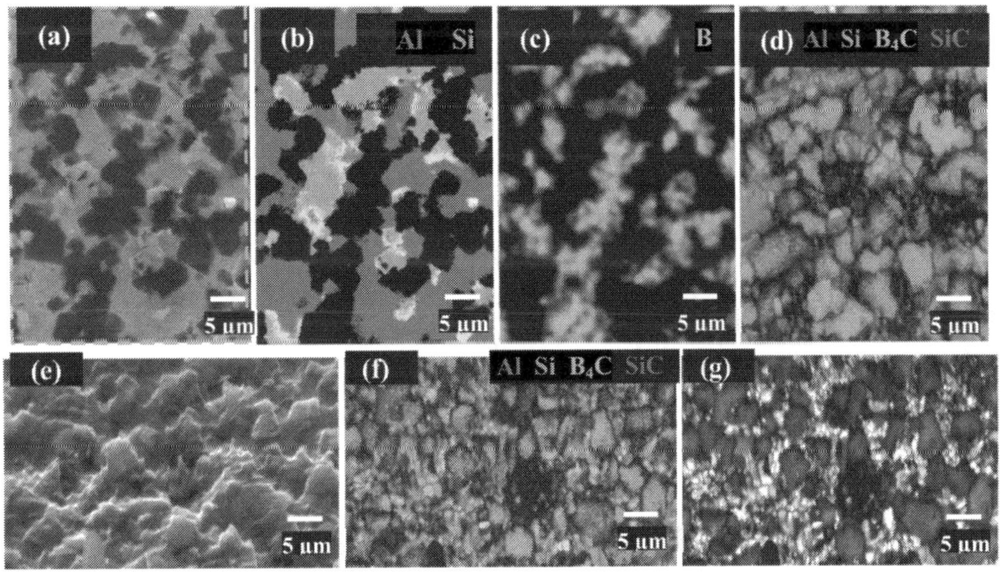

Figure 5. Multi-carbide B_4C-SiC/(Al, Si) composite. (a) - (d) polished section of with chemical etching: (a) SEM microstructure; (b) - (c) Elemental maps from EDS to Al+Si and B; (d) EBSD analysis for Al, Si and B_4C, with area of 46.5 x 33.9 μm^2 and 0.082 µm step size; (e) - (g) polished section with ion beam polishing: (e) SEM microstructure; (f) EBSD phase map for Al, Si and B_4C; and (g) the pattern quality map, with area of 47.5 x 40.6 μm^2 and 0.082 µm step size.

The constrains of the mechanical polishing technique with chemical polishing to improve surface quality to the level needed for EBSD analysis of a four-phase composite may seemingly be lessened by using ion beam polishing (or direct cross section) for the same purpose [11, 27].

The quality of EBSD data of the composite sample was also improved by adjusting the main controls of operation of the SEM: accelerating voltage (V_{acc}) and current emission (I_e). By increasing V_{acc}, the intensity of available signal is larger, but the size of the interaction volume between the electron beam and sample is also enlarged, decreasing spatial resolution of EBSD in conflict with the need to index the small (often sub-micrometre sized) grains of the Al and Si phases in the narrow spaces between B_4C particles. In SEM-EDS, due to high capability of X-rays to penetrate matter, the calculated depth of the interaction volume for 20 kV electron beam is 3.1 µm while for 15 kV it decreases to 1.9 µm, as calculated for applied EBSD conditions (incident beam at 70°, 27 mm work distance) and composition of the composite determined by EDS in the area of figure 5e. Differently, backscattered electrons are generated in the near surface layers of the sample, decreasing therefore the depth of backscattering to 10 - 20 nm, the large fraction of useful EBSD signal being emitted from this volume [28].

A larger overlap with blur of Kikuchi lines at high V_{acc} may prevent the correct indexing of phases with similar crystalline structures. Otherwise, I_e can be increased by changing the objective lens aperture to a larger one thus intensifying the available signal. The effect of varying the two SEM operation controls, V_{acc} and I_e, on EBSD image quality for Al phase is illustrated in the images of figure 6.

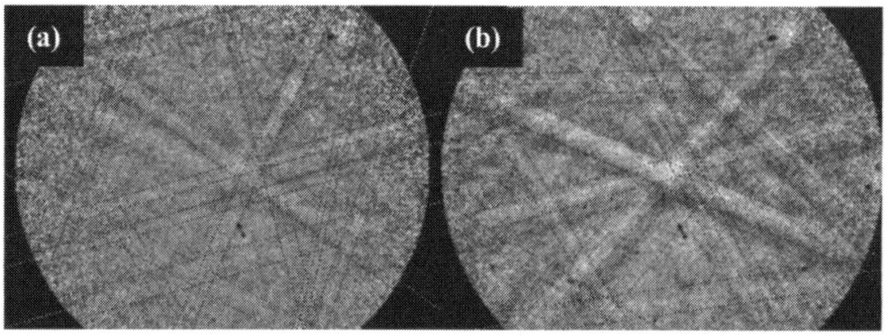

Figure 6. Kikuchi diffraction pattern for the same micrometre-sized crystal of the Al phase and approximately fixed position inside the grain with different SEM operation conditions: (a) V_{acc} = 20 kV and 0.086 nA probe current (50 μm objective lens aperture); (b) V_{acc} = 20 kV and 0.194 nA probe current (100 μm objective lens aperture).

Due to the unresolved issues, it was not possible to make the identification if Al-B-C rims or ternary phases formed on B_4C cores by using EBSD in SEM up to the present stage of this study. Two directions for future work emerge from the present study. The improved spatial resolution down to nanometre range of the SEM-TKD technique, transmission Kikuchi diffraction in the SEM microscope, as compared to SEM-EBSD [24, 29], will expectedly aid the automated recognition of electron diffraction patterns of the Al and Si phases with sub-grain structures of submicron sizes. But, for SEM-TKD the sample has to be prepared as a thin section of transmission electron microscopy.

The Al phase plays a key role in toughening of the B_4C-SiC/(Al, Si) composites [1]. Aligned with the broad objectives of the study, the assessment of fracture toughness of the composite by indentation fracture and demonstration by EBSD techniques of the extent of plastic deformation in crack-bridging Al particles contributing to the fracture energy demands at the same time samples with dimensions large enough for the indentation fracture tests but with surfaces rectified and polished to EBSD quality.

4. Conclusions

Multi-carbide B_4C-SiC/(Al, Si) composites were prepared by the reactive melt infiltration method from the B_4C porous preforms, with Al-Si alloy at temperatures between 1200 and 1350 °C under Ar atmosphere. The study of microstructure and composition of the Si-Al alloy and the composites by XRD and SEM/EDS with EBSD analysis lead to the following conclusions:

a) Present results corroborate other studies on polishing of composite materials by conventional mechanical methods for EBSD analysis; the method applied to the Al-Si alloy and to composites of this study gave quality enough for EBSD characterisation after chemical polishing. Limited by the constraints of surface relief and resemblances of crystallography of main phases of the composite, namely of the Al and Si crystalline lattices, the automated software was unable to select clearly the right Al phase in many points, or conversely the Si phase, and the reliable quantification of phases by EBSD especially of the Al phase was not feasible in spite of much better EBSD spatial resolution than the EDS one.
b) Plastic smearing more than passivation of the Al surface layer, small (sub-micrometre) sub-grain structures combined with surface relief can explain the difficulty of indexing this phase in the

composites by the automated software of phase recognition of EBSD. Chemical etching of the polished surface for up to 1 minute improved in part the indexing of Al by EDSD.

c) Ion beam polishing also improved the finishing of the cross-sections of the composite for the use in EBSD analysis. But, the noticeable increasing of relief stand for further study particularly directed to find proper conditions for preparation of low relieve polished surfaces from multi-phase materials with a large span of hardness of their phases.

Acknowledgments
The authors would like to acknowledge the RNME - Pole University of Aveiro (FCT Project REDE/1509/RME/2005) for access to SEM microscopy, the scientific and technical assistance and one of the authors the EMMS consortium funded by the European Commission for the M.Sc.-scholarship.

References

[1] Wu H, Zhang S, Gao M, Zhu D, Pan Y, Liu Y, Pan H, Oliveira F J and Vieira J M 2012 *Mater. Sci. Engng. A* **551** 200-208
[2] Domnich V, Reynaud S, Haber R A and Chhowalla M 2011 *J. Am. Ceram. Soc.* **94** 3605-3628
[3] Lipp A 1996 *Tech. Rundsch.* **58** [7] 1-47
[4] Thavenot F 1990 *J. Eur. Ceram. Soc.* **6** 205-225
[5] Vast N, Sjakste J and Betranhandy E 2009 *J. Phys. Conf. Ser.* **176** 012002
[6] Suri A K, Subramanian C, Sonber J K and Murthy T 2010 *Int. Mater. Rev.* **55** 4-11
[7] Werheit H, Filipov V, Kuhlman U, Schwarz U, Armbruster M, Lei-the-Jasper A, Tanaka T, Higashi T, Lundstrom T, Gurin V N and Korsukova M M 2010 *Sci. Technol. Adv. Mat.* **11** 023001
[8] Morosin B, Kwei G H, Lawson A C, Aselage T L and Emin D 1995 *J. Alloy. Compound.* **226** 121-125
[9] Jiménez I, Sutherland D G J, van Buuren T, Carlisle J A, Terminello L J and Himpsel F J 1998 *Phys. Rev. B* **57** 13167-13174
[10] Schwetz K A and Karduck P 1991 *Boron-rich solids.* Vol. 231 (Emin D, Aselage T, Beckel C L, Switendick A C and Morosin B; Eds.) (New York: American Institute of Physics)
[11] Guo J, Amira S, Gougeon P, Chen X-G 2011 *Mater. Charact.* **62** 865-877
[12] Li A J, Zhen Y H, Yin Q, Ma L P, Yin Y S 2006 *Ceram. Int.* **32** 849-856
[13] Zorzi J E, Perottoni C A and Da Jornada J A H 2005 *Mater. Lett.* **59** 2932-2935
[14] Mashhadi M, Taheri-Nassaj E, Sglavo V M, Sarpoolaky H and Ehsani N 2009 *Ceram. Int.* **35** 831-837
[15] Miyazaki H, Zhou Y, Hyuga H, Yoshizawa Y and Kumazawa T 2010 *J. Eur. Ceram. Soc.* **30** 999-1005
[16] Hayun S, Rittel D, Frage N and Dariel M P 2008 *Mater. Sci. Eng. A* **487** 405-409
[17] Mizrahi I, Raviv A, Dilman H, Aizenshtein M, Dariel M P and Frage N 2007 *J. Mater. Sci.* **42** 6923-6928
[18] Lin Q L, Shen P, Qiu F, Zhang D and Jiang Q H 2009 *Scripta Mater.* **60** 960-963
[19] Toptan F, Kilicarslan A, Karaaslan A, Cigdem M and Kerti I 2010 *Mater. Design* **31** S87-S91
[20] Kennedy A R and Brampton B 2011 *Scripta Mater.* **44** 1077-1082
[21] Halverson D C, Pyzik A J, Aksay I A and Snowden W E 1989 *J. Am. Ceram. Soc.* **72** 775-780
[22] Warmuzek M 2004 *Aluminum-silicon casting alloys - Atlas of microfractographs.* (ASM International, 2
[23] Ünlü N 2008 *Mater. Charact.* **59** 547-553
[24] Eisenreich N, Fietzek H, Juez-Lorenzo M M, Kolarik V, Koleczko A and Weiser V 2004 *Propellants, Explosives, Pyrotechnics* **29** 137-145
[25] Keller R and Geiss R H 2012 *J. Microsc.* **245** 245-251

[26] Humphreys F J, Huang Y, Brough I and Harris C 1999 *J. Microsc.* **195** 212-216
[27] Schwartz A J, Kumar M, Adams B L and Field D P 2009 *Electron backscatter diffraction in materials science - 2nd Edition.* (Springer)
[28] Winkelmann A 2010 *J. Microsc.* **239** 32-45
[29] Trimby P W 2012 *Ultramicrosc.* **120** 16-24

IOP Conf. Series: Materials Science and Engineering 55 (2014) 012002 doi:10.1088/1757-899X/55/1/012002

Measurement and Monte Carlo simulation of the spatial resolution in element analysis with the FEG-EPMA JEOL JXA-8530F

D Berger and J Nissen

Technische Universität Berlin, Zentraleinrichtung Elektronenmikroskopie ZELMI, Straße des 17. Juni 135, DE-10623 Berlin, Germany

E-mail: dirk.berger@tu-berlin.de, joerg.nissen@tu-berlin.de

Abstract. To get a reliable estimation of the spatial analytical resolution of electron probe microanalysers with field emission electron sources (FEG-EPMA) it is necessary firstly to know the probe diameter, especially for low acceleration voltages, and secondly to calculate the X-ray emitting volume by Monte Carlo (MC) simulations. In the first step of the investigations presented here the probe diameter was determined by recording SE- and BSE-images of a gold-on-carbon high resolution specimen at various acceleration voltages and probe currents. From the rise of the contrast profiles of the gold insulars edges the probe diameter was derived by using the 1 % - 99 % criterion. The obtained probe diameter ranges from 10 nm at 15 kV/10 pA and 20 nm at 5 kV/10 pA to 65 nm @ 5 kV/10 nA. In the next step line scans measured with the FEG-EPMA are compared to corresponding MC-simulations on especially crafted thin layer specimens. The agreement is not well in all cases but MC simulations are suitable to get an approximation of the expected spatial analytical resolution in dependence of the used acceleration voltage, the probe current and the detected X-ray transition.

To get useful numbers it is necessary to distinguish between the quantitative and the qualitative spatial analytical resolution. The former refers to applications requiring the quantification of the elements, i.e., the elemental composition is homogenous within the X-ray emitting volume. Therefore, the quantitative resolution is derived from the edge steepness of the line scans using the 1 % - 99 % criterion. At our specimens resolutions suitable for the quantitative analysis down to about 100 nm were observed. The qualitative spatial analytical resolution is evaluated by means of the 16 % - 84 % criterion and applies to the pure recognition of elemental changes in line scans across precipitates or layers whose sizes might be smaller than the X-ray emitting volume. We obtained qualitative spatial analytical resolutions down to several ten nanometres.

1. Introduction

Element analysis with high spatial resolution becomes more important for the investigation of nanoparticles, nanolayers and core-shell-structures in geology, material science and chemistry. The commonly used transmission electron microscopy (TEM) is not suitable in numerous cases where areas with small differences in Z should be imaged or for the precise quantification of medium to heavy weighted elements [1]. In these cases the use of wavelength-dispersive X-ray spectrometers (WDS) in an electron probe microanalyser (EPMA) with their high spectral resolution and sensitivity is advantageous over TEM-analysis. This paper regards bulk EPMA-specimens only, since the specimen preparation then requires less effort compared to the TEM-specimen preparation.

Content from this work may be used under the terms of the Creative Commons Attribution 3.0 licence. Any further distribution of this work must maintain attribution to the author(s) and the title of the work, journal citation and DOI.
Published under licence by IOP Publishing Ltd

So far the limited spatial analytical resolution is the main drawback of conventional EPMA. In this paper first attempts and measurements are reported to understand the mechanisms that are relevant for the spatial analytical resolution. The main task is to decrease the interaction volume of the emitted and detected signal. This might be done by the decrease of the penetration volume of the electrons by considerably decreasing the acceleration voltage (so called low voltage analysis) or by the decrease of the analytical volume by particularly decreasing the acceleration voltage with respect to the ionisation energy of the detected X-ray transition (so called low overvoltage analysis) [2, 3]. The low overvoltage analysis increases the spatial resolution by the effect that the electrons that still have enough energy to excite the X-ray transition have a small beam spread only. Applying the low overvoltage analysis one accepts that the overvoltage might not have the optimal value to excite the X-ray transition with the highest probability. The focus lies on the optimisation of the spatial resolution.

To compare both methods, the sizes of the penetration volume of all electrons and of the electrons with energy larger than the ionisation energy are determined with Monte Carlo simulations [4]. For the comparison of these simulations with measurements several variables need to be considered additionally. First of all fluorescence has to be regarded unless it is not included in the simulations. Secondly the interaction volume is enlarged by the diameter of the probe. Therefore, the analytical volume is the convolution of the beam diameter with the interaction volume.

The measurements are made using an EPMA with a field emission gun (FEG-EPMA) since only the FEG emits an electron beam with a high brightness allowing a strong focussing even for low acceleration voltages. EPMA with a thermionic electron gun are not suitable for the low voltage analysis at high spatial resolution (but might be used for the low overvoltage–analysis). However, the beam size depends inversely proportional to the acceleration voltage and proportional to the beam currents even for the FEG. Therefore, the probe diameter needs to be determined for all acceleration voltages and currents.

This paper focusses on the optimisation of the analytical resolution in EPMA and leaves problems that might arise for the quantification (by using L- and M-lines) out. Great efforts in understanding the element analysis using L- and M-lines and in the development of solutions have already been made [5-13].

2. Experimental and results

2.1. Measurement of the probe diameter

SE- and BSE-images of a gold-on-carbon high resolution specimen were recorded at various accelerating voltages (3, 5, 8, und 15 kV) and probe currents in steps from 10 pA to 100 nA using a FEG-EPMA JEOL JXA8530F (figure 1). The probe diameter is derived from these images in correspondence with ISO/TS 24597 by the analysis of the contrast profile of the gold insulars. The programme "Image" by FEI (figure 2) has been used since it allows freely choosing the thresholds and averaging over many grains and orientations. The programme „Image" determines the beam diameter from the image sharpness which is calculated following the derivative method specified in the standard ISO/TS 24597. The derivative method fits error functions to edge profiles extracted from the SEM image. The image needs to contain of as many Au particles, so that at least 2000 edge profiles with a homogenous distribution of orientations contribute to the calculated mean value. In our case images with about 30000 edge profiles were analysed.

The analytical resolution is determined by the convolution of the probe diameter with the interaction volume. Since Monte Carlo simulations of the interaction volume trace the electron path until the complete stopping of the electrons, the probe diameter is determined from the line profile with a 1 % to 99 % criterion as well. Usually resolution measurements using profile scans are evaluated using the 16 % - 84 % criterion corresponding to the full-width-half-maximum (FWHM) of

IOP Conf. Series: Materials Science and Engineering 55 (2014) 012002 doi:10.1088/1757-899X/55/1/012002

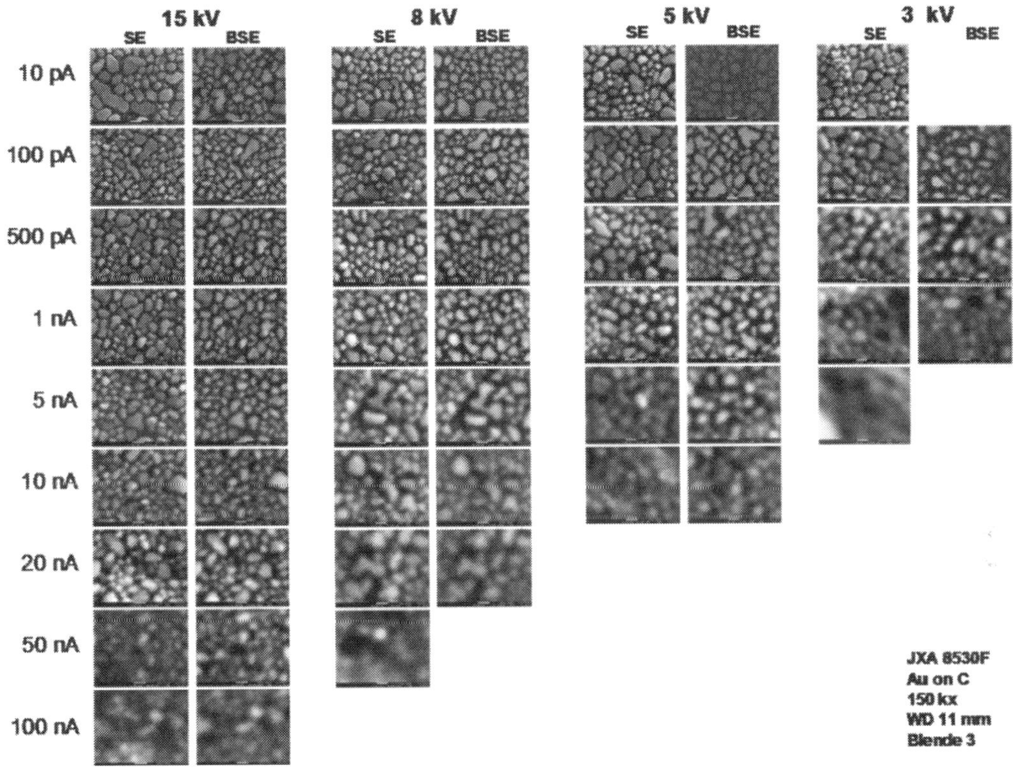

Figure 1. SE- and BSE-images at various accelerating voltages and probe currents used for the measurements of probe diameters.

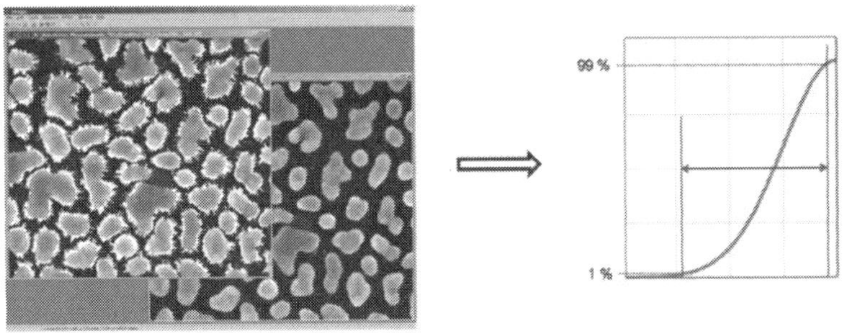

Figure 2. Contrast profile of the gold insulars edges in the SE- and BSE-images (analysed with the programme "Image" by FEI).

of a Gaussian function convoluted with an abrupt edge, i.e., the Gaussian beam diameter [5]. In figure 3 the different criterions are compared with the occasionally used 30 % - 70 % criterion. In figure 4 the measured edge steepness's are plotted for different acceleration voltages, e.g., 15 nm for 15 kV and 10 pA or 31 nm for 8 kV and 1 nA. For high analytical resolution low beam currents and high acceleration voltages are required. These data correspond with the specifications as provided by the JEOL Company [14].

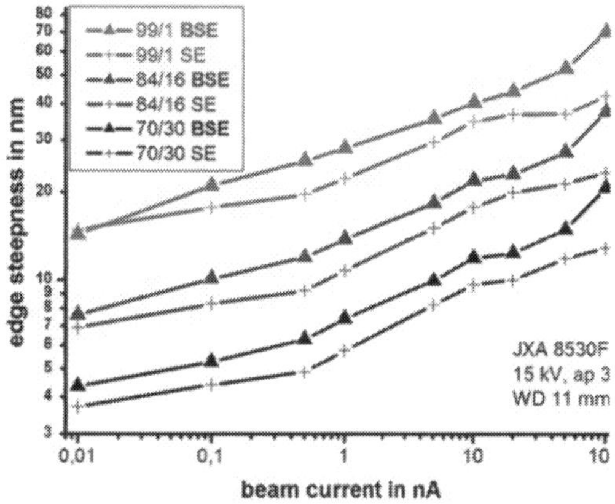

Figure 3. Measured edge steepness from line scans over contrast edges for different threshold criterions. For the 16 % - 84 % criterion the measured edge steepness corresponds to the Gaussian FWHM beam diameter and to the qualitative spatial analytical resolution (sect. 2.2).

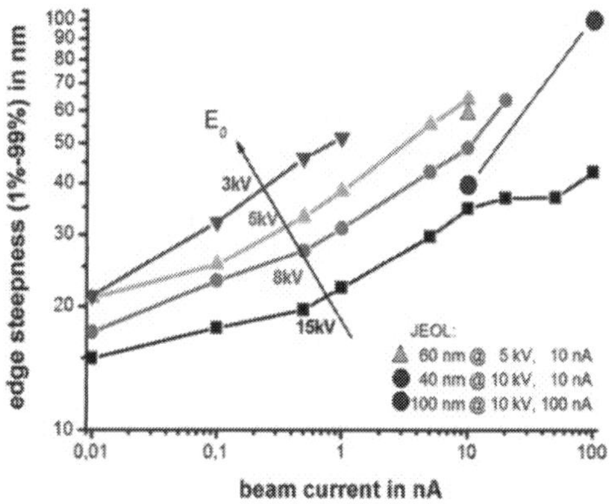

Figure 4. Measured edge steepness for different currents and acceleration voltages as an estimate for the probe diameter with the 1 % - 99 % criterion which is used to measure the quantitative spatial analytical resolution (sect. 2.3).

2.2. Monte Carlo simulation and measurements of the quantitative spatial analytical resolution

In the next step measurements of the analytical resolution are compared to predictions from Monte Carlo (MC) simulations obtained from the programme CASINO (version 2.48) [15]. Three specimens were specially made for this purpose consisting of thin layers or sharp transitions of pure elements. The specimens are designed to allow both low voltage and low overvoltage measurements. Cross-sections are prepared by classical metallography or by use of a focused Ga-ion beam (FIB, FEI Helios NanoLab600). WDX line scans of particular X-ray transitions are recorded across the element interfaces (FEG-EPMA, JEOL JXA 8530F). Assuming a Gaussian beam profile and a Gaussian distribution of emitted X-rays, the line scans have a sigmoidal shape corresponding to the integral of the Gaussian function. Unfortunately the real beam profiles and the X-ray distribution have more complex shapes and might not be easily fitted at least for the 1 % - 99 % criterion to analyze the edge steepness. Therefore, the measured line scans are smoothed over 5 points and the analytical resolution is estimated as the distance between points having 1 % and 99 % of the maximum signal respectively. This 1 % - 99 %-criterion is necessary to determine the analytical resolution for a quantitative analysis. In this case it is guaranteed that all detected photons are emitted from a homogenous volume of the sample. We call this the quantitative analytical resolution (see sections 2.2.1 - 2.2.3). Additionally it is interesting to know the minimum size of features (having another element distribution) that can be detected by WDX line scans. In this case the source area of emitted photons might be larger compared to the size of the feature. We call this the qualitative analytical resolution (see section 2.3). It is determined by analysing the steepness of line scans with a 16 % - 84 % criterion which corresponds to the full width half maximum (FWHM) of a Gaussian function convoluted with a step function.

The MC simulations are performed for the same specimen and measurement conditions, i.e., the same X-ray transitions, element configuration, acceleration voltages and beam currents. Furthermore, the simulations consider the particular probe diameters (as determined in sect. 2.1) as a parameter. The simulated line scans are analysed with the 1 % - 99 %-criterion to apply for comparisons with the quantitative analytical resolution. Moreover the 1 % - 99 %-criterion offers the possibility to compare the measurements with MC simulations that consider a complete stopping of the electrons.

2.2.1. First example: Fe-Ni interface. At the metallographic prepared cross-section of an electroplated Fe-Ni interface (figure 5) we measured line scans of the Ni-Lα transition with accelerating voltages of 15, 8 and 5 kV and a beam current of 5 nA. 450 points were analysed with a step size of 10 nm and a dwell time per point of 3 s. The Ni Lα-line (0.851 keV) was measured with the TAP-crystal.

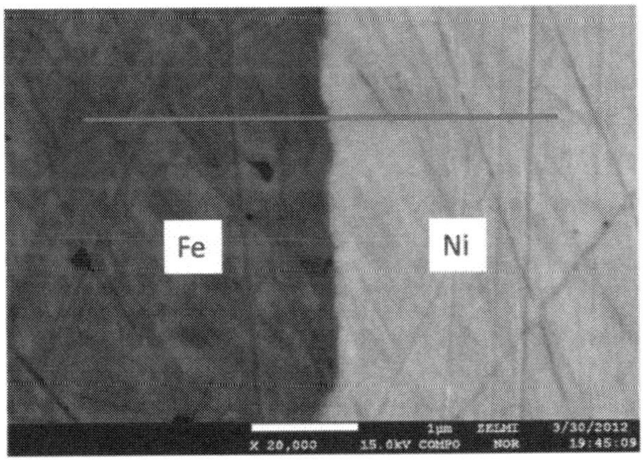

Figure 5. Backscattered electrons image of the Fe-Ni interface and direction of the line scan.

Figure 6 compares simulated and measured line scans for the three acceleration voltages. As expected the analytical resolution increases as the acceleration voltage is decreased. Considering the measurements the resolution is getting better from 800 nm at 15 kV to 200 nm at 5 kV. At 15 kV simulations and measurements coincide quite well. But for lower accelerating voltages simulation and measurement deviate from each other and the underestimation of the simulation increases with the decrease of the voltage. The reason for this discrepancy is not clear yet. Maybe the simulation is not optimized for low energies since radial distributions of emitted X-rays, which were simulated with CASINO showed a maximum several nanometre out of the centre but not at the centre. The simulation at 5kV would coincide with the measurement if a beam diameter of 120 nm is postulated. According to the measurements (figure 4) the beam diameter has a value of 56 nm and there is no reason to assume a deviance of more than a factor of 2. In any case fluorescence is not responsible for the effect, since the energy of all Fe L-lines (0.60 - 0.72 keV) is well below the Ni-L ionisation energy (0.853 keV).

2.2.2. Second example: Ag-Au-Ag interface. This second specimen is suitable for the comparison of low voltage and low overvoltage analysis. Two silver (500 nm) layers with an intermediate gold layer (75 nm) were deposited on a silicon wafer with a sputter coater. Afterwards a 15 μm wide cross-section was prepared using the FIB. The top layer visible in figure 7 is a platinum layer necessary during the FIB-preparation. On the layers cross-section, a line scan with 15 kV and 5 kV and 1 nA was measured using the beam scan with a step size of 3 nm and a dwell time of 20s. The Au Mα- and the Au Lα- signals were recorded with the TAP and the LIFH crystal respectively.

First an acceleration voltage of 15 kV was used allowing the excitation of the Au Mα- and the Au Lα-transitions. The later transition is excited with a low overvoltage. From Monte Carlo simulation curves (figure 8) we get an estimated resolution of 375 nm for the Au Mα-signal and 144 nm for the Au Lα-signal. Figure 9 plots the measured line scans with evaluated resolutions of 200 nm and 360 nm for comparison. We found a quite good accordance for the Au Mα-signal but a

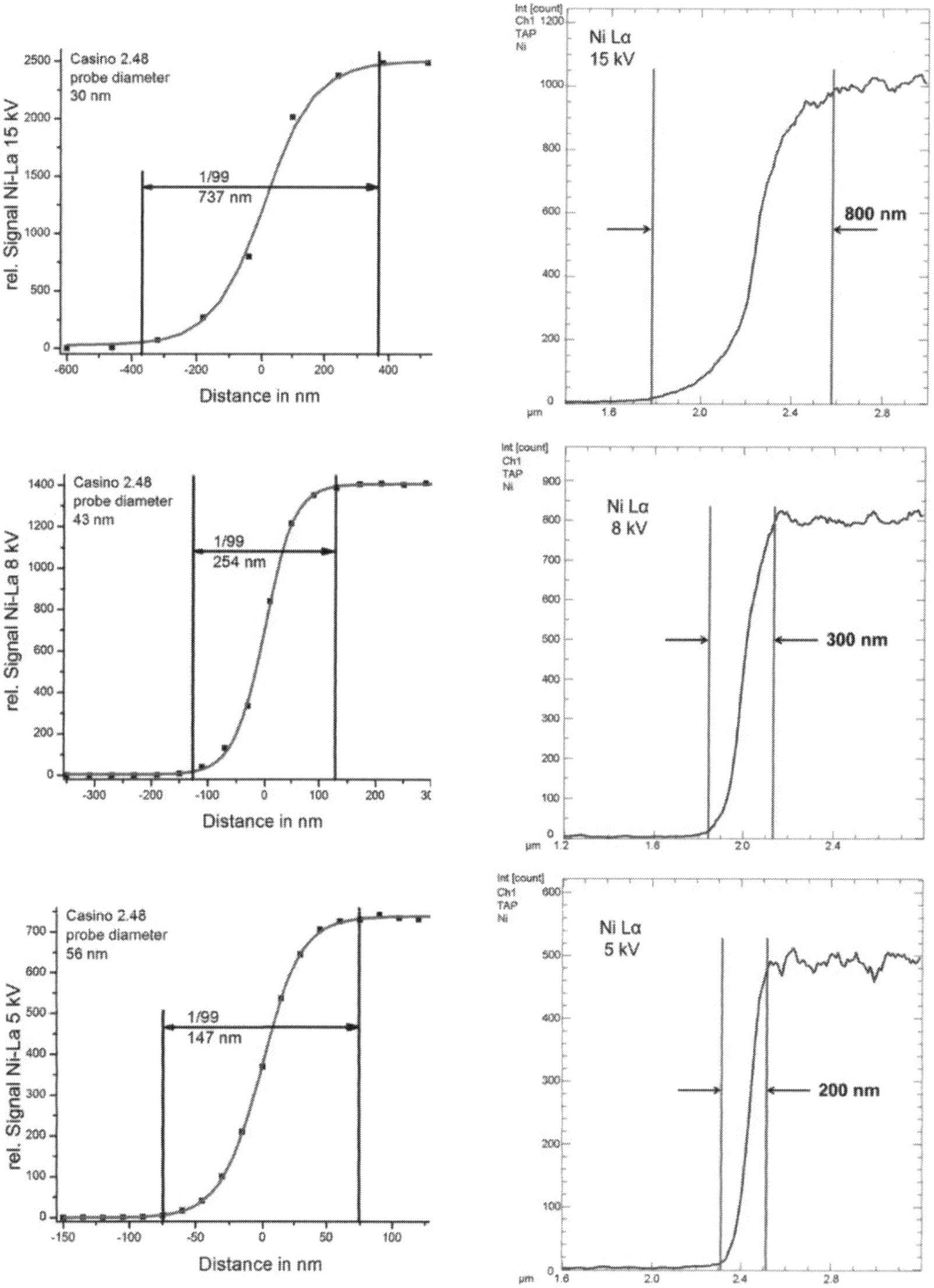

Figure 6. Comparison of MC simulated and measured line scans of Ni-Lα across the Fe-Ni interface at 15, 8 and 5 kV.

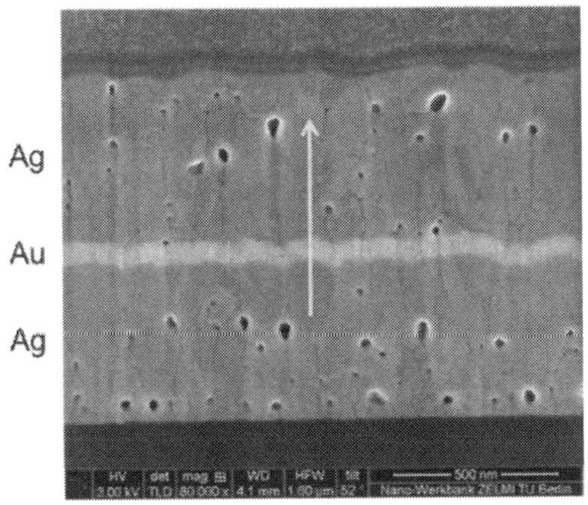

Figure 7. In-lens SE-image of the Ag-Au-Ag interface and direction of the line scan.

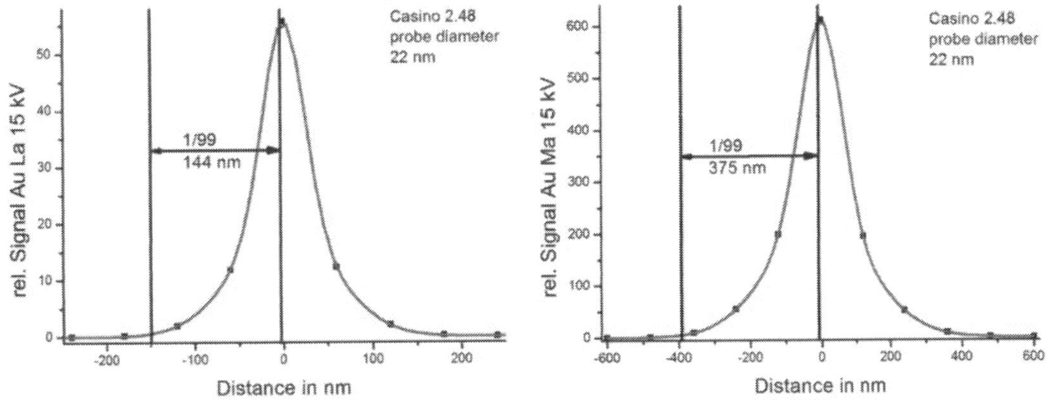

Figure 8. MC simulated line scans of Au-Mα and Au-Lα across the Ag-Au-Ag interface at 15 kV.

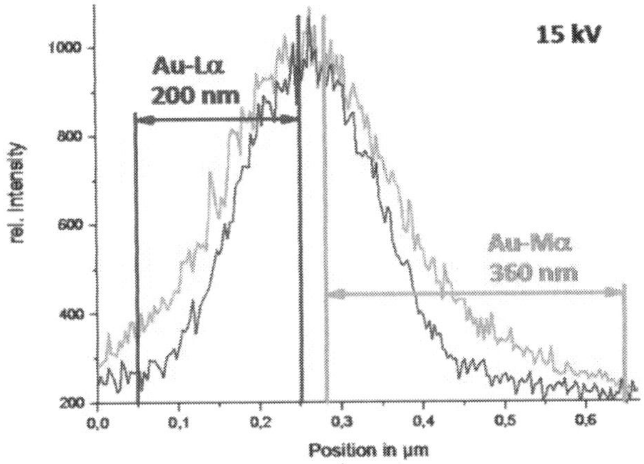

Figure 9. Measured line scans of Au-Mα and Au-Lα across the Ag-Au-Ag interface at 15 kV

clear 25 %-underestimation for the Au Lα-signal. Again this discrepancy is not understood yet. It might be caused by the roughness of the interface which worsens the resolution. This effect has a different influence on Au Mα- and Au Lα-photons. Apparently, the measured Au signals are additionally not in saturation even in the middle of the gold layer, i.e., that the volume of the emitted photons exceeds the thickness of the gold layer. Therefore, the line scans were repeated at an acceleration voltage of 5 kV. Now only the Au Mα-signal occurs and it is excited with a low voltage.

The reduction of the acceleration voltage to 5 kV improves the analytical resolution by decreasing the range of electrons. Additionally it might be used for the comparison of low voltage and low overvoltage analysis. From figures 8 and 9 at 15 kV it is already visible that the detection of the Au Lα-signal enables a better resolution. The reason for this is that only few (sharply localized) electrons have enough energy to excite L-transitions (ionisation energy 11.92 keV) while many electrons from a wide excitation area might excite the M-transitions (ionisation energy 2.21 keV).

The question is: How does the Au Lα-line scan recorded at 15 kV compares to the Au Mα-line scan at 5 kV? The MC simulation in figure 10 shows that the range of electrons nearly has the same size as the thickness of the gold layer since the Au Mα-signal is in saturation at the centre. Unfortunately simulation and measurements do not coincide very well again. Probably CASINO includes the fluorescence of Au-Mα (2.21 keV) excited by Ag-Lα (2.98 keV) not in a suitable way since most matrix correction models consider fluorescence ratios smaller 1.25 only. In our case it has a value of 1.35. Furthermore, it is not clear whether the measured Au-signal has reached the saturation.

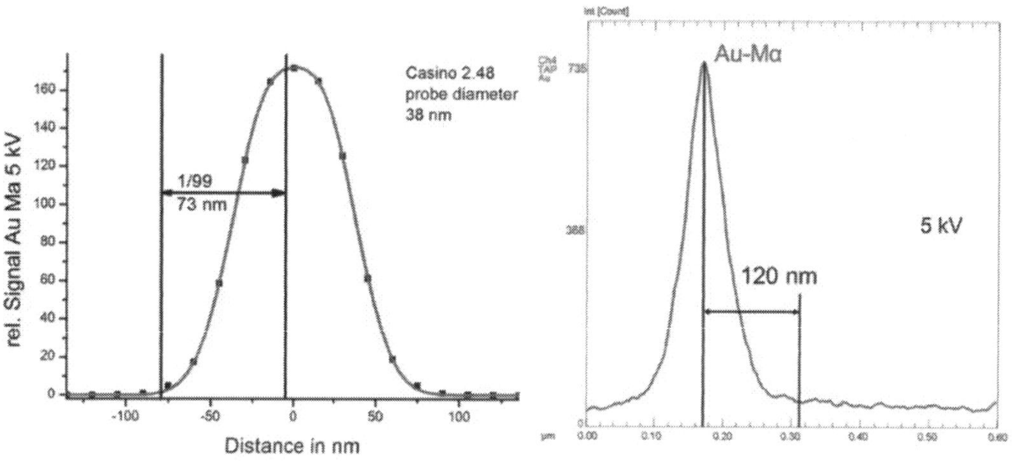

Figure 10. Comparison of MC simulated and measured line scans of Au-Mα and Ag-Lα across the Ag-Au-Ag interface at 5 kV.

Obviously, the low voltage analysis delivers in both cases a better resolution compared with the low overvoltage analysis. The further decrease of the overvoltage in figures 8 and 9 will increase the resolution but the detected signal will further decrease. In addition it should be mentioned that the low overvoltage analysis might not be optimal for the later quantification of the revealed small structures. Using an acceleration voltage of 15 kV to measure the Au Lα-line gives an overvoltage ration of 1.25 only whereas it should have a value of 2.5 … 3 to maximize the excitation. From this point of view it is more suitable to use the Au Mα-line at 5 kV (overvoltage ration of 2.6), i.e., the low voltage analysis is preferable due to the better spatial resolution and due to the better quantification.

2.2.3. Third example: Si-Au-C interface. In the next step the analytical resolution is examined in the presence of light elements. A 75 nm thick gold layer was deposited by sputter coating on a silicon wafer substrate. A 275 nm thick carbon layer was then added by electron beam evaporation on top. The line scan (200 points with step size 3 nm and dwell time 20 s) was measured at a cleaved cross-section of the specimen (figure 11) with 5 kV (1 nA). Here, the Au-M_α signal was recorded with the PETH crystal, but similar results were obtained with the TAP crystal.

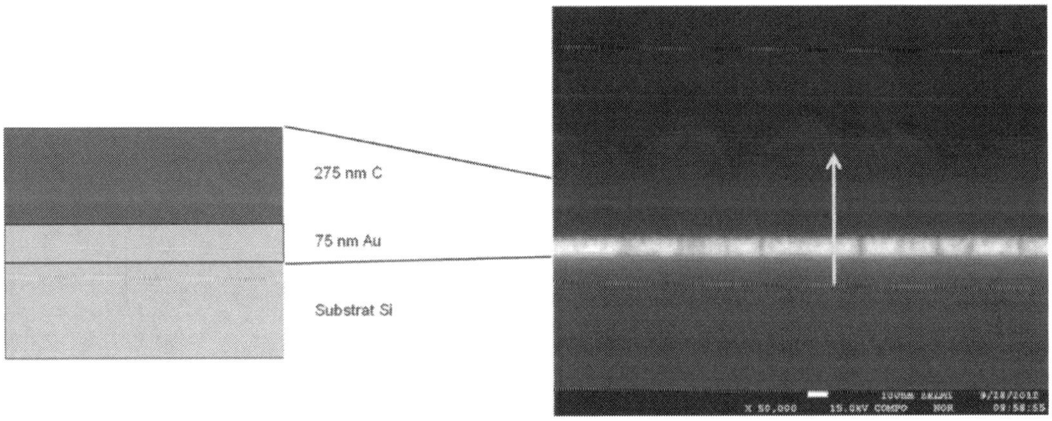

Figure 11. Layout and backscattered electrons image of the Si-Au-C interface and direction of the line scan.

By comparing both line scans in figure 12 it is clear that there is a poor agreement between simulation and measurement. The measured Au Mα-line scan shows a strong but reasonable asymmetry since the gold layer is embedded in silicon (left) and carbon (right). In contrast, the simulation is nearly symmetric. Obviously the density of the electron beam evaporated carbon film differs too much from the density of bulk carbon used by the simulation. In addition the simulation might include the absorption of Au-Mα in Si which is ten times higher compared to the absorption in C not in a suitable way.

2.3. Measurements of the qualitative spatial analytical resolution

In some applications it is not necessary to quantify the elements but only to trace small areas with another elemental composition. Then, it is sufficient that the WDX-signals change along the line scan well above noise [16]. Therefore, it is reasonable here to define a qualitative analytical resolution. This value is a measure for the size of structures that might be detected using the WDX-signal. Then it is meaningless to use the 1 % - 99 % criterion to analyze the steepness in line scans. In figure 13 the Au-Mα and the Si-Kα line scans from the third specimen are re-analysed by fitting Gaussian and sigmoidal functions respectively. From these fitting functions the edge steepness is measured using the 16 % - 84 % criterion corresponding to the FWHM of the Gaussian beam convoluted with a sharp transition of elements. With this commonly used criterion a (qualitative) analytical resolution of 26 nm for the Au Mα-line and 45 nm for the Si Kα-line is obtained. From the clear contrast of the Au Mα-signal it might be derived that even thinner layers with a better analytical resolution are detectable.

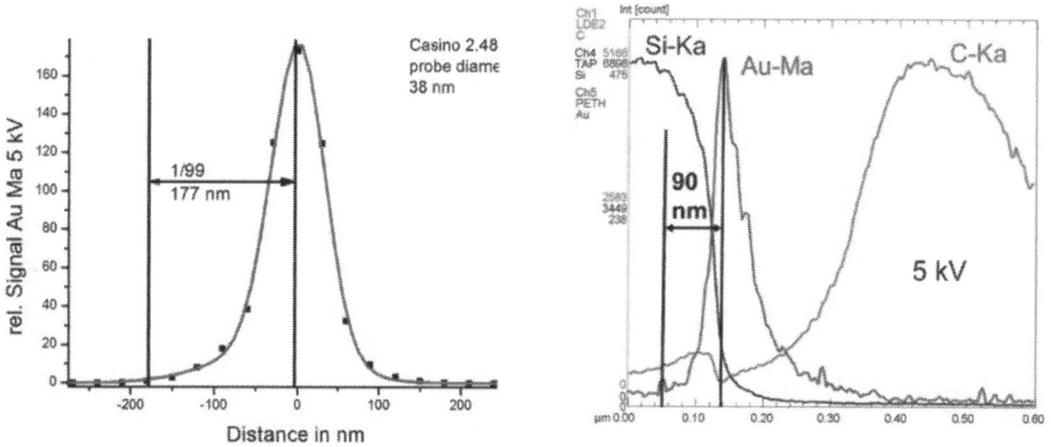

Figure 12. Comparison of MC simulated and measured line scans of Au-Mα, Si-Kα and C-Kα across the Si-Au-C interface at 5 kV. Right: The measured line scans are smoothed over 5 points and the markers for the edge steepness's are set by hand.

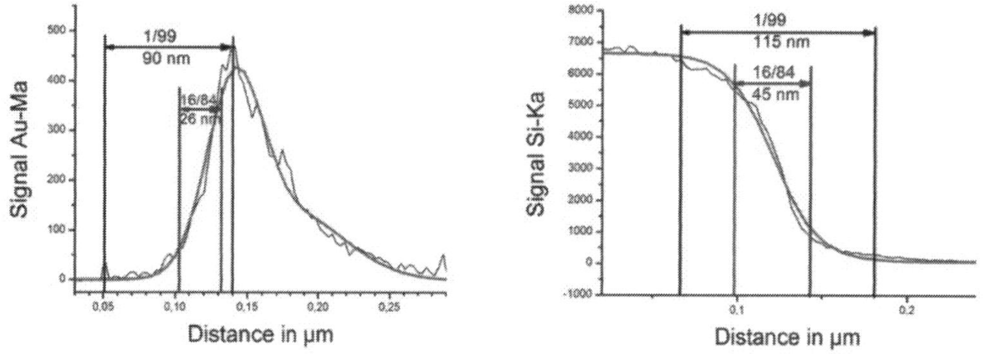

Figure 13. Measured line scans of Au-Mα and Si-Kα across the Si-Au-C interface at 5 kV and indicated quantitative and qualitative analytical resolution. The Au Mα-signal is fitted by two Gaussian functions considering the two interfaces. The Si Kα-signal is fitted with a sigmoidal function. The 16 % - 84 % analysis is obtained from the fitting function while the 1 % - 99 % analysis is obtained from the smoothed measurements figure 12 right.

3. Conclusions

The FEG-EPMA has a significantly increased analytical resolution compared to EPMAs with thermionic electron emitter reaching values down to several ten nanometres. When measuring the analytical resolution two applications should be distinguished. On the one hand, the quantitative analytical resolution is considered as a measure allowing the quantification of the elements since the analysed structures are larger than the area of emitted photons. The quantitative analytical resolution is determined by steepness of line scans with a 1 % - 99 % criterion. Here we measured values of about 100 nm. On the other hand, for some applications it is enough just to detect small elemental changes using the WDX-signal. Then the qualitative analytical resolution is regarded which is derived from line scans with a 16 % - 84 % criterion. This resolution value was found to be in the order of tens of nanometres. At particular samples the recognition of even 10 nm thin layers seems to be feasible.

For the prediction of both kinds of analytical resolutions, MC simulations are helpful. Appropriate simulations consider the actual probe diameter which needs to be measured for all individual acceleration voltages and probe currents as an input parameter for the simulation. Commonly Au-on-C reference specimens are used for these measurements. If the quantitative analytical resolution is required then also the probe diameters are determined with the 1 % - 99 % criterion.

At three specially made specimens we compared resolutions derived from MC simulations with measurements under different measurement conditions. In most cases, the MC simulations are suitable to get an approximation of the expected resolution but often they differ by 25 % to 33 % from the measurements. The discrepancy seems to be larger for small accelerations voltages and large differences in the atomic number. In the future further attempts to understand the differences are necessary.

Acknowledgement
The financial support by the Europäische Fonds für Regionale Entwicklung (EFRE) within the project "Nano-Analytiklabor" is gratefully acknowledged. We appreciate fruitful discussion on the presented topic with colleagues during the great EMAS meetings, especially with J. Börder, P.T. Pinard, J. F. Thiot and P. McSwiggen.

References

[1] Kubo Y *et al* 2013 *Ultramicroscopy* **135** 64-70
[2] McSwiggen P 2013 Book of Tutorials and Abstracts of the EMAS 2013 Workshop (Dresden: European Microbeam Analysis Society), 57-71; *IOP Conf. Ser.: Mater. Sci. Engng.*, this volume
[3] Newbury D 2002 *J. Res. Natl. Inst. Stand. Technol.* **107** 605-619
[4] Berger D and Nissen J 2012 *JEOL User-Meeting*, Göttingen, Germany
[5] Hombourger C, Outrequin M 2013 Microscopy Today, Vol. 21 No. 03 10-15
[6] Fialin M *et al* 1999 *Am. Mineralogist* **84** 70-77
[7] Llovet X *et al* 2012 *IOP Conf. Ser.: Mat. Sci. Engng.* **32** 012014
[8] Newbury D 2002 *J. Res. Natl. Inst. Stand. Technol.* **107** 605-619
[9] Barkshire I *et al* 2000 *Mikrochim. Acta* **132** 113-128
[10] Merlet C and Llovet X 2012 *IOP Conf. Ser.: Mat. Sci. Engng.* **32** 012016
[11] Gopon P *et al* 2013 Book of Tutorials and Abstracts of the EMAS 2013 Workshop (Dresden: European Microbeam Analysis Society), 323
[12] Duncumb P and Statham P J 2002 *Microchim. Acta* **138** 249-258
[13] Assmann A and Wendt M 2003 *Spectrochim. Acta B* **58** 711-716
[14] Mori N 2010 *JEOL News* **45** 42-46
[15] Drouin D *et al* 2007 *Scanning* **29** 92-101
[16] Kimura T *et al* 2006 *Microchim. Acta* **155** 175-178

Measuring conditions for second order X-ray Bragg-spectrometry

J Dellith, A Scheffel and M Wendt

Institute of Photonic Technology, Albert-Einstein-Strasse 9, DE-07745 Jena, Germany

E-mail: jan.dellith@ipht-jena.de

Abstract. The $KL_{2,3}$ $(\alpha)_{1,2}$-lines of ^{19}K, the $L_3M_{4,5}$ $(\alpha)_{1,2}$-lines of ^{48}Cd, and the $M_5N_{6,7}$ $(\alpha)_{1,2}$-lines of ^{92}U are lines of comparable energy in the region of approximately 3 keV. In none of these cases were we able to resolve the three doublets when recording the spectra in first order Bragg spectrometry using a PET crystal as the dispersing element. For the purpose of enhancing the resolving power of the spectrometer, the three α spectra were recorded in second order reflection, thereby transferring the lines into another spectral region dominated by X-ray quanta of half the energy. In order to achieve high net peak intensities as well as a high peak-to-background ratio and, consequently, a high level of detection capability, the discriminator settings should be optimized quite carefully. In this manner, we were able to resolve the three α doublets and estimate α_2/α_1 intensity ratios. Inexplicably, current monographs, e.g., by Goldstein *et al*, do not contain any indications about the rational use of high order spectrometry. Only a few rather old monographs contain some hints in this regard.

1. Introduction

Compared to energy-dispersive X-ray spectrometry (EDS), wavelength-dispersive X-ray spectrometry (WDS) provides a better energy resolution and, consequently, a higher sensitivity for the detection of trace elements. However, WDS measurements are more time-consuming and also more complicated than EDS measurements because the measured WD intensities depend on a number of circumstances which can be neglected in ED measurements. For example, EDS spectra are completely free of influences caused by high order reflections. Most of the aspects described in connection with high order reflections focus on their disturbing effects. Here, the interference possibility of a given characteristic X-ray line with one originating from high order reflections is discussed initially. In fact, this aspect is problematic, particularly in the field of light element analysis, because often only one usable X-ray line is available. To give an example: In the analysis of boron doped silica, the boron KL_3 (α)-line at 183 eV interferes with the third order reflection of the oxygen KL_3 (α)-line with an energy of 525 eV. Normally, the high order reflections can be efficiently discriminated by using the pulse height analysis mode. However, especially in the example mentioned above, it is sometimes tricky because of the spread pulse height spectrum of the soft boron K-radiation [1].

On the other hand, there are quite useful applications of high order reflections. For instance, a precise calibration of wavelength-dispersive X-ray spectrometers can be achieved using one and the same element and measuring the spectra in first, second, and third order reflection. Furthermore, in the simultaneous quantitative analysis of trace and major elements, the strong line intensities and subsequent high count rates of the major elements can be reduced by falling back on high order spectra, the intensity of which is only a fraction of those of the first order. But perhaps the most

Content from this work may be used under the terms of the Creative Commons Attribution 3.0 licence. Any further distribution of this work must maintain attribution to the author(s) and the title of the work, journal citation and DOI.
Published under licence by IOP Publishing Ltd

important reason for using high order reflections is the possibility of improving the resolving power of a given WD spectrometer and consequently enhancing the line separation.

Unfortunately, information about useful aspects of high order WD spectrometry can only be found in a few rather old specialized books, e.g., by Müller [2] or Jenkins [3]. In the current monographs by Reed [4], Scott *et al* [5], or Goldstein *et al* [6] this topic is completely lacking.

In two previous works, we studied the M-spectra of ^{68}Er [7] and ^{92}U [8], and have already briefly reported on the potential of high order spectrometry. In the present work, we aim to illustrate the resolution enhancement in particular. We will propose a strategy to determine optimized discriminator settings, which are a basic requirement to achieving high peak-to-background ratios and a level of detection capability which is comparable to that provided by first order spectrometry.

2. Experimental method

The measurements were carried out using a JEOL JXA-8800L microprobe analyzer (Jeol Ltd., Tokyo, Japan). The take-off angle of this instrument is 40°. All spectra were recorded using a PET crystal with a periodicity of 2d = 8.742 Å as the dispersing element. The L-value range of the WD spectrometer used is 55 mm ≤ L ≤ 255 mm. Thus, the detectable energy with the PET crystal ranges from 1550 eV to 7200 eV according to equation (1).

$$E = \frac{12398 \cdot R}{d \cdot L} \qquad (1)$$

The numerical equation (1) was derived from Bragg's law and the Rowland circle conditions for the WD spectrometer used. Here, E [eV] is the radiation energy, L [mm] is the L value, R [mm] is the diameter of the Rowland circle, and d [Å] is the periodicity of the dispersing element. For the spectrometer used, the diameter of the Rowland circle is 140 mm.

The energy of the exciting electrons, E_0, was set to 20 keV in all measurements. The probe current used was, depending on the sample and the reflecting order, in the range between 10 nA and 1 μA. In order to reduce the sample strain, the probe diameter was widened to 100 μm. The effect of beam widening concerning potential spectra deterioration was carefully proofed previously by a series of measurements with different probe diameters. Here, spectral changes are vanishing for probe diameters below 200 μm.

If the PET crystal is used, the X-rays in our microprobe are detected by a sealed Xe-filled proportional counter (XPC). Nearly all WDS spectra were recorded in the differential pulse height analysis mode. But in order to determine optimized measuring conditions, spectra were initially recorded in the integral mode. The XPC voltage and the gain of the pulse amplifier were selected in such a manner that the mean pulse height of the detected X-ray quanta was 4 V.

As standard samples we have selected the three elements potassium (^{19}K), cadmium (^{48}Cd), and uranium (^{92}U), which exhibit intensive (K, L, M)$\alpha_{1,2}$-doublets of nearly comparable energy. The relevant line energies, taken from the compilation tabulated by Bearden [9], are summarized in Table 1.

With the exception of potassium, pure element metal standards were used. Nearly all metals have a more or less strong tendency to oxidize. By grinding the samples immediately before the measurement, we were able to strongly reduce the ^{8}O Kα-intensity to a negligible level. For the potassium measurements a polished KBr sample was used. This sample was, in order to avoid charging effects, coated with approximately 20 nm of carbon.

Table 1. $\alpha_{1,2}$-line energies for the K-radiation of K, the L-radiation of Cd, and the M-radiation of U (values are rounded to full eV; data taken from [9]).

Transition/Line	[19]Potassium	[48]Cadmium	[92]Uranium
KL_3 (α_1)	3314 eV	-	-
KL_2 (α_2)	3311 eV	-	-
L_3M_5 (α_1)	-	3134 eV	-
L_3M_4 (α_2)	-	3127 eV	-
M_5N_7 (α_1)	-	-	3171 eV
M_5N_6 (α_2)	-	-	3160 eV

3. Results and discussion

3.1. Resolution enhancement

In figure 1 the energy difference $\Delta E = E(\alpha_1) - E(\alpha_2)$ for K-, L-, and M-radiation versus the atomic number Z is shown. In the classic compilation by Bearden [9], separated energies for α_1- and α_2-lines are listed first for $Z > 12$ (K-radiation), $Z > 36$ (L-radiation), and $Z > 73$ (M-radiation). As illustrated in figure 1, ΔE rises rapidly with increasing Z. The elements used in this work are marked by dotted lines. Here, the energy difference ranges from 3 eV for potassium up to 11 eV in the case of uranium.

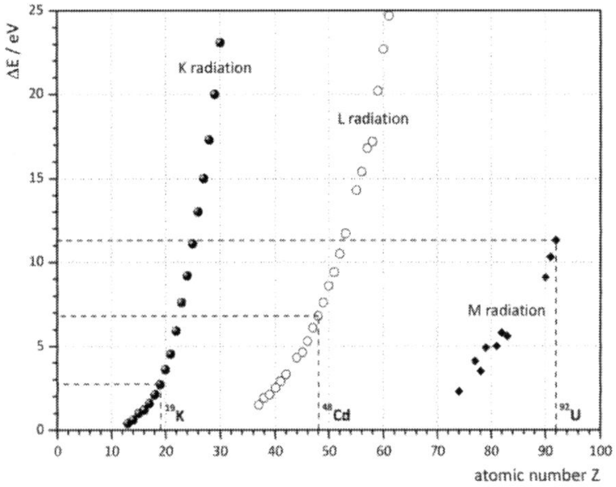

Figure 1. Energy difference $\Delta E = E(\alpha_1) - E(\alpha_2)$ versus atomic number Z. ΔE rises rapidly with increasing Z. The line energies were taken from reference [9].

In none of these cases were we able to resolve the (K, L, M)$\alpha_{1,2}$-doublets into their components when recording X-ray spectra in first order reflection with our setup. The resolving power of a given WD spectrometer is mainly determined by the crystallographic perfection of the dispersing element used. Furthermore, a high dispersion, $D = \Delta\theta/\Delta\lambda$, of this element is crucial for sufficient line separation. More details in this regard can be found e.g., in the monograph written by Reed [4].

Nevertheless, using a given setup, it is possible to enhance the spectrometer resolution significantly. For instance, one can narrow the entrance slit of the detector in order to exclude reflections originating from the parts of the crystal which are outside of the Rowland circle plane. Another, more powerful way involves the use of the nonlinear resolving power of Johannson-type

WD spectrometers. In order to illustrate this point initially, one has to consider the spectrometer position of the radiation of interest. The energy of the (K, L, M)$\alpha_{1,2}$-line doublets of the three elements used in this work is comparable in the range slightly above 3 keV. According to figure 2, the spectrometer position of the radiation of this energy is – using our spectrometer setup – between 120 mm ≤ L ≤ 125 mm.

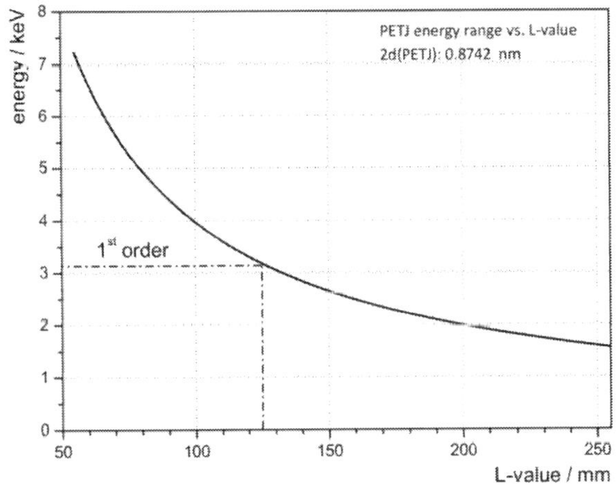

Figure 2. Dependency of the detectable energy range versus the spectrometer position. The curve was calculated using equation 1. The approximate spectrometer position of the lines of interest in first order reflection is marked.

By recording the X-ray spectra in second order reflection, they will be transferred to twice the initial L value of the first order into a region in which X-ray quanta of half the energy are normally detected. This means that we will find the second order reflection of the three line doublets between 240 mm ≤ L ≤ 250 mm. In order to better understand the advantageous consequences of this, one has to differentiate Eq. 1.

$$\frac{\Delta E}{\Delta L} = f(L) = \frac{-397099}{L^2} \qquad (2)$$

In this manner, one is able to calculate the energy difference, ΔE, for one fixed increment, ΔL. In figure 3, this has been done for a typical interval of ΔL = 100 μm.

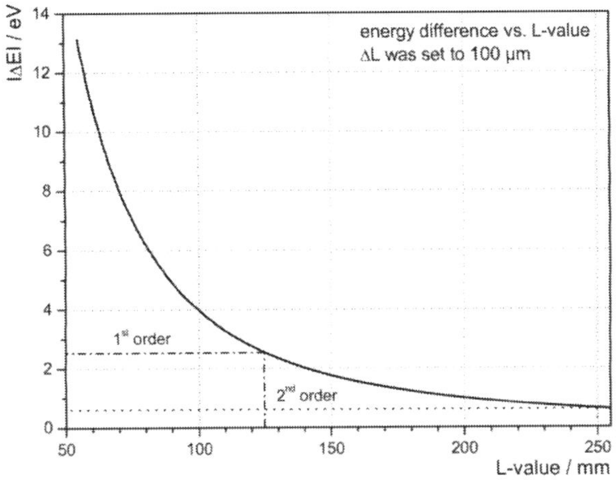

Figure 3. Dependency of the energy difference, ΔE, from the spectrometer position, calculated for a fixed interval of ΔL = 100 μm. The lines of interest in first (dashed-dotted) and second order (dotted) reflection are marked.

The curve shown in figure 3 makes it clear that the theoretical energy resolution of the lines of interest in the frame of the measurement conditions in first order reflection cannot be better than nearly 3 eV. In contrast, by changing to second order spectrometry and thus shifting the spectra towards larger L-values, the resolution limitation can be lowered to approximately 0.5 eV. Consequently, in second order reflection a given X-ray spectrum extends a much larger L-value range than in first order reflection and the line separation is greatly improved. In practice, the measured energy resolution is – as already mentioned – determined by several additional aspects. It is generally much worse than the theoretical considerations. One option for an experimental approach to estimating the real spectrometer resolution as a function of the L-value is to measure one and the same element in first, second, and third order reflection. Then, one can determine the full width at half maximum (FWHM) of a given X-ray line as a measure of the resolving power. Suitable lines for this purpose are K-lines initially because they are narrower than both L-lines and M-lines. Figure 4 shows the natural KL_3 (α_1)-line widths (FWHM) of the elements $10 \leq Z \leq 30$. These values were taken from reference [10].

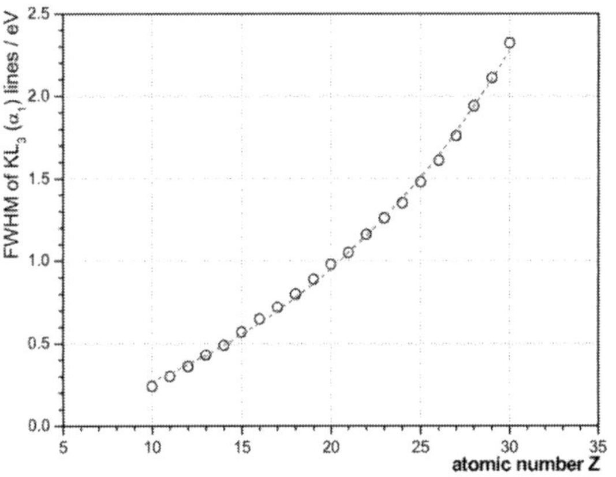

Figure 4. Natural KL_3 (α_1)-line width (FWHM) for the elements $10 \leq Z \leq 30$. The data were taken from Krause *et al* [10].

One element usually used to specify the resolving power of energy-dispersive X-ray spectrometers is ^{25}Mn. Mn is also suitable for the specification of a WD spectrometer. The energy of Mn-Kα_1 is 5899 eV, and that of Kα_2 is 5888 eV. Using the PET crystal our spectrometer is able to detect the Mn $KL_{2,3}$ ($\alpha_{1,2}$)-radiation in first, second, and third order reflection. In this connection, it is assumed that the actual spectrometer resolution is always worse than the natural ^{25}Mn $KL_{2,3}$ ($\alpha_{1,2}$)-line width. Figure 5 shows the experimentally estimated FWHM values for the three reflection orders versus the L-value.

The natural line widths of the ^{25}Mn K-lines were given by Krause to 1.48 eV and 1.50 eV, respectively. The energy difference, $\Delta E = E(\alpha_1) - E(\alpha_2)$, is about 11 eV, and the relative line intensity, α_2/α_1, is shown in White and Johnson [11] to be about 50 %. Consequently, the overall FWHM of the unresolved K$\alpha_{1,2}$-line is expected to be significantly broader than the values for each single line.

In first and second order reflections we were unable to resolve the ^{25}Mn K$\alpha_{1,2}$-doublet. Here, we have estimated the FWHM to 75 eV in the first order and to 10 eV in the second order. In third order reflection we were able to separate the Kα_1- and Kα_2-line; thus, we have estimated the Kα_1 FWHM to be approximately 3 eV. Here, one can directly compare the measured and the "true" line width and thus estimate the resolving power at that spectrometer position. For the first and second order measurement, the uncertainty is greater because of the unknown line width. Nevertheless, the second order measurement, in which the Mn K$\alpha_{1,2}$-doublet is still unresolved, indicates an overall FWHM of

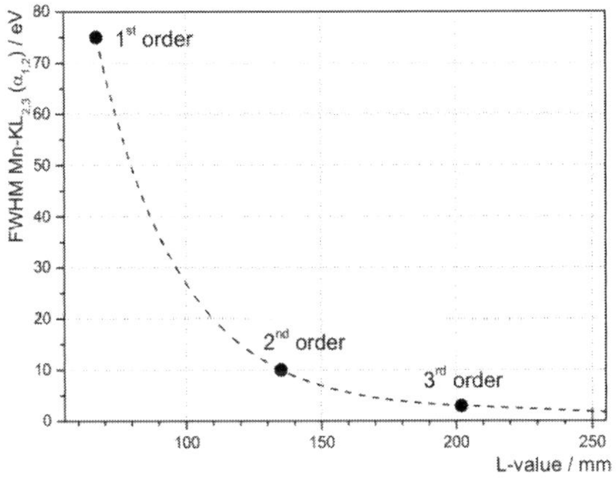

Figure 5. Measured [25]Mn $KL_{2,3}$ ($\alpha_{1,2}$)-line widths (FWHM) in first and second order reflection. In the third order, the $\alpha_{1,2}$-doublet is resolved, and therefore, only the α_1 FWHM is shown.

no more than 10 eV and this value seems to be mainly determined by the energy difference $\Delta E = E(\alpha_1) - E(\alpha_2)$

3.2. Discriminator settings

By changing from first to second order spectrometry, the lines are transferred to twice of the initial spectrometer position. Here, normally X-ray quanta of half the energy are detected. In the case of our standard measurements this concerns solely continuous bremsstrahlung and consequently the second order spectrum involves now bremsstrahlung reflected in both the first and the second order. This fact leads to a worsening of the peak-to-background ratio.

In order to enhance the detection capability of the second order lines, the counter settings must be optimized quite carefully. For this purpose, we initially choose a sufficient gain according to the energy of the detected X-ray quanta. Then, a so called high-voltage scan was performed to determine the optimal detector high voltage. Thereby, a lower gain requires a higher voltage and consequently a higher gain a lower voltage to achieve one and the same pulse height of the detected X-ray quanta. An optimal combination of gain and high voltage for a given X-ray quantum energy leads in our microprobe always to a mean pulse height of 4 V. A corresponding base level scan is shown in figure 6. Here, the base line is set to 3 V, and the window width amounts to 2 V. Consequently, only pulse energies within this window will contribute to the differential mode measurement.

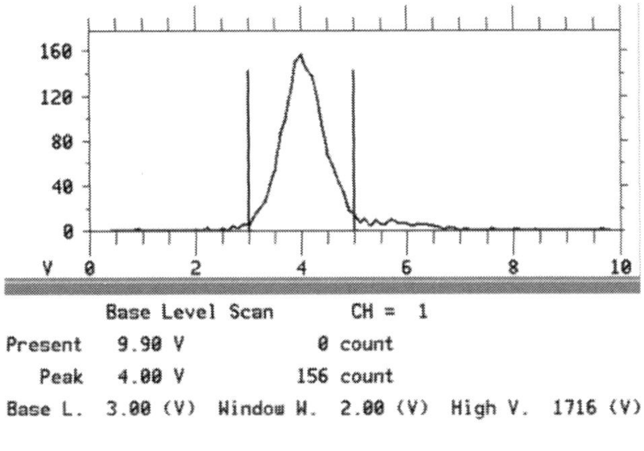

Figure 6. Base level scan for second order [19]K KL_3 (α_1)-quanta. The gain and the high voltage of the XPC were selected so that the mean pulse height was 4 V.

In order to determine the optimized discriminator settings in the first step, we used the so-called integral detection mode. In this step, a base line is set and all pulses above this base line contribute to the measured X-ray spectrum. In our experiments we started at a low base level of 1 V. Consequently, the resulting (K, L, M)$\alpha_{1,2}$ second order spectrum exhibited a poor peak-to-background (P/B) ratio because a broad bremsstrahlung spectrum (reflected in both the first and second order) contributes to the measurement. Then, we increased the base level step by step and recorded X-ray spectra after each single step. By doing so, we excluded radiative contributions step by step starting at the low energy side of the incoming spectral range. This applies in particular to background radiation of half the energy detected in first order reflection.

As a measure of optimisation, the product of P/B times net peak intensity (PBNP) was used. The dependency of this value as a function of the base level setting is shown for the potassium KL_3 (α_1)-radiation in figure 7a. In addition, figure 7b shows the corresponding detection limit (DL) for ^{19}K KL_3 (α_1)-radiation in potassium bromide (KBr).

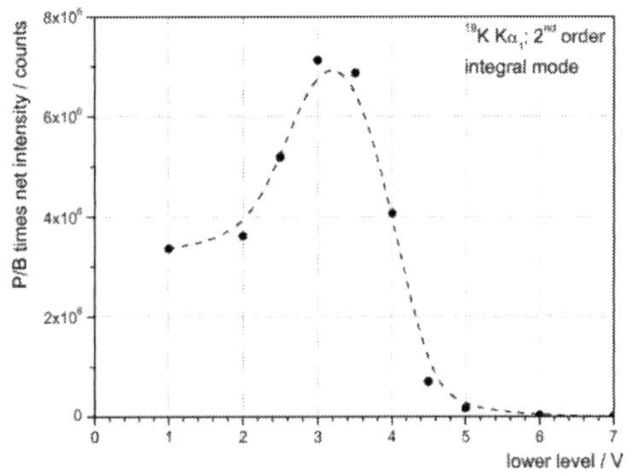

Figure 7a. P/B times net peak intensity (PBNP) versus the base level setting. The chart exhibits a maximum at a lower level of 3 V.

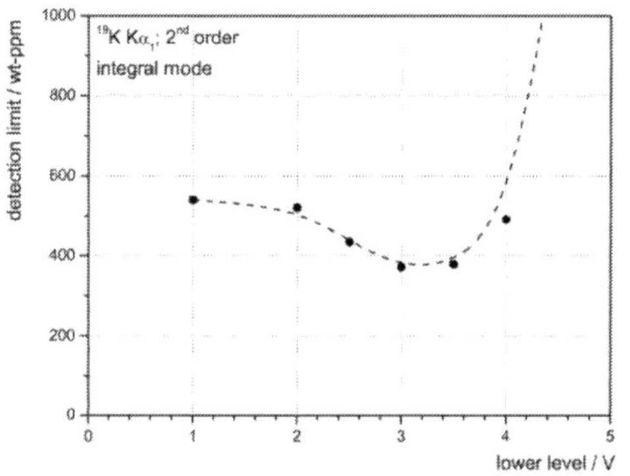

Figure 7b. Corresponding detection limit of K-Kα_1 in KBr. Also here, an optimized setting around 3 V can be observed.

The detection limits for each single step and for each single spectrum were estimated using equation 3.

$$DL = 3 \cdot \sqrt{BG} \tag{3}$$

Here, BG is the background intensity [counts], and DL is the calculated detection limit [counts] which is then converted into a concentration. To do this, the K-Kα_1 net intensity is measured in relation to it and normalized concerning the KBr stoichiometry. Both the PBNP chart and the DL curve indicate an optimum lower level around 3 V. By further increasing the base level, a progressive discrimination of the actual peak signal is carried out. Consequently, the PBNP course drops, whereas the detection capability gets worse and worse.

In the second step of optimisation, the base level was set to the ascertained value of 3 V as done in the first experimental step. Then, the window width was reduced step by step, and again here spectra were recorded after each step. This strategy leads to a step-by-step suppression of the high energy spectral part of the incoming radiation. The subsequent effect is expected to be much smaller than the base level adaption implemented in step one because the high energy contributions belong mainly to the second order spectrum. The results of this series of measurements are shown in figure 8a. At a decrease in window width, an increasing tendency can be observed in the PBNP value. This behaviour can mainly be explained by a lowering of the background intensity and thus an enhancement of the P/B ratio. When the window width becomes even narrower than 1.5 V, the PBNP value drops immediately due to the progressive discrimination of the actual peak signal. Here, too, detection limits were estimated for the K Kα_1-radiation in the KBr compound. The course of the curve mirrors inversely those of the PBNP value as shown in figure 8b.

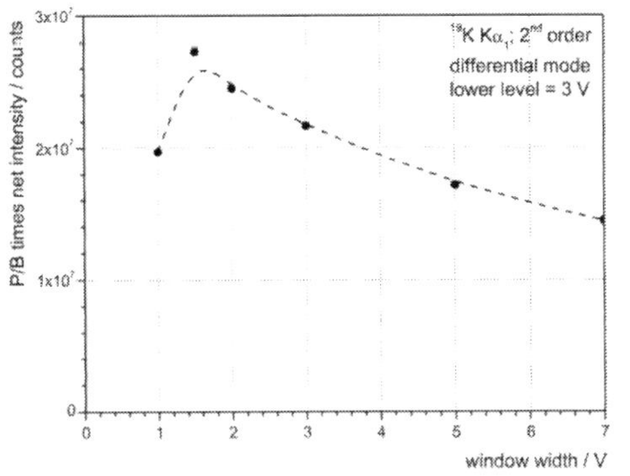

Figure 8a. P/B times net peak intensity (PBNP) versus window width. The course exhibits a maximum at 1.5 V.

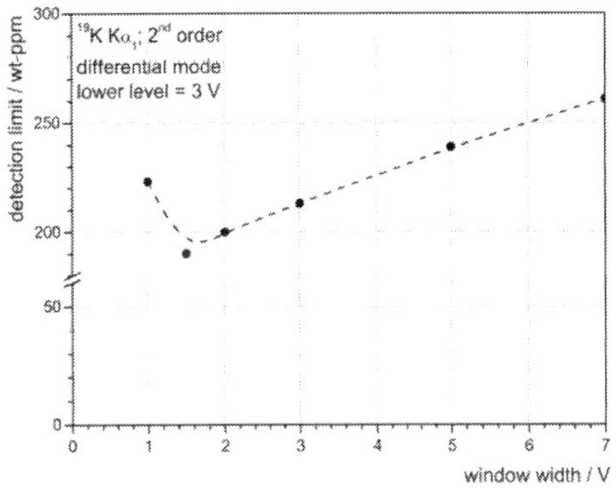

Figure 8b. Corresponding detection limit for K-Kα_1 in KBr. Again, the best detection capability is achieved at a window width of 1.5 V.

3.3. Measurements

The $KL_{2,3}$ $(\alpha)_{1,2}$-lines of ^{19}K, the $L_3M_{4,5}$ $(\alpha)_{1,2}$-lines of ^{48}Cd, and the $M_5N_{6,7}$ $(\alpha)_{1,2}$-lines of ^{92}U recorded in first (a) and second (b) order reflection are shown in figures 9, 10, and 11. The second order measurements were carried out using the optimized discriminator setting procedure described in section 3.2.

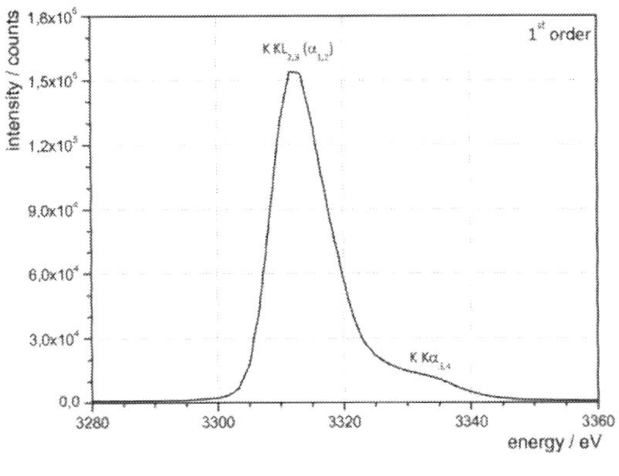

Figure 9a. Unresolved $KL_{2,3}$ $(\alpha)_{1,2}$-lines of ^{19}K recorded in first order reflection. On the high energy side of K$\alpha_{1,2}$ the weak satellite line K$\alpha_{3,4}$ is observable.

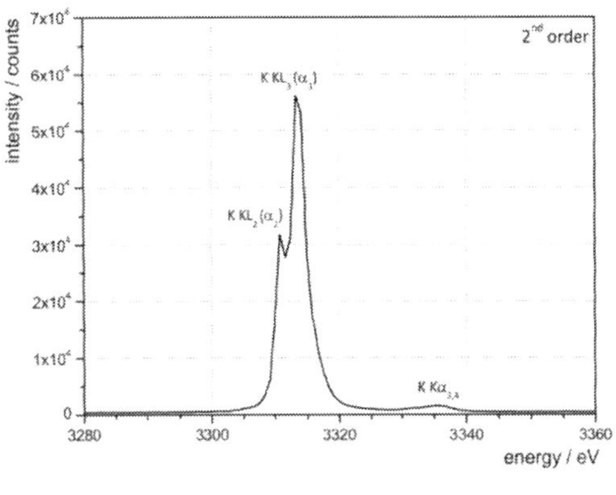

Figure 9b. $KL_{2,3}$ $(\alpha)_{1,2}$-lines of ^{19}K recorded in second order reflection. Here, the Kα_2 line can be clearly observed.

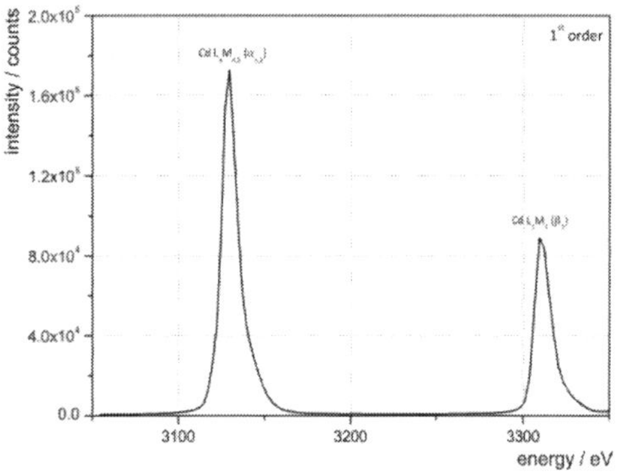

Figure 10a. Unresolved $L_3M_{4,5}$ $(\alpha)_{1,2}$-lines and L_2M_4 (β) of ^{48}Cd recorded in first order reflection.

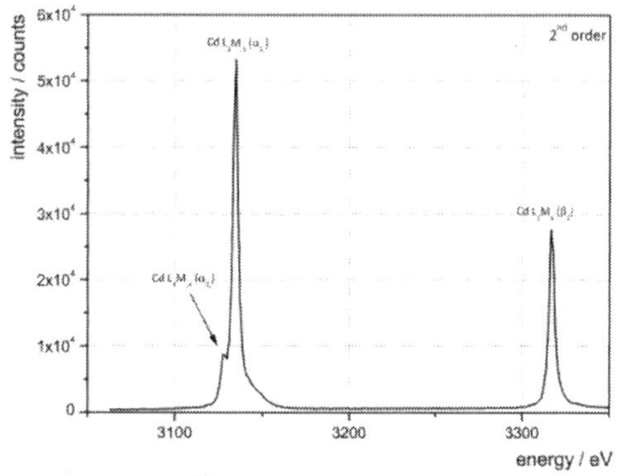

Figure 10b. ^{48}Cd L-lines recorded in second order reflection. Here, the L_3M_4 (α_2) -line can be clearly observed.

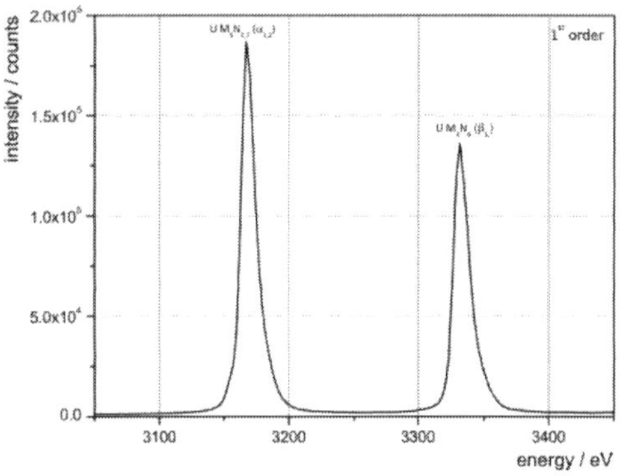

Figure 11a. Unresolved $M_5N_{6,7}$ $(\alpha)_{1,2}$-lines and M_4N_6 (β) of ^{92}U recorded in first order reflection.

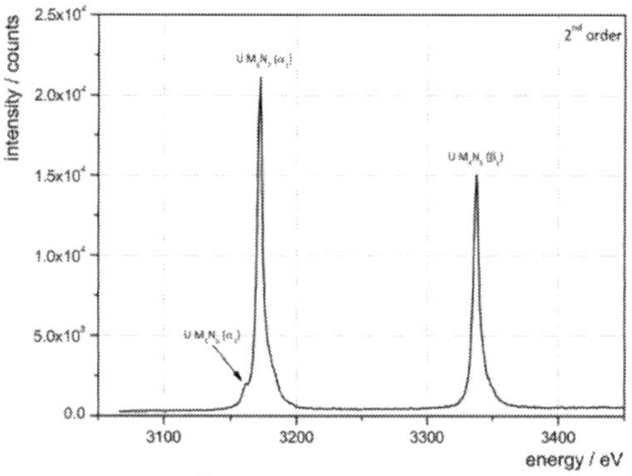

Figure 11b. ^{92}U M-lines recorded in second order reflection. Here, the M_5N_6 (α_2)-line is visible as a low energy shoulder of M_5N_7 (α_1).

In the case of the second order measurements, we were able to resolve the (K, L ,M)$\alpha_{1,2}$-doublets of all three elements into their components α_1 and α_2. This enabled us to estimate α_2/α_1 intensity ratios and to compare these values with the established literature. Furthermore, we estimated FWHM data for the unresolved $\alpha_{1,2}$-doublets in the case of the first order measurements and for the separated α_1-line in the case of the second order measurements. All results are summarized in Table 2.

The FWHM data compiled in Table 2 show the resolution enhancement connected with the transition from first to second order spectrometry. By shifting the spectrometer position to twice the first order L value, we observed a FWHM decrease by a factor of approximately 3. The estimated FWHM values agree for potassium and cadmium qualitatively well with the spectrometer performance curve shown for Mn K-radiation in figure 5. For these two elements as well as the manganese, the natural line widths are in a comparable range. Some data available in the literature are shown in Table 3. Due to the lack of reliable data for Cd L-lines, we have used those of the neighbouring element silver.

Table 2. $\alpha_{1,2}$-spectrometer position for first and second order, line widths (FWHM) and relative intensities for the K-radiation of K, the L-radiation of Cd, and the M-radiation of U.

X-ray line	Order	L value [mm][a]	FWHM [a,b] [eV]	α_2/α_1 intensity ratio [%]	
				Present work	Other literature[c]
K-$KL_{2,3}$ ($\alpha_{1,2}$)	1	120	10	-	
K-$KL_{2,3}$ ($\alpha_{1,2}$)	2	240	3	50	50
Cd-$L_3M_{4,5}$ ($\alpha_{1,2}$)	1	127	10	-	
Cd-$L_3M_{4,5}$ ($\alpha_{1,2}$)	2	254	3	10	10
U-$M_5N_{6,7}$ ($\alpha_{1,2}$)	1	125	14	-	
U-$M_5N_{6,7}$ ($\alpha_{1,2}$)	2	250	6	5	100!

[a] For second order spectra, the α_1-data are shown; [b] values are rounded to full eV; [c] relative intensities shown in [11].

Table 3. Natural line widths for selected K- and L-lines found in the literature.

Element/line	Natural line width [eV]	Reference
K-KL_3 (α_1)	0.89	[10]
K-KL_2 (α_2)	0.89	[10]
Mn-KL_3 (α_1)	1.48	[10]
Mn-KL_2 (α_2)	1.50	[10]
[a]Ag-L_3M_5 (α_1)	2.34	[12]
[a]Ag-L_3M_4 (α_2)	2.20	[12]
U-M_5N_7 (α_1)	-	-
U-M_5N_6 (α_2)	3.40	[13]

[a] Instead of the Cd L-line widths, those of the neighbouring element silver are shown.

In the case of K and Cd we can conclude that the spectrometer resolution limits the measured FWHM values because the natural line widths are significantly smaller. It has to be said that in the case of the first order spectrum of potassium, the relatively intensive KL_2 (α_2)-line contributes to the line width of the unresolved doublet; therefore, it is expected to be somewhat broader than the single α_1-line. Because of the low relative intensity of Lα_2 and, in particular, Mα_2 the influence of these lines on the FWHM of the unresolved $\alpha_{1,2}$-doublets can be neglected.

In the case of U M-lines we have determined slightly broader FWHM values, which can mainly be explained by the generally broader line width of M-lines. Unfortunately, we were unable to find a reliable line width for the U M_5N_7 (α_1)-line in the literature. If one uses the energetic spread of the participating subshells as shown in [13], one can estimate the FWHM for U-Mα_1 to be also approximately 3.5 eV. For U M_5N_6 (α_2) a FWHM value of 3.4 eV is shown in the same reference.

Through the evaluation of the second order spectra we have estimated α_2/α_1 intensity ratios which are also shown in Table 2. However, according to the close energy proximity the determination of the α_2 net intensity can only be a rough approximation. Nevertheless, our values for potassium and cadmium correspond well with literature references. In contrast, a great discrepancy was observed in the case of uranium. We reported in a previous work about the possible causes of this [8]. A rough prediction of relative intensities of characteristic X-ray lines can be made applying the Burger-Dorgelo rule (Burger and Dorgelo, 1924). This approximation leads to α_2/α_1 intensity ratios of 50 % for ^{19}K, 11 % for ^{48}Cd, and 5 % for ^{92}U, which corresponds well with our experimental results. A short but instructive example of such a calculation was presented in Reed (p. 300) in his monograph [4].

4. Concluding remarks

The resolving power of a given wavelength-dispersive X-ray spectrometer exhibits a strong L-value dependency. If an X-ray line is detectable in first and second order reflection as well, one can reach a significantly improved energy resolution by choosing the second order. It was possible in this manner only to resolve the (K, L, M)$\alpha_{1,2}$-doublets of the elements potassium, cadmium, and uranium into their components and to estimate α_2/α_1 intensity ratios. Except for the M$\alpha_{1,2}$-radiation of uranium, our α_2/α_1-intensity values correspond well with the data shown in the literature.

A significant loss of intensity is connected with the transition to the second order reflection, which can, at least partially, be compensated by increasing the probe current. In addition, the background intensity is increased because the bremsstrahlung reflected in first as well as in second order reflection contributes to the measurement. Therefore, it is of great importance to optimize the discriminator settings in order to maintain sufficient peak-to-background ratios and, consequently, sufficient detection limits. By doing so, we were able to reach nearly the same detection limit for ^{19}K Kα_1-radiation in KBr in the second order spectra as in first order reflection. Consequently, it can be concluded that the strategy of second order spectrometry can be a helpful performance gain for the solution of tricky analytical tasks.

References

[1] Kirchhof J, Unger S, Dellith J, Scheffel A and Teichmann C 2012 Diffusion coefficients of boron in vitreous silica at high temperatures. *Optical Materials Express* **2** 534
[2] Müller R 1967 *Spektrochemische Analysen mit Röntgenfluoreszenz.* (Munich & Vienna: R. Oldenburg Verlag)
[3] Jenkins R 1977 *An introduction to X-ray spectrometry.* (Heyden & Son; ISBN 0-85501-035-5)
[4] Reed S J B 1993 *Electron probe microanalysis, 2nd edition.* (Cambridge: Cambridge University Press; ISBN 0-521-41956-5)
[5] Scott V D, Love G and Reed S J B 1995 *Quantitative electron probe microanalysis, 2nd edition.* (Ellis Horwood; ISBN 0-13-104050-2)
[6] Goldstein J *et al* 2003 *Scanning electron microscopy and X-ray microanalysis, 3rd edition.* (Springer Verlag; ISBN 0-306-47292-9)
[7] Dellith J and Wendt M 2007 The M emission spectrum of 68erbium. *Microsc. Microanal.* **13** 191-195
[8] Dellith J, Scheffel A, Terborg R and Wendt M 2011 Reinvestigation of the M emission spectrum of 92uranium. *Microsc. Microanal.* **17** 296-301
[9] Bearden J A 1967 X-ray wavelengths. *Rev. Mod. Phys.* **39** 78-124
[10] Krause M O *et al* 1979 Natural line widths of atomic K- and L-levels, Kα X-ray lines and several KLL Auger lines. *J. Phys. Chem. Ref. Data* **8** no. 2
[11] White E W and Johnson G G 1970 Emission and absorption wavelengths and two-theta tables. *ASTM Data Series DS37A.* (Philadelphia)
[12] Parratt L G 1938 The silver L series X-ray spectrum: line widths, wavelengths, relative intensities, satellites, and widths of energy levels. *Phys. Rev.* **54** 99-114
[13] Zschornack G 2007 *Handbook of X-ray data.* (Springer Verlag; ISBN 978-3-540-28618-9)

IOP Conf. Series: Materials Science and Engineering 55 (2014) 012004 doi:10.1088/1757-899X/55/1/012004

Specimen homogeneity requirements for EDS in high count rate mode

F Eggert, M Schleifer and F Reinauer

EDAX Inc., AMETEK Materials Analysis Division, 91 McKee Drive, US-07430 Mahwah, New Jersey, U.S.A.

E-mail: frank.eggert@ametek.de

Abstract. This work describes limits with pile-up correction considering EDS spectra evaluation in the case of processing high count rate spectra from inhomogeneous specimen areas. The pulse processor pile-up rejection has some limitations with detection of X-ray event coincidences. Therefore, remaining pile-up artefacts are always present in measured spectra and it is necessary to consider these artefacts to avoid mistakes in qualitative and quantitative results. Any pile-up calculation is based on probabilities for coincidences of X-rays, some of which cannot be resolved by pile-up rejection due to the constantly limited time resolution. A fundamental assumption with all calculation approaches is the time-related random distribution of emitted X-rays. This rule is violated with spectra from inhomogeneous specimen areas. Examples are shown and opportunities to avoid the fundamental shortcoming are discussed.

1. Introduction

One of the major benefits of 'silicon drift detectors' (SDD) in electron microscopes is their ability to process very high count rates during collection. Thus it is possible to collect spectra at each point while scanning the electron beam at high speed. As previously described, 'Smart PileUp Consideration' (SPUC) [1] can evaluate spectra collected at rates greater than 200 kcps, considering the pile-up distribution in the spectrum during all qualitative and quantitative evaluation processes. Therefore, the published automated qualitative analysis with EXpertID [2] which also incorporates SPUC can also deal with very high count rate spectra.

Often one adds single pixel spectra from selected specimen areas into one "sum" spectrum for evaluation. However, if the specimens are heterogeneous, the sum-spectrum from areas of different elemental composition is not the same as a spectrum with the same elements from a homogenous specimen. The well-known differences are due to variation in excitation and absorption of X-rays. The combination of high count rates and spectra summed from heterogeneous samples presents a substantial additional challenge for quantitative analysis. SPUC and other pile-up corrections assume the detected X-rays are emitted randomly within a given time interval and therefore follow probability estimations about X-ray detection and processing [1]. But this assumption is not the case for inhomogeneous specimen surfaces with sum-spectrum from area with different element composition.

Content from this work may be used under the terms of the Creative Commons Attribution 3.0 licence. Any further distribution of this work must maintain attribution to the author(s) and the title of the work, journal citation and DOI.
Published under licence by IOP Publishing Ltd

IOP Conf. Series: Materials Science and Engineering **55** (2014) 012004 doi:10.1088/1757-899X/55/1/012004

2. Problem

The basic theory is that the probability of an un-rejected pile-up pulse P^{pu}_{C12} (pile-up event in channel $C1+C2$, at sum of both energies) is calculated in the following way:

$$P_{C12}^{pu} = P_{C1} \times P_{C2} \times CR \times PPR \tag{1}$$

where P_{C1} is the probability of an event in energy channel $C1$ considering all X-ray counts being measured, P_{C2} is the same for events in channel $C2$, CR is count rate, and PPR is the average pulse pair resolving time (determined by the fast channel). Therefore, CR x PPR describes the probability that another measured X-ray quantum is detected before the fast channel is open to recognize additional events. Following [1] eq. (1) is possible to transform into (2):

$$N_{C12}^{pu} = N_{C1} \times N_{C2} \times CR \times PPR / TC \tag{2}$$

Actually this is a calculation of the contribution of the pile-up distribution to the measured spectra, based on the spectrum itself, where N_{C1} and N_{C2} are the counts actually measured in channel $C1$ and $C2$ of the spectrum and TC is total number of counts in the spectrum. But all these calculations are based on the probabilities of X-ray emission being randomly based. This assumption is not satisfied if the electron beam excitation involves areas of different specimen constituent (composition). Sometimes X-rays from two different elements will never be detected at the same time, because the elements are in one phase but not in the other. Then these sum energy peaks will never be present in the inhomogeneous specimen spectrum. More generally, the probabilities for emission are not constant, but depend on the concentration of the element in different phases. This is not in agreement with the basic approach (eq. 1). Moreover, it is not possible to cover the variable probabilities with a pile-up distribution calculation. This results in spectra evaluation issues, even if SPUC is applied.

3. Example with simulation of Al_2O_3 and SiO_2 spectra

Al_2O_3 and SiO_2 spectra were simulated [3] with a 50 kcps count rate assumed (figure 1). A simple addition of both spectra includes no Al+Si sum peak in the summed spectrum (figure 2), because both element characteristic lines will never be detected by EDS at same time. But the pile-up approximation based on this sum-spectrum does not fit. The fundamental reason is the calculated pile-up distribution assumes wrongly that the probability for X-ray emission is constant for each type of radiation.

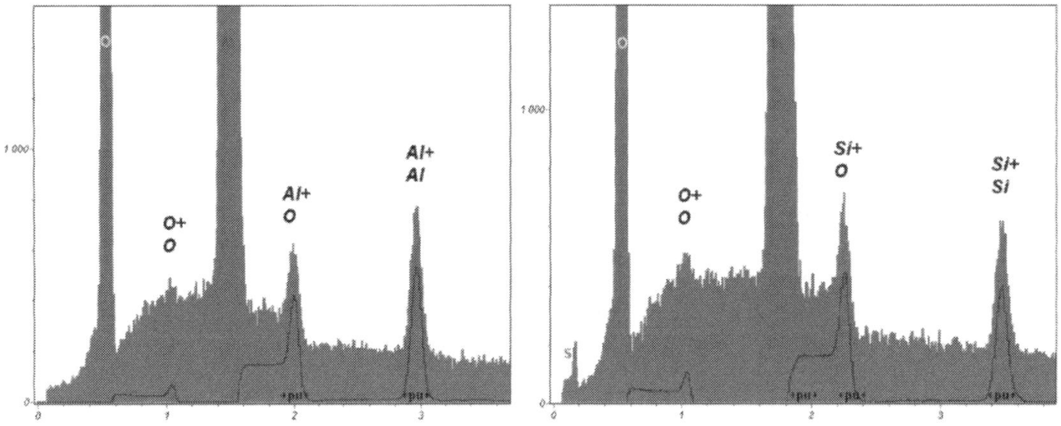

Figure 1. Simulated Al_2O_3 and SiO_2 spectra (50 kcps) displayed together with pile-up (blue).

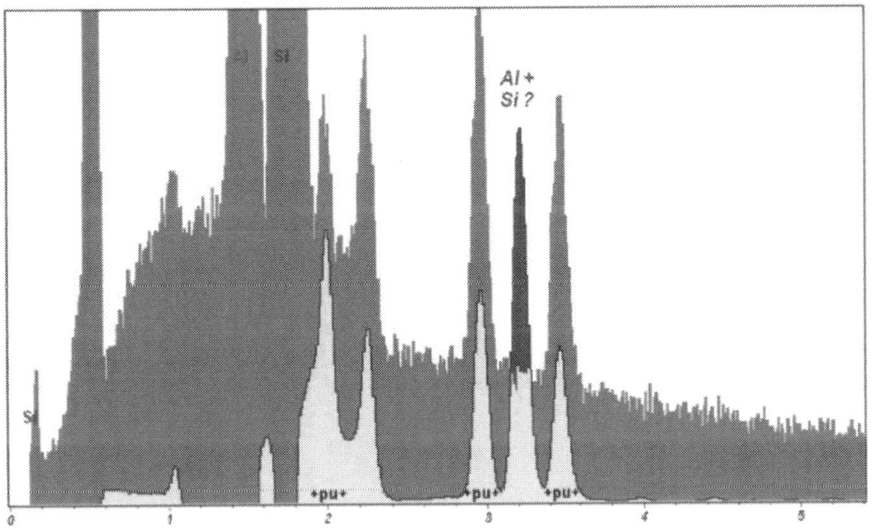

Figure 2. Sum-spectrum of the two spectra from figure1displayed together with calculated pile-up.

Therefore, any pile-up correction will fail to correctly evaluate this spectrum (as a model for adding spectra from areas in specimen of different composition). Al+Al, Si+Si and Si+O are underestimated. Moreover, the Al+Si sum is overcorrected with a common pile-up correction or could be displayed, if SPUC [1] is used, as an anomalous feature in reconstruction which is not really present in the spectrum (figure 3).

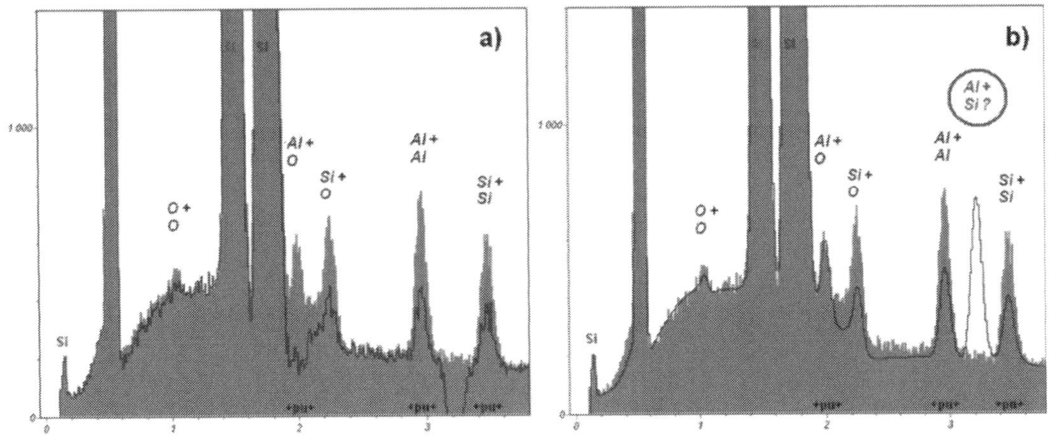

Figure 3. Original sum-spectrum compared: to pile-up corrected (a); to SPUC reconstruction [1] (b).

The pile-up was properly calculated in the low energy region of the spectrum up to about 1.5 keV, because both individual spectra are similar in this energy region. Thus all calculated pile-up fits the sum-spectrum up to the O+O sum peak energy, but not at higher energies where the sum events with Al and Si X-rays are incorrectly considered.

4. Example with real measurements of a Kiruna mineral stone

Another example is with real measured spectra from different phases of a Kiruna stone, see phase map image (figure 4) below. Each phase is distinguished by colour. The sum-spectrum examples were obtained from pure phase pixel spectra summation. The evaluation of these high count rate spectra (~ 150 kcps) uses a proper consideration of the pile-up effects (figure 6).

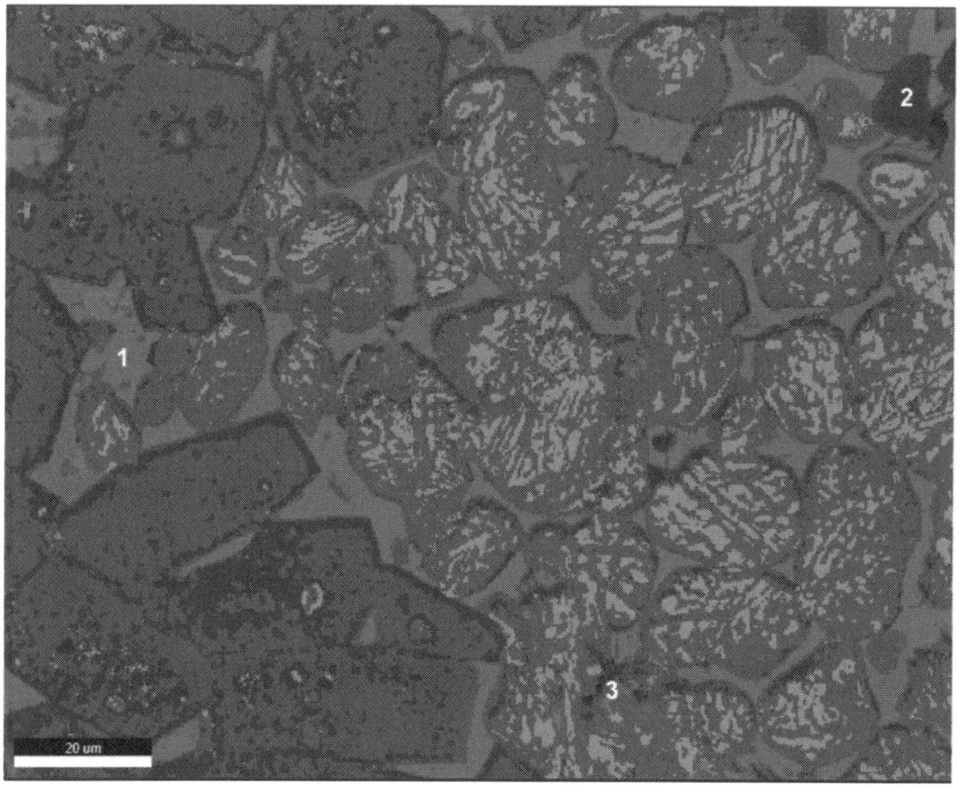

Figure 4. Phase map of Kiruna mineral with mean count rate about 150 kcps. Three different phase areas were selected as examples for sum-spectrum of homogeneous phases, marked with number 1, 2 and 3.

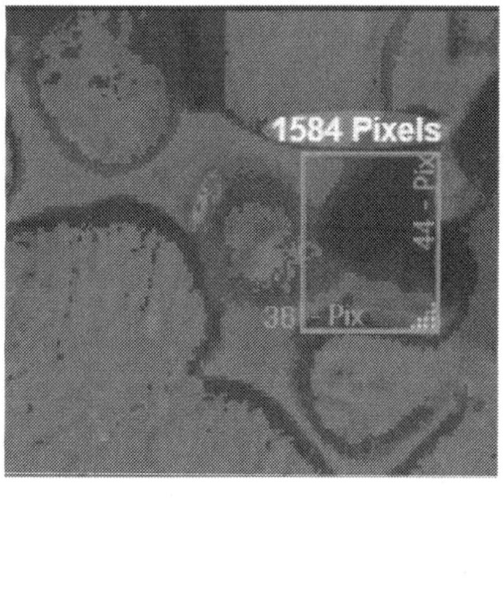

Figure 5. Enlarged image of a specimen area near mark 2 in figure 4. The inhomogeneous area outlined in green encloses the pixels from which the sum-spectrum shown in figure 7.

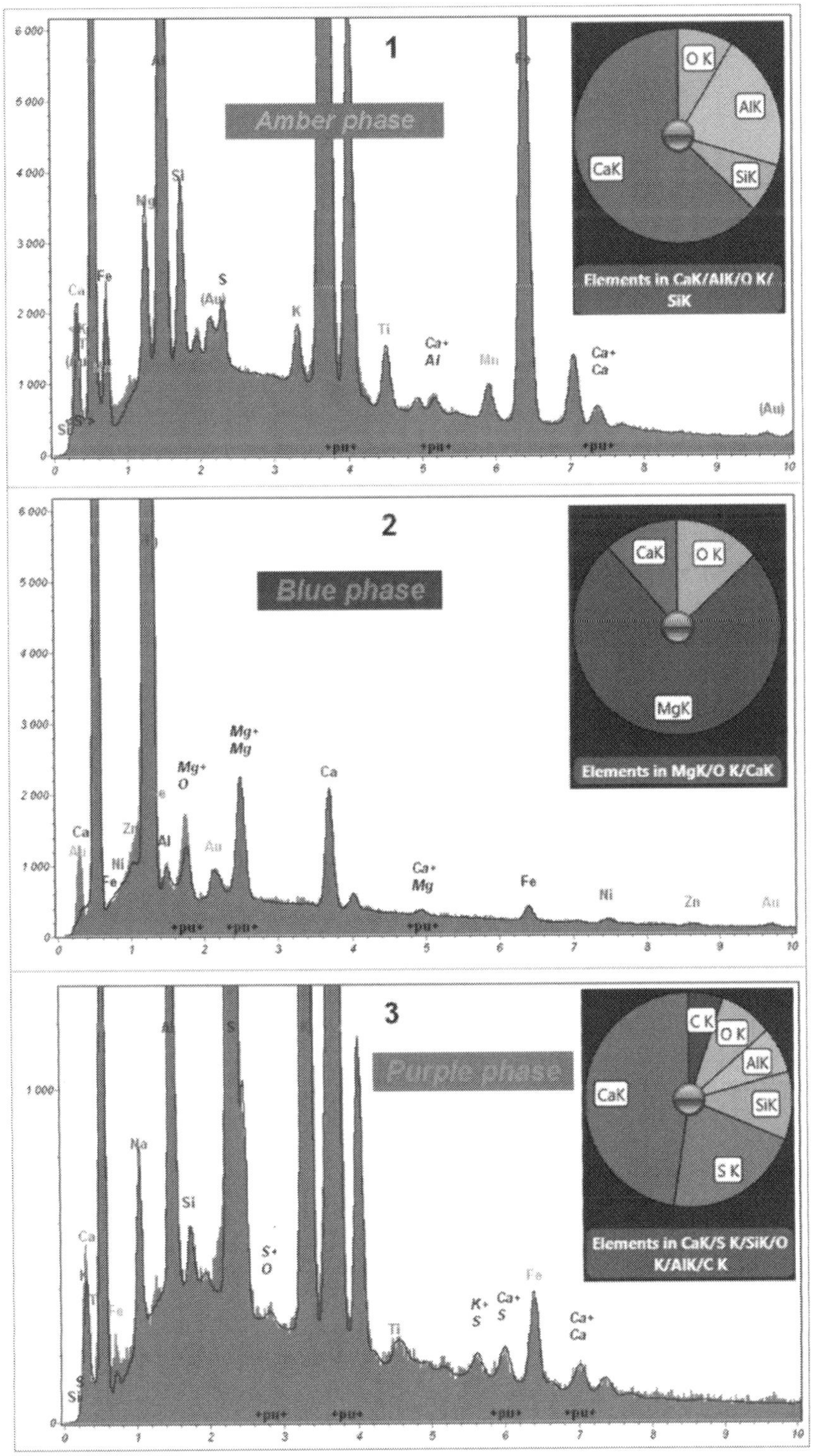

Figure 6. Examples of the evaluation of three spectra from mineral phases in Kiruna. The blue line reconstruction includes pile-up consideration. The most severe pile-up regions are labelled +pu+.

There is a different result when collecting all pixel spectra from a sample area which is not a homogeneous, such as that marked in figure 5. The pile-up consideration in the evaluation has uncertainties for element identification due to violations the pile-up calculation pre-assumptions (see spectra evaluation of this sum-spectrum in figure 7).

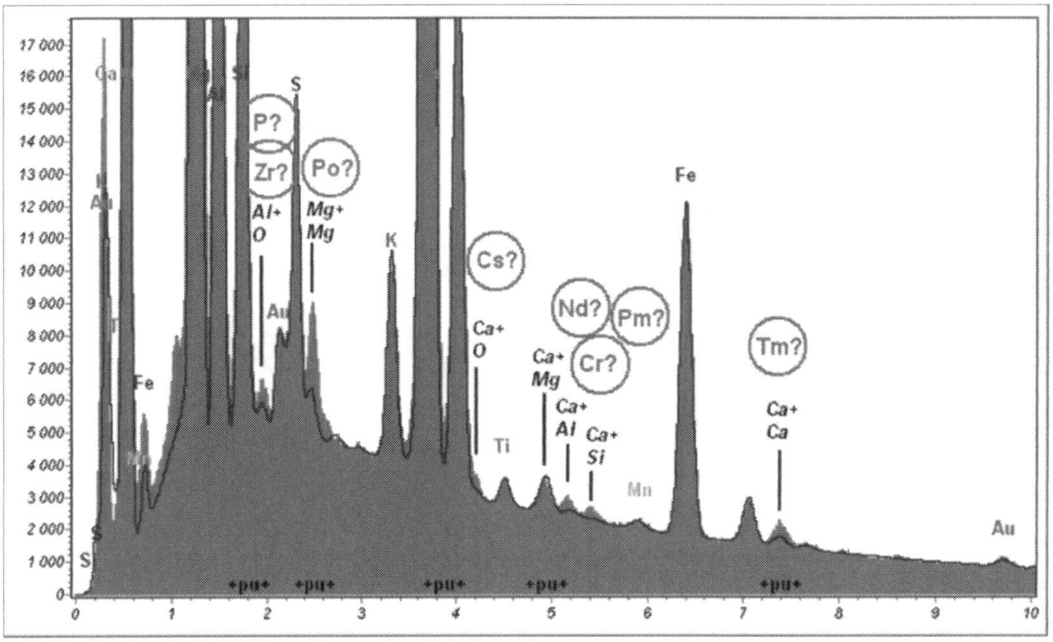

Figure 7. Evaluation of the Kiruna sum-spectrum from the area of inhomogeneous composition delineated in figure 5 using SPUC to consider the pile-up counts. There are remaining questions, also for element identification, due to violation of pile-up calculation pre-assumptions (X-ray emission not constant).

False positive element identification is one of the results. Also, incorrect consideration of the pile-up will have additional effects for quantification. This is further to the incorrect calculation of all excitation and absorption effects in the ZAF correction. The quantitative results will not only depend on element distribution and area fractions of single phases, but also on the count rate.

Although SPUC is not able to handle the inhomogeneous specimen case, this is much better than other commonly used pile-up corrections (see figure 8), where overcorrection makes the entire spectrum unusable for any further evaluation. This advantage is due to SPUC self adjustment, which finally in most cases produces some underestimation of the correction, rather than overcorrection.

Conclusions

EDAX's Smart Phase Mapping [4] or other methods, based on PCA (principle component analysis) [5], supports the operator by selecting regions of uniform composition to avoid acquiring a sum-spectrum from inhomogeneous area. However, even if the selected area is inhomogeneous, the SPUC [1] still addresses the issue reasonably and generally avoids over-correction, which occur with other pile-up correction solutions and helps to provide the user with the best possible results. But basically, the usual pile-up detector artefacts are not possible to take exactly into account for

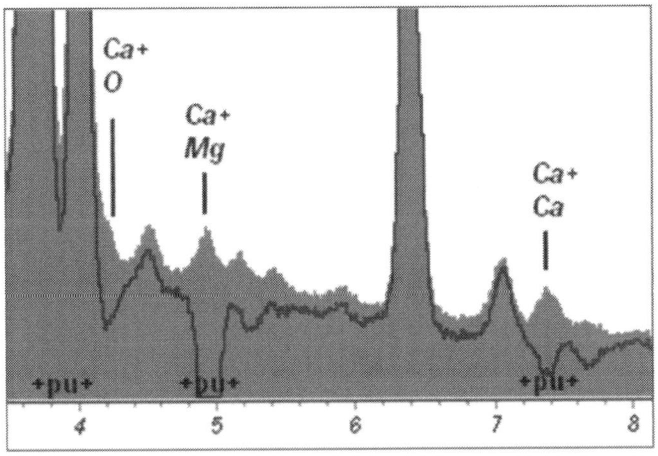

Figure 8. Evaluation of the Kiruna sum-spectrum from the area of inhomogeneous composition delineated in figure 5 using common pile-up correction. Overcorrection occurs at several points due to violation of pile-up calculation pre-assumptions (X-ray emission not constant).

qualitative and quantitative analysis with a spectrum from inhomogeneous specimen area. Therefore this should be avoided.

Conclusions

EDAX's Smart Phase Mapping [4] or other methods, based on PCA (principle component analysis) [5], supports the operator by selecting regions of uniform composition to avoid acquiring a sum-spectrum from inhomogeneous area. However, even if the selected area is inhomogeneous, the SPUC [1] still addresses the issue reasonably and generally avoids over-correction, which occur with other pile-up correction solutions and helps to provide the user with the best possible results. But basically, the usual pile-up detector artefacts are not possible to take exactly into account for qualitative and quantitative analysis with a spectrum from inhomogeneous specimen area. Therefore this should be avoided.

References

[1] Eggert F, Elam T, Anderhalt R and Nicolosi J 2012 *IOP Conf. Ser.: Mater. Sci. Engng.* **32** 012008
[2] Eggert F 2010 *IOP Conf. Ser.: Mater. Sci. Engng.* **7** 012007
[3] Eggert F 2006 *Microchim. Acta* **155** 129
[4] Nowell M M and Wright S I 2004 *J. Microsc.* **213** 296
[5] Kotula P G, Keenan M R and Michael J R 2003 *Microsc. Microanal.* **9** 1

IOP Conf. Series: Materials Science and Engineering **55** (2014) 012005 doi:10.1088/1757-899X/55/1/012005

Understanding the structure of nanocatalysts with high resolution scanning/transmission electron microscopy

L D Francis[1], J Rivas[1] and M José-Yacamán[2]

1 International Iberian Nanotechnology Laboratory, Avenida Mestre Jose Veiga, PT-4715-330, Braga, Portugal
2 University of Texas at San Antonio, Department of Physics and Astronomy, One UTSA Circle, US-78249 San Antonio, Texas, U.S.A

E-mail: leonard.francis@inl.int

Abstract. Nanomaterials including nanoparticles, nanowires and nanotubes play an important role in heterogeneous catalysis. Thanks to the rapid improvement of the electron microscopic techniques and with the advent of aberration corrected electron microscopy as well as theoretical methodologies, the potential effects induced by nanocatalysts are better understood than before by unravelling their atomic structure. A brief introduction to advanced electron microscopic techniques namely aberration corrected scanning transmission electron microscopy (Cs-STEM) is presented and subsequently two examples of nanocatalysts are considered in the present review. The first example will focus on the study of bimetallic/core-shell nanoalloys. In heterogeneous catalysis, catalysts containing two or more metals might show significantly different catalytic properties compared to the parent metals and thus are widely utilized in several catalytic reactions. Atom-by-atom insights of the nanoalloy based catalysts ex: Au-Pd will be described in the present review using a combination of advanced electron microscopic and spectroscopic techniques. A related example on the understanding of bimetallic clusters by HAADF-STEM will also be presented in addition to nanoparticles. In the second case understanding the structure of transition metal chalcogenide based nanocatalysts by HRTEM and aberration corrected STEM, for the case of MoS_2 will be discussed. MoS_2-based catalysts serve as model catalysts and are employed in the hydrodesulphurisations (HDS) reactions in the removal of sulphur from gasoline and related petrochemical products. They have been studied in various forms including nanowires, nanotubes and nanoplates. Their structure, atomic insights and as a consequence elucidation of their corresponding catalytic activity are thus important.

1. Introduction

Electron microscopic techniques present a very powerful approach to the systematic elucidation of structure of nanocatalysts and thereby enable correlation of heir corresponding catalytic activity. In the case scanning/transmission electron microscopy (S/TEM) one employs different imaging modes to obtain different and yet complementary information of nanostructured catalysts. The common imaging modes include bright-field (BF), dark-field (DF) and phase contrast imaging [1]. In comparison to all these imaging modes the high-angle annular dark field (HAADF) imaging or Z-contrast imaging provides a significant advantage since in this case it is possible to differentiate alloys of two different materials from the difference related to their atomic number (intensity $I \alpha$ atomic number $Z^{1.46}$). In the case of aberration corrected electron microscopes HAADF-STEM

Content from this work may be used under the terms of the Creative Commons Attribution 3.0 licence. Any further distribution of this work must maintain attribution to the author(s) and the title of the work, journal citation and DOI.
Published under licence by IOP Publishing Ltd

imaging technique offers extreme resolution limits and thus single atom imaging based on the Z-contrast is now routinely achieved. In combination with spectroscopic techniques such as energy-dispersive X-ray analysis (EDS) and electron energy loss spectroscopy (EELS) one can obtain single atom information combined with their corresponding chemical information [2]. Figure 1 shows a Z-contrast STEM simulation of single atoms of different elements, arranged in a 4x4 matrix. The line scan through the centre of the atomic positions is shown at the right of the simulated image. The trend follows approximately as $I \alpha Z^{1.46}$ relation as shown by Pennycook [2].

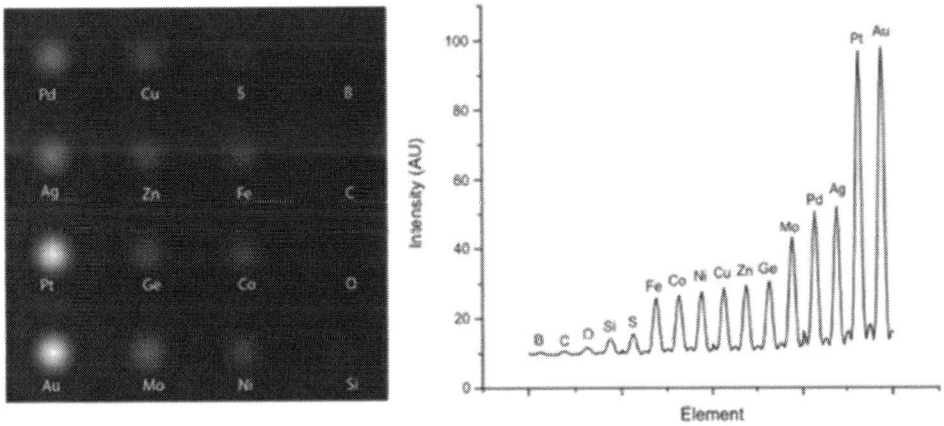

Figure 1. Profiles of intensity obtained from a series of HAADF-STEM simulated images (at a defocus of -41 nm). Note the intensity variation with the different elements. In each case the columns of elements have the same number of atoms.

In the field of catalysis, nanoalloys of transition metals have generated tremendous interest since they display structures and properties which are quite distinct and different from those of the pure elemental clusters: the structures of binary clusters may be quite different from the structures of the corresponding pure clusters of the same size. When two metals combine to form a nanoparticle there are two situations that can be produced: (i) either the particles form strictly what can be thought of as an alloy, or (ii) a core-shell structure is produced. In the former case, the two atomic species are distributed homogeneously on the particle; in the latter, a core of one of the metals is surrounded by a thin shell of the other metal forming a complex shape. Usually, the metal with larger size and lower surface energy will be on the outside layer, but the opposite may happen depending on the synthesis conditions. The core-shell structure is believed to be a very important factor affecting the electronic properties of the cluster. In reality, bimetallic particles are more complex. A nanoparticle might have an incomplete core or it may have a eutectic type of structure, where the two metals form separated particles joined in an interface. Understanding such an atomic structure is also not trivial [3]. Nanoscale transition-metal (TM) nanoparticles exhibit enormous complexity, with their structure and properties intimately connected and, at times, varying dramatically. Alloyed nanoparticles have been shown to improve catalytic, magnetic, optical, and biomedical properties needed in a wide range of applications. For example oxygen reduction reaction (ORR) electrocatalysts, consisting of Pt and another late transition metal (M, where M = Ru, Rh, Pd, Au), have a very high activity compared to that of pure Pt ORR catalysts [4]. Real-world bimetallic catalysts, usually composed of nanoscale bimetallic transition metal clusters well-dispersed on non-catalytic supports, lies at the heart of numerous important industrial applications. For example, late transition metal alloy catalysts are the major working horses in the petroleum industry for the production of lead-free gasoline with high octane numbers and are indispensable ingredients in three-way catalysts used in combustion cars for substantially cutting down the environmentally-detrimental emission of NO_x and CO [4]. In general, bimetallic systems offer many more possibilities than their single-metallic counterparts in tailoring surface physical and chemical properties, and therefore, represent a very promising field in which a rational design of novel and cost-efficient bimetallic catalysts and structure elucidation need to be achieved – a demanding challenge of the discipline in meeting our society's ever increasing demands for more environmentally benign and energy efficient technologies. The lack of detailed structural information at the atomic scale in real-world bimetallic catalysts under working conditions remains one of the major hurdles for an ultimate rational design of bimetallic catalysts for industrial applications [5].

In the present review a combination of aberration corrected STEM along with HRTEM imaging is presented to understand the nature and structure of nanocatalysts. Thus in the first example of nanocatalysts, bimetallic nanoparticles obtained by chemical routes are presented. Chemical routes can enable the production of metal nanoparticles with a well-controlled size and shape distribution. Bimetallic nanoparticles of Au-Pd find important applications in catalysis. Their catalytic performance is directly related to the structure, alloy formation and variation of composition in the structure. Although the standard idea is that bimetallic nanoparticles can be either an alloy or a core shell structure, in the present example a more complex structure has been shown. Thus the structure and composition of complex three-layered Pd-Au nanoparticles by using aberration corrected high angle annular dark field scanning transmission electron microscopy (HAADF-STEM) has been investigated. The investigations reveal that the nanoparticles are composed of an evenly alloyed inner core, an Au-rich intermediate layer, and a Pd-rich outer shell. The structure is more complicated than what simple models can predict. Additional studies of this system were also carried out to identify and understand the interface structure better. This was done by carrying out spectral and chemical analysis (STEM-EDS, STEM-EELS). Apart from the three-layered core-shell structures it has also been possible to observe in some cases a four-layered core-shell structure as well. The entire core-shell structure is not rigid and there is indeed intercalation of Au-Pd into the other layers as well. In addition stacking faults present in the nanoparticles have been located. The problem of the interface structure between the layers is also investigated and the results are presented [6, 7].

In the second example transition metal chalcogenide nanocatalysts are presented. Molybdenum disulphide/tungsten disulphide (MoS_2/WS_2) is a compound very useful for its properties; it is used as lubricant, catalyst in hydrodesulphuration, in hydrogen fuel storage, etc. [5]. Transition metal chalcogenides like MoS_2 or WS_2 are quasi two dimensional (2D) compounds. Atoms within a layer are bound by strong covalent forces, while individual layers are held together by van der Waals (vdW) interactions. Although there are a large number of polytypes observed in the case of the transition metal chalcogenides the most notable experimentally observed polytypes in the case of MoS_2 and WS_2 based on the stacking sequence of the layers are either a hexagonal polymorph with two layers in the unit cell – 2H (*P63/mmc*) symmetry, rhombohedral with three molecular layers in the unit cell layers – 3R (*R3m*) symmetry and the less common trigonal with one layer – 1T. MoS_2 nanoplates that find important applications in catalysis have been investigated by using advanced electron microscopic techniques namely aberration corrected STEM and HRTEM [8].

For the electron microscopy analysis, the samples were dispersed in ethanol and a drop of this suspension was deposited onto a holey carbon grid. The samples were characterized using the aberration (Cs) corrected JEOL JEM-ARM200F (ARM) 200 kV FEG-STEM/TEM, equipped with a CEOS Cs corrector on the illumination system. The probe correction was performed through a dodecapole corrector (CEOS GmbH) aligned through the CESCOR software, to finally obtain a twelve-fold Ronchigram with a flat area of 50 mrad. The pixel spacing was calibrated using Si [110] lattice images in the HAADF mode, and confirmed by using gold standard particles. Images were commonly recorded for 10 to 16 s. The HAADF as well as the BF images were acquired using a digiscan camera. In order to reduce the noise of the images and to obtain clearer images the raw data was filtered using the 2D Wiener filter and the Richardson-Lucy / maximum entropy algorithm implemented by Ishizuka. The EDS analysis was performed using EDAX instrumentation attached to the JEOL-ARM microscope. Spectra, line scans as well as chemical maps for the various elements were obtained using the EDAX Genesis software. One- and two-dimensional electron energy loss spectroscopy (EELS) analysis was performed with a GIF-Tridiem spectrometer.

2. Bimetallic nanoparticles

2.1. Nanoparticles obtained by chemical methods – The case of the three-layered Pd-Au-Pd nanoparticles

HAADF and BF images for nanoparticles with composition Au1Pd5 are shown in figures 2a and b, respectively. The three- layer structure observed in these nanoparticles was found to be dependent on the diameter. It was observed that 25 % of the nanoparticles had a diameter of 5.5 ± 1.3 nm, and only displayed two clear layers. The remaining 75 % of the nanoparticles showed a size of 12.7 ± 0.8 nm, with three layers being evident.

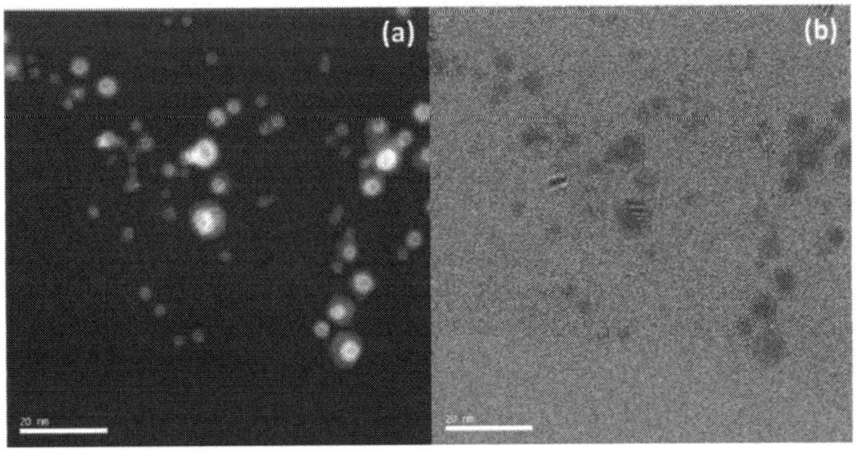

Figure 2. (a) Low magnification HAADF, and (b) BF-STEM images of Au-Pd nanoparticles obtained by successive reduction of monometallic ions.

A closer look at the nanoparticles is provided by the ultra-high resolution HAADF images of the nanoparticles in figure 3. From the image (figure 3a) it can be seen that formation of the three-layers in the nanoparticle is clearly evident. The diameter of the nanoparticles is ~ 8 nm. This consists of a Pd core (A ~ 2 nm) – Au-rich first shell (B ~ 3 nm) – Pd-rich outer shell (C ~ 3 nm) as can be seen from the contrasts in the HAADF images in figure 3a. The shape of this particle corresponds to a truncated octahedron, which has (111) and (100) facets. Figure 3b shows a three layer particle with the (100) orientation in which one can see roughness on the surface, and distortion in the core of the nanoparticle. In some nanoparticles it may also be seen that the Au atoms are indeed diffused into the Pd-rich outermost shell as can be seen from the bright atom contrast in the Pd-layer. Figures 3c and d show the case of the atom by atom intercalation of Au in the Pd-layer. This shows how gold can be present in the Pd-layers and is not entirely rigid. So indeed the layers are alloys of different concentrations [9].

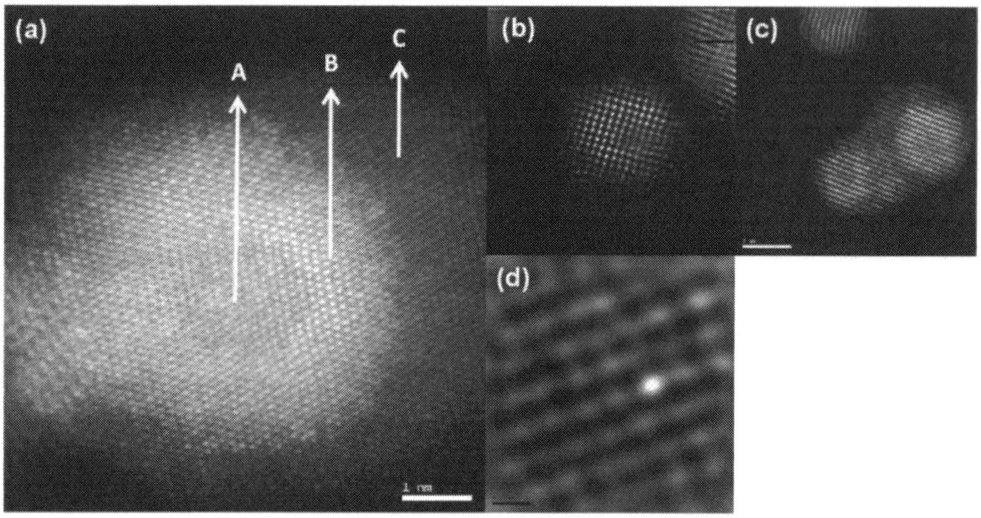

Figure 3. Aberration-corrected STEM images of the three layer Pd-Au nanoparticles: (a) Atomic-resolution HAADF-STEM image of a truncated cuboctahedral Au/Pd nanoparticle. The contrast of the three distinct regions (marked A-C) can be clearly seen. (b) Nanoparticles with surface roughness and distortions at the core. (c, d) Au intercalation into the Pd-layer. (Reprinted with permission from Ref. 6, Copyright (2011) Elsevier).

EDS line scan in the case of the three-layer Pd-Au-Pd nanoparticles reveal the following interesting features as can be seen in figure 4. The Pd L- and the Au L- and M-signals can be clearly traced across the region of the individual Pd- and Au-layers with the maximum intensity of the signals varying between the respective layers (figure 4b). In addition we have also carried out EDS elemental mapping for the respective elements. Figure 5 shows the chemical maps of the Au-Pd nanoparticles. It can be seen from the figure that the Pd (K, L) and Au (L, M) maps clearly reveal the presence of the individual elemental contrasts in the three-layer core-shell structure of the nanoparticles. With the information obtained from the STEM-EDS line scans and elemental maps of the individual nanoparticles the presence of the core-Pd – first shell-Au – outer shell-Pd is clearly confirmed without any ambiguity.

The distribution of the chemical elements of the three-layer nanoparticle (figure 6a) was investigated by EELS, and the result for the EELS line profile of Pd-$M_{4,5}$ is shown in figure 6b. For purposes of comparison, the figure also shows the high-angle annular dark field (HAADF) intensity line profile of the same particle (figure 6d). From the EELS profile (figure 6c) it can be noted that, the distribution of Pd appear to be in accordance with the presence of Pd in the core and in the outermost shell as is the case for a tri-layer particle. However the EELS Au-$M_{4,5}$ signal at 2206 eV turns out to be weak, as expected for a heavy element, but high enough to show the presence of Au in the particle. From the comparison between the Pd EELS profile and the HAADF profile, it can be inferred that there exists a region close to the centre of the particle as well as in the outermost shell which is rich in Pd. The corresponding EELS map of Pd reveals clearly the presence of Pd in the core and in the outermost shell of the tri-layered nanoparticles as is shown in figure 7.

It is worthy to mention that in case of these nanoparticles occasionally four-layered structure is seen. Shown in figure 8 is the case of the four layer nanoparticles wherein it can be seen that near the surface there is an extra layer of atoms of gold and this suggests a 4 layered structure. This is a different case from the previously mentioned three-layered nanoparticles and has not been reported till date.

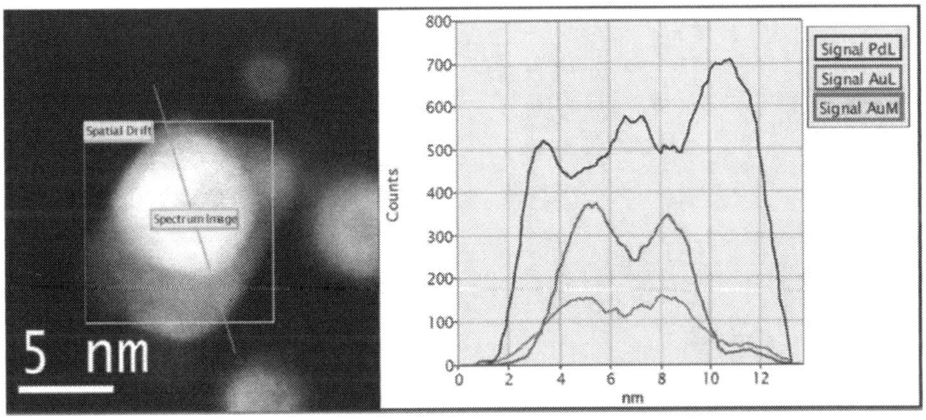

Figure 4. (a) and (b) Characterisation of the Au and Pd elemental distribution across the nanoparticle by STEM-EDAX line-scanning technique across the individual three-layer nanoparticle. The Pd L- and the Au L- and M-line scan signals can be clearly seen varying in intensity along the different regions of the nanoparticles. (a) Shows the analysed area and the direction of analysis. (Reprinted with permission from Ref. 6, Copyright (2011) Elsevier).

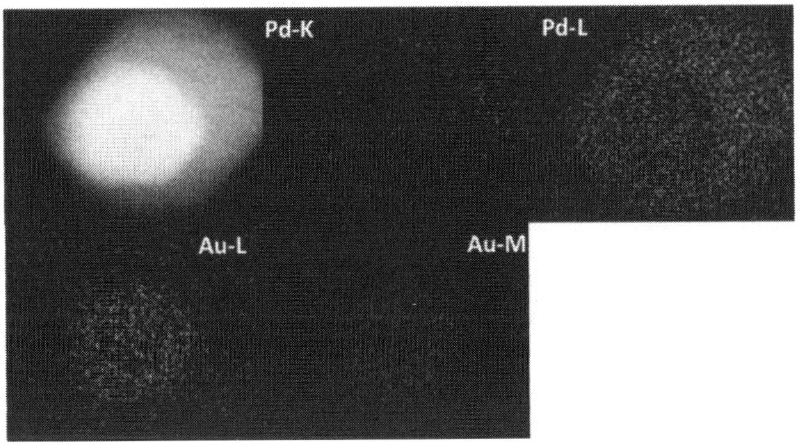

Figure 5. Elemental maps of the nanoparticles – Pd (K, L) and Au (L, M) maps.

A particle in its <110> orientation is shown in figures 9a and b in the HAADF and BF-STEM modes respectively. Profiles of the first and second layer of the particle are shown in figures 9c and d respectively. The profiles show a very rough surface; this can only be explained if we consider two factors. One, atoms of gold migrated to the surface causing an increase in signal in those particular columns. Two, there are incomplete columns of atoms (e.g., steps) on the surface.

The BF-STEM images of the three layer nanoparticle (figures 10a and b) along with the HAADF image (figure 10c) are displayed (Au_1Pd_3). The incident electron beam direction is along [110]. The side surfaces of the particles can be indexed as {111} and {100} planes, as labelled in figure 10a and d. The centre of the nanoparticles is round but the exterior surface is in the shape of a square. The real shape of these particles can be associated with the truncated-octahedron. This is extremely interesting since the individual metals themselves will tend to form icosahedral or decahedral shapes. On the other hand we have not observed these shapes in bimetallic Au/Pd [22]. This is a very

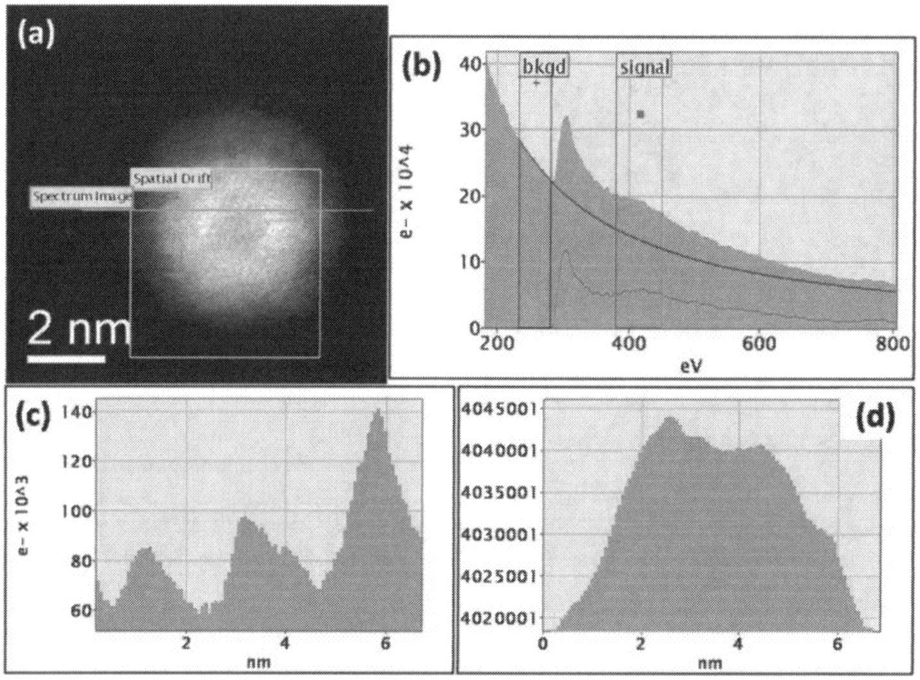

Figure 6. The distribution of the chemical elements of the three-layer nanoparticles (a) investigated by EELS, and the result for the EELS line profile of Pd-$M_{4,5}$ is shown (b). Linear Pd EELS profile performed on the particles (c), and its corresponding HAADF intensity profile (d). (Reprinted with permission from Ref. 6, Copyright (2011) Elsevier).

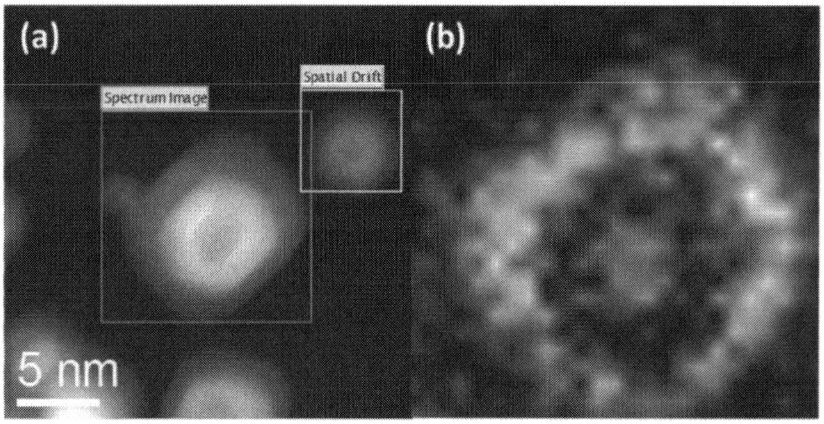

Figure 7. EELS map of Pd reveals clearly the presence of Pd in the core and in the outermost shell of the tri-layered nanoparticles.

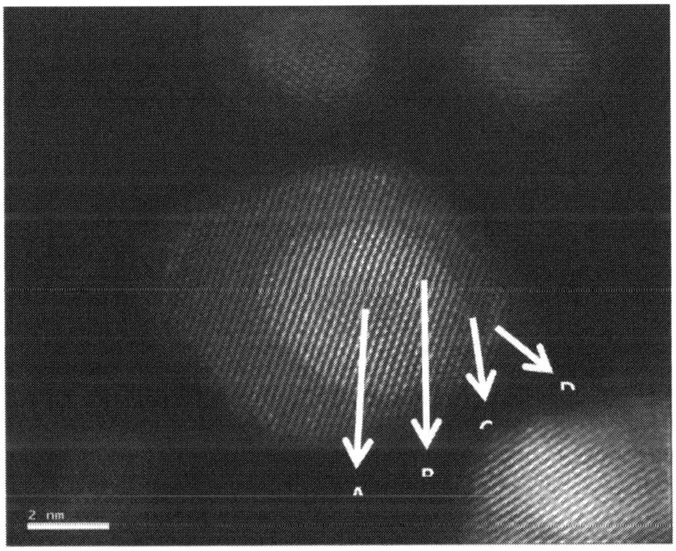

Figure 8. Aberration-corrected STEM images of the four layer Au/Pd nanoparticles. The contrast of the four distinct regions (marked A, B, C and D) can be clearly seen.

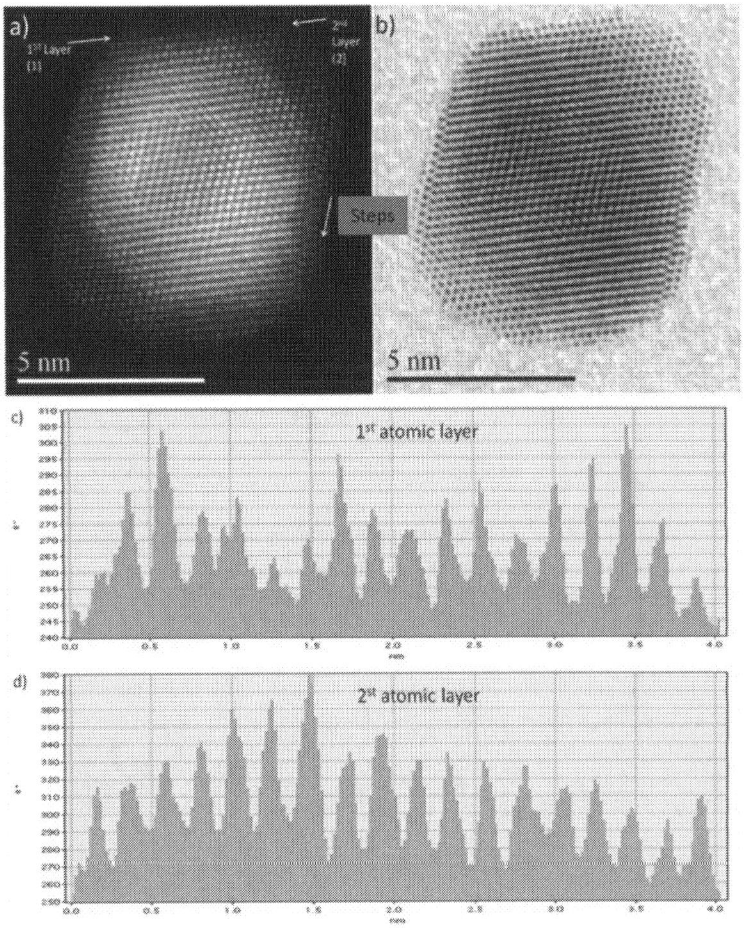

Figure 9. (a) and (b) HAADF and BF-STEM images of Au_1Pd_4 nanoparticle. c) and d) shows the intensity profile of the first two atomic layers of the nanoparticle. (Reprinted with permission from Ref. 6, Copyright (2011) Elsevier).

IOP Conf. Series: Materials Science and Engineering **55** (2014) 012005 doi:10.1088/1757-899X/55/1/012005

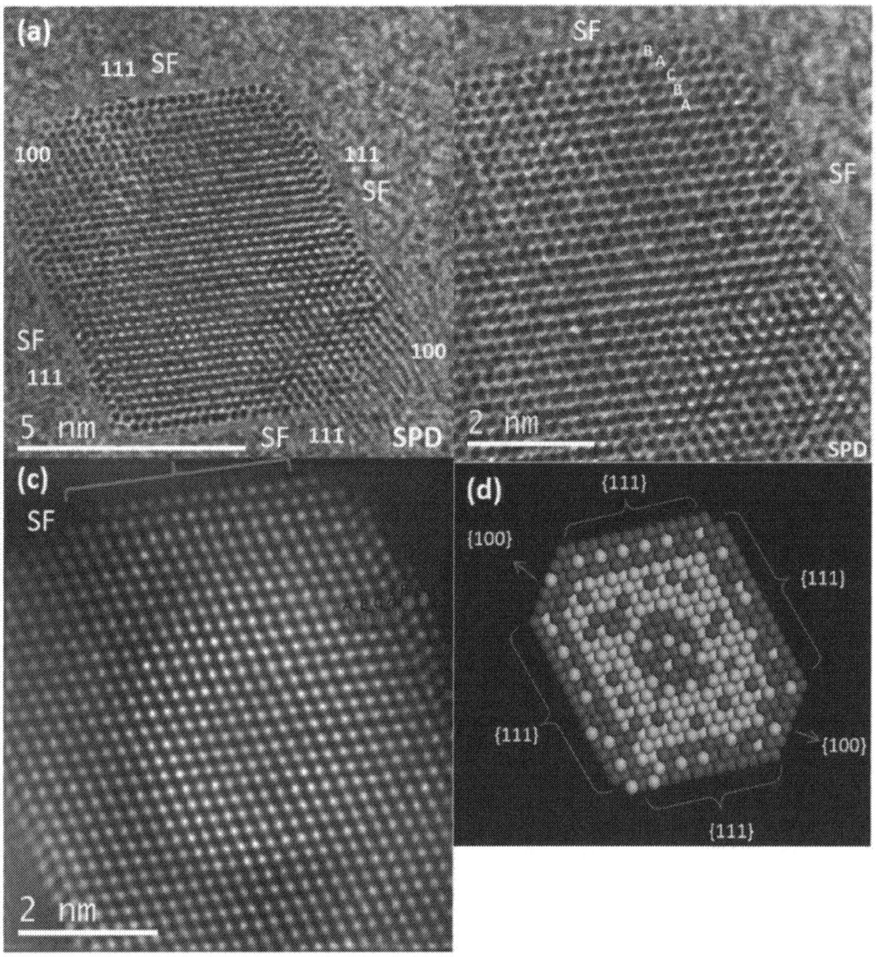

Figure 10. The BF-STEM images of the three layer nanoparticle (a and b) along with the HAADF image (c) are displayed. A higher magnification of the BF-STEM image of (a) is shown in (b) and the HAADF image shown in (c) clearly gives indication of the variation in the stacking sequence leading to the stacking faults which is in combination with SPDs. (d) Atomistic model of Au/Pd modified cuboctahedron with the three layers and showing the {100} and {111} facets: a Pd-rich inner-core, Au-rich intermediate-shell, and a Pd-rich exterior shell. (Reprinted with permission from Ref. 6, Copyright (2011) Elsevier).

significant fact that shows how the structure can be changed with the chemical composition. The HAADF intensity is taken into account considering the variation in intensities that change significantly when one moves from the Pd-rich layer to the Au-rich layer. In this case the profile of the layers show the same thickness within a particular layer attributing uniformity in the thicknesses of the layers and variation in intensity arising only as a consequence of the difference in their atomic numbers between Pd and Au.

The nanoparticles have defects at the {111} surface in all the directions (figure 10 a). In order to see the defects clearly, a higher magnification of the BF-STEM image of figure 10a is shown in figure 10b and along with the HAADF image shown in figure 10c clearly gives indication of the variation in the stacking sequence leading to the stacking faults. Although the standard stacking sequence is ABCABC... in face-centred cubic (FCC) structures, in the regions marked in the images (figures 10b and c) this sequence is not followed. Instead, the sequence becomes ABCAB... (FCC-HCP structure), as can be seen in the figure 10b and ABCAB... in figure 10c. This corresponds to stacking faults. In addition the presence of Shockley partial dislocations (SPDs) can be observed (figure 10a). The formation of stacking faults in Au-Pd core-shell structures has been shown to be linked to SPDs [6]. A partial dislocation is accompanied by the presence of a stacking fault. It may be noted here that the large strain involved in the formation of the three-layer structures may be alleviated to some extent by the formation of such stacking faults and related defects. The stacking faults are also evident in the case of the Au_1Pd_4 tri-layer particles. Figures 11a and b are the HAADF STEM and the BF-STEM images which shows a variation in the stacking sequence from the previously mentioned case. Here the stacking sequence of ABCBA is produced by twinning and the layer C corresponds to the twinning mirror plane [6].

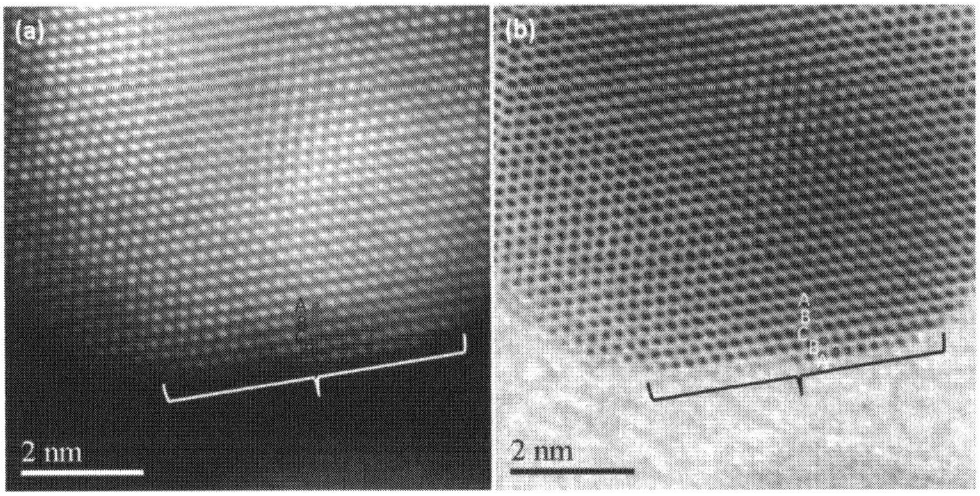

Figure 11. HAADF and BF-STEM images of Au_1Pd_4 nanoparticle showing the presence of twinning. (Reprinted with permission from Ref. 6, Copyright (2011) Elsevier).

2.2. Clusters
Metal atomic clusters consist of groups of atoms with well-defined compositions and one or very few stable geometric structures. They represent the most elemental building blocks in nature – after atoms – and are characterized by their size, comparable to the Fermi wavelength of an electron, which makes them a bridge between atoms and nanoparticles or bulk metals, with properties very different from both of them.

2.3. Catalysis
Small clusters have been found to own a catalytic activity not observed in their bulk analogues or even nanoparticles, which makes them very attractive as new catalytic materials. Quantum chemical calculations indicate that such high reactivity is due to under coordination of the metal atoms forming

the cluster. Several families of metal clusters have been found to possess surprisingly high and selective catalytic activity when immobilized on a support. For example, platinum clusters with 8 – 10 atoms can be used as catalysts for the oxidative dehydrogenation of propane, while gold clusters with 6 - 10 atoms have been shown to be highly active for catalysing propene epoxidation. The control and tunability of the catalytic oxidation of CO by Au_{20} clusters deposited on MgO surfaces has also been studied. Thus it has been found that the active site on the cluster is characterized by enhanced electron density, which activates the adsorbed O_2 molecule and promotes the bonding of CO. Small metal clusters have been also found to display electrocatalytic activities different from the material in bulk or as nanoparticles. Although this is a field not too much explored so far, these electrocatalytic properties make metal clusters to be promising materials in fuel-cell applications [10, 11].

It is indeed interesting to note that the ultimate resolution of observing individual atoms/small clusters have been achieved using probe aberration corrected STEM. Shown in figure 12, are the very small bimetallic clusters of Pd-Ir (inside the circle). The size of the cluster is just about ~ 2 nm. It is interesting to note that based on the Z-contrast, the two different atoms can be distinguished very clearly (Pd, Z = 46, Ir = 77). In addition in the vicinity of these small clusters it is also possible to locate individual atoms of Ir/Pd (outside the circle) [7].

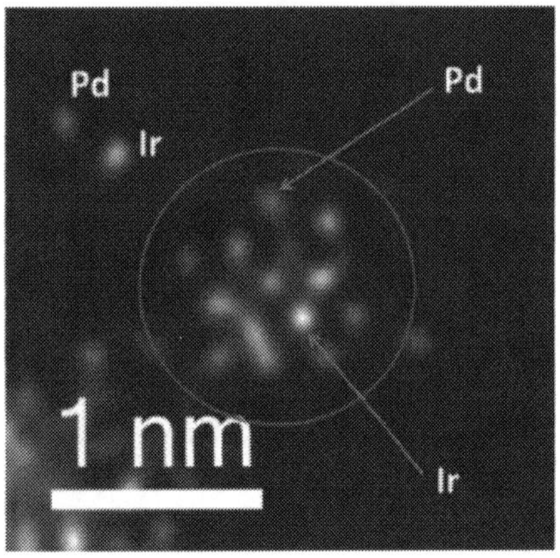

Figure 12. Clusters of Pd-Ir are shown (inside the circle). It is possible to distinguish each element based on the Z-contrast (Pd = 46, Ir = 77). Notice also the presence of individual atoms (outside and close to the circle).

3. MoS_2 nanoplates

Figures 13a and b show the SEM micrograph of the MoS_2-plates synthesized at 700 °C and annealed at 900 °C and 1000 °C, respectively starting from MoO_3 and S powder. Both the images show the presence of hexagonal plates with nanometric size in the sample. The size of the plates is approximately 30 - 35 nm of diameter and the plates are hexagonal in shape. In the figures the hexagonal plates are stacked in agglomerates consisting of the plates, which can be distinguished individually. These agglomerates can be broken down by ultrasonication.

Figure 14a shows the BF-STEM image of a MoS_2-plate, revealing in detail the stacks of Mo; with this image it was possible to measure the Mo–Mo distances in the a_1-, a_2- and a_3-directions. Distances of 2.76 Å in the a_1-direction, 2.8 Å in the a_2-direction and 2.87 Å in the a_3-direction are obtained; these values correspond to Mo–Mo interatomic distances. A line profile is shown in the inset of figure 14a. Thus the hexagonal atomic structure of MoS_2 is observed with a resolution sufficient enough to resolve the Mo–Mo interatomic distance. Figure 14b shows the diffraction pattern wherein, the a_1-, a_2- and a_3-directions are indicated in the figure. Thus it is possible to clearly see the crystal array of MoS_2, corresponding to a hexagonal crystal system.

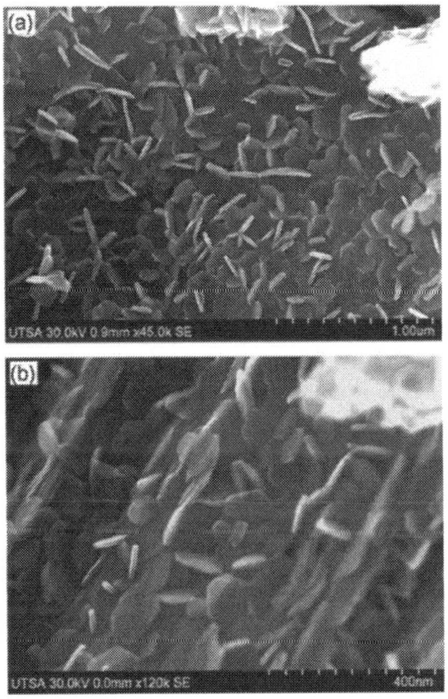

Figure 13. SEM micrographs of the MoS_2-plates synthesized and annealed at: (a) 700 °C - 900 °C; (b) 700 °C - 1000 °C. (Reproduced by permission of The Royal Society of Chemistry).

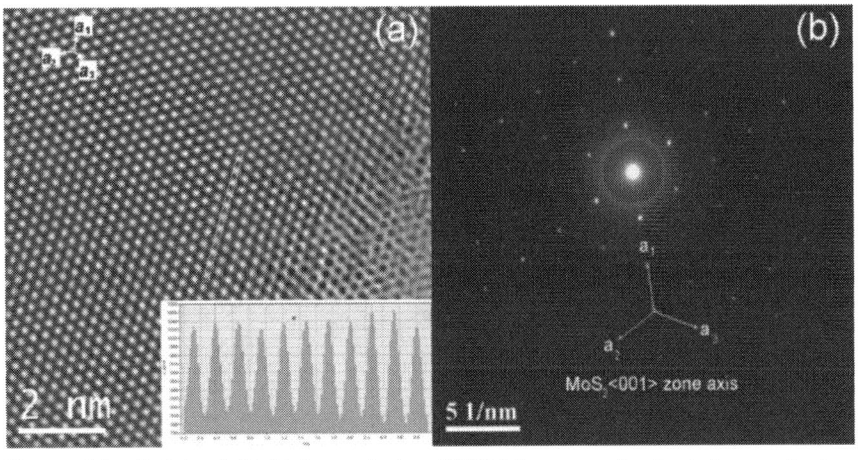

Figure 14. Bright field high resolution STEM image and selected area electron diffraction pattern from the MoS_2-plates. (a) BF-STEM image showing the stacks of Mo-atoms, the inset a_1, a_2 and a_3 in the figure shows the directions in the six fold zone axis and the measurements of the Mo–Mo interatomic distances, (b) electron diffraction pattern shows the hexagonal symmetry in the <001> direction. (Reproduced by permission of The Royal Society of Chemistry).

Figure 15 shows a three-layered MoS_2 plate, forming a Moiré pattern composed by the three individual layers of MoS_2. The upper inset of figure 15 shows a HRTEM micrograph of a Moiré pattern, a Moiré lattice is observed; the lower inset shows the fast Fourier transform (FFT) of the layers, three sets of hexagonal spots can be seen, corresponding to the three layers; the angles between the layers are measured in the figure, the angle between the first two layers is 19°, and the angle between the next two layers is 16°. These are the rotational angles of the layers, here three layers were found and their superimposition causes the Moiré pattern. Some of the plates observed contained Moiré patterns of three layers and some of them with two layers. There are also cases of single plates. Figure 16 shows a HRTEM micrograph of a MoS_2 plate seen in the lateral view. The inset in the left shows the diffraction pattern of the interior, it was possible to obtain the direction of the crystal, that results to be [000n], where n is even. The inset in the right shows an amplification of the main image, several bands of different size are observed, the broad lines are molybdenum, as it has a high atomic number and a high atomic weight; the thin lines are sulphur. The lines are regular and are alternating in the image revealing the stacks of S–Mo–S forming sandwiches, three of these layers form a unit cell in MoS_2, one of which is marked in the inset of figure 16. The measured distances between three atoms of Mo in a cell unit was found to be c = 0.12 nm or 1.2 Å. These results agree with the parameters obtained with the XRD, and with the reported values in literature of c/2= 6.53 Å (shown in the right inset of figure 16).

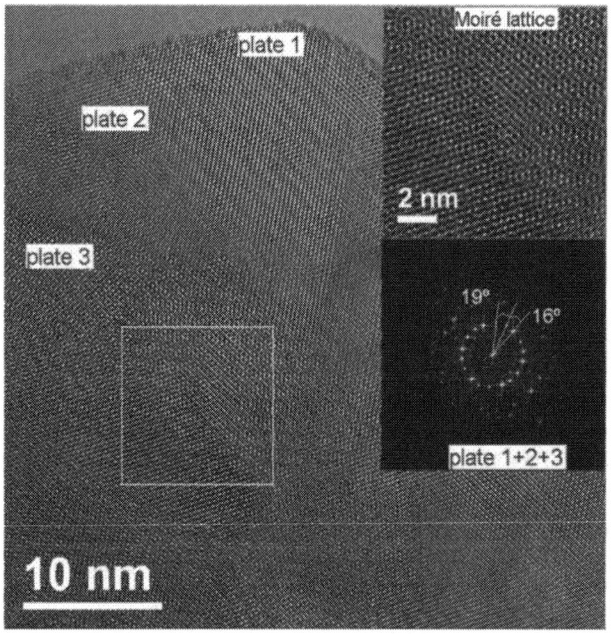

Figure 15. Image showing the superimposition of three plates of MoS_2 and the formation of a Moire pattern. The upper inset shows a HRTEM micrograph of the Moire pattern and the lattice. The lower inset shows the electron diffraction pattern with three sets of points representing the three plates, the angles between the plates are 19° and 16°. (Reproduced by permission of The Royal Society of Chemistry).

It must be noted that from the measured catalytic activity it is found that the selectivity and activity of these MoS_2-nanoplate catalyst are very high, corresponding not to an unpromoted MoS_2 catalyst but almost that of a doped MoS_2 with cobalt. Theoretical analysis performed on unrotated as well as rotated 2H-MoS_2 and monitoring the total average energy in both cases yielded indication that the Moiré patterns observed in the rotated case are in stable states. On the other hand, energy bands on both systems indicate that for the rotated case the system indicates a trend toward the metallicity of the system, this is not observed when the system is unrotated, which could explain the enhancement in the catalytic properties. It is hoped that such improved catalysts will lead to a cleaner gasoline and to a cheaper HDS process.

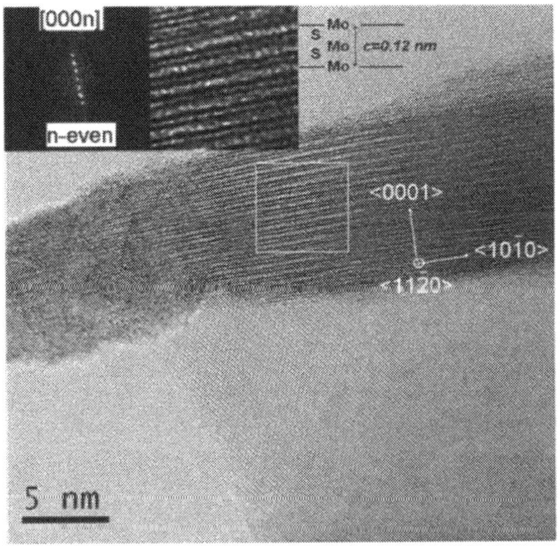

Figure 16. HRTEM image of a plate with the self-assembled MoS_2-layers. The inset in the left shows the FFT in the <11-20> zone axis and the inset in the right shows a zoom of the self-assembled unit cell (c = 1.2 Å). The stripes consist of alternating stacks of Mo and of S. (Reproduced by permission of The Royal Society of Chemistry).

4. Conclusions

Understanding the structure of nanomaterial catalysts has tremendously improved in recent years with the advent of aberration corrected electron microscopes. Two different examples have been described to showcase the recent developments in this field of nanocatalysis. In the first example, the case of bimetallic three-layered Au-Pd nanoparticles is presented. Unravelling the atomistic structure of nanoalloys is vital towards providing the complete understanding towards the way catalysts behave under different situations. This has been accomplished by a combination of atomic scale imaging and simultaneous spectral analysis. In the second example, the structure and intricate details of MoS_2-nanoplates have been deciphered using aberration corrected STEM and HRTEM. These catalysts find important applications in the removal of sulphur compounds from gasoline and hence it is critical to understand their structure and morphology at the atomic level (especially when promoters are incorporated into their structure). This thereby provides to enhance their activity and selectivity, towards specific catalytic processes. Thus it is clear that the current state of the art aberration corrected electron microscopic techniques provide to enhance our understanding of nanocatalysts in an unprecedented manner [12-14].

Acknowledgements

We would like to thank all our collaborators involved in this work, especially Dr. Alvaro Mayoral, Dr. Sergio Meijia-Rosales, Dr. Rodrigo Esparza, Dr. Xochitl Lopez-Lozano, Dr. Arturo Ponce-Pedraza, Dr. Carlos Fernando Castro-Guerrero, Dr. Belsay Borges, Dr. Gilberto Casillas, and Hector Barron.

References

[1] Pyrz W D and Buttrey D J 2008 *Langmuir* **24** 11350-11360
[2] Pennycook S J 2012 *MRS Bulletin* **37** 943-951
[3] Mayoral A, Barron H, Estrada-Salas R, Vazquez-Duran A and Jose-Yacaman M 2010 *Nanoscale* **2** 335-342
[4] Debe M K 2012 *Nature* **486** 43-51
[5] José-Yacamán M, Ponce A, Mejía-Rosales, S and Deepak F L 2013 *Advances in imaging and electron physics*. (London: Academic Press)

IOP Conf. Series: Materials Science and Engineering **55** (2014) 012005 doi:10.1088/1757-899X/55/1/012005

[6] Deepak F L, Casillas G, Esparza R, Barron H and José-Yacamán M 2011 *J. Crystal Growth* **325** 60-67
[7] Mayoral A, Esparza R, Deepak F L, Casillas G, Mejía-Rosales S, Ponce A and José-Yacamán M 2011 *Jeol News* **46** 1-5
[8] Castro-Guerrero C F, Deepak F L, Ponce A, Cruz-Reyes J, Valle-Granados M D, Fuentes-Moyado S, Galvan D H and José-Yacamán M 2011 *Cat. Sci. Tech.* **1** 1024-1031
[9] Mayoral A, Deepak F L, Esparza R, Casillas G, Magen C, Perez-Tijerina E and José-Yacamán M 2012 *Micron* **43** 557-564
[10] Calvo J, Rivas J and López-Quintela M A 2012 Synthesis of subnanometric nanoparticles. *Encyclopedia of nanotechnology*. (Bhushan B; Ed.). (Heidelberg: Springer Verlag) 2639-2648
[11] Bahena D, Bhattarai N, Santiago U, Tlahuice A, Ponce A, Bach S B H, Yoon B, Whetten R L, Landman U and José-Yacamán M 2013 *J. Phys. Chem. Lett.* **4** 975-981
[12] Zhang H, Watanabe T, Okumura M, Haruta M and Toshima N 2012 *Nat. Mater.* **11** 49-52
[13] Zhou Z-Y, Tian Na, Li J-T, Broadwell I and Sun S-G 2011 *Chem. Soc. Rev.* **40** 4167-4185
[14] Henning A M, Watt J, Miedziak P J, Cheong S, Santonastaso M, Song M, Takeda Y, Kirkland A I, Taylor S H and Tilley R D 2013 *Angew. Chem. Int. Ed.* **52** 1477-1480

Electron microprobe analysis of cryolite

F Guimarães[1], P Bravo Silva[1], J Ferreira[1], A P Piedade[2] and M T F Vieira[2]

[1] LNEG - National Laboratory of Energy and Geology, Rua da Amieira, Ap. 1089, PT-4466-901 S. Mamede de Infesta, Portugal

[2] University of Coimbra, Faculty of Sciences and Technology, Department of Mechanical Engineering, Rua Silvio Lima, PT-3030-790 Coimbra, Portugal

E-mail: fernanda.guimaraes@lneg.pt

Abstract. A sample of cryolite was studied with a JEOL JXA 8500-F electron microprobe under several operating conditions. A TAP crystal was used to analyse Na and Al and a LDE1 crystal to analyse F. As F and Na are both highly "volatile" elements, special care must be taken during analysis. The measurement order of Na, F and Al is not irrelevant and optimum conditions may also result in different combinations of accelerating voltage, beam current, beam size or counting times. Relevant X-ray signals were recorded in order to investigate the behaviour of the Na Kα and F Kα counts with elapsed time. The incident beam current was also recorded at the same time. In a clear contrast to what has normally been reported in the EPMA analysis of aluminosilicates and silicate glasses, we found that the Na X-ray counts increase with time. This increment of X-rays intensities for sodium in cryolite depends on the operating conditions and is accompanied by a strong migration of fluorine from the beam excitation volume, leading to a decrease in F X-ray counting rates. It was also observed that higher incident beam currents induce higher radiation damage in the mineral. The current instability is consistent with possible electron induced dissociation in the cryolite structure. An analytical protocol was achieved for 6 kV and 15kV accelerating voltage for the correct EPMA analysis of cryolite.

1. Introduction

Cryolite (Na_3AlF_6) is an uncommon mineral, although it is the most frequent aluminofluoride in nature [1]. Its chemical composition is extremely rich in fluorine and sodium (F: 53 – 54 %; Na: 32 – 34 %). Historically, natural cryolite has been used as an aluminium ore and, more recently, as a solvent for bauxite in the electrolytic refining of aluminium, and also as a pesticide. Natural cryolite was only extracted in large quantities from a world-class deposit within a granite stock in Ivigtut, Greenland [1]. After the exhaustion of the Ivigtut deposit, another outstanding deposit of cryolite was found in the Pitinga mining district, Brazil [2, 3]. Minor occurrences of cryolite and other aluminofluorides occur around the world, mostly in peralkaline granites and pegmatites, but its rarity precludes their economic application. The unavailability of natural cryolite in economic quantities led to the production of synthetic cryolite, which is widely used in the electrolytic processing of alumina. In the last few years, there has been a growing interest on cryolite as a preferred component for thermodynamic models of peralkaline granitic and rhyolitic systems [4-6].

References to electron microprobe WDS analysis of aluminofluorides are rare in the literature, as accurate fluorine determination is still a difficult task, especially in F-rich minerals. In the case of cryolite, this analytical problem gets worse due to the coexistence of two highly "volatile" elements,

Content from this work may be used under the terms of the Creative Commons Attribution 3.0 licence. Any further distribution of this work must maintain attribution to the author(s) and the title of the work, journal citation and DOI.
Published under licence by IOP Publishing Ltd

(F and Na). In this study, several WDS analyses on a cryolite sample from the Academia das Ciências de Lisboa Museum were performed under different operating conditions. Relevant X-ray count rates were recorded in order to investigate the behaviour of the Na Kα and F Kα X-ray with elapsed time. The incident beam current was also recorded at the same time in cryolite, and also in the calibrating standards used for Na (albite) and F (fluorite). The measurement order of the elements Na, F and Al was also taken into account. Both sequences F-Na-Al and Na-F-Al were used for analyses. In order to determine the effect of temperature on the cryolite structure, some heating tests followed by XRD spectra acquisition were performed at the Faculty of Sciences and Technology of Coimbra on another cryolite sample.

2. Analytical methods

The EPMA analyses were performed with a JEOL JXA 8500-F electron microprobe, at LNEG, under several operating conditions. The following standards were used for all the quantitative determinations: fluorite (F Kα), albite (Na Kα) and orthoclase (Al Kα). One WDS spectrometer, equipped with a TAP crystal, was used to analyse Na and Al and a LDE1 crystal was used to analyse F. All elements were analysed with 20 s counting times. The sample and standards were carbon coated (approx. 30 nm thickness).

The XRD spectra were obtained using a Philips X'Pert powder diffractometer, at 40 kV and 35 mA, with Co radiation (λα1 = 0.178896 nm and λα2 = 0.179285 nm), equipped with collimator and Bragg-Brentano geometry, at University of Coimbra. A cryolite was heated from room temperature to 300 °C, at 5 $°C·min^{-1}$ and 150 mbar against a blank of pure alumina with equivalent weight. All the tests were performed with a step size of 0.025° and a time per step of 0.5 s, in 20° - 80° range. The heating was performed at 5 $°C·min^{-1}$ and, after a 5 minute wait at each of the pre-defined temperatures, the spectra collection begun. The temperatures were: room temperature (RT), 60 °C, 100 °C, 205 °C, 300 °C and room temperature after cooling. The diffraction spectra were compared with ICDD (International Center for Diffraction Data) pattern no. 82-0216, which corresponds to monoclinical cryolite.

3. Results

Time-dependent X-rays intensity variation of Na Kα and F Kα in cryolite is presented in figure 1. The shaded areas show the 10s intervals of F and Na peak measurements during cryolite analysis.

At 15 kV and 1 nA, F and Na counts/s (cps) have opposite behaviours. In general, there is a decrease of F cps and an increase of Na cps with time (figures 1a and b). Furthermore, the slopes of F count rates decrease and Na count rates increase are higher for a beam diameter of 5 μm. However, in the first 10 s an increase on F count rates was recorded before it starts to drop off (figure 1a). The same phenomenon of F decrease and Na increase was observed at higher current beam (10 nA), but, for larger beam diameters, there was an increase in the F cps with time (figure 1c).

When using a lower accelerating voltage (6 kV), again the F count rate decreases with time as opposed to the increase of Na count rate, although the slopes of these variations are less steep (figure 2). At this accelerating voltage, when Na is the second element to be analysed and the beam diameter is 5 μm, the Na Kα intensity is comparatively too high and leads to an excessive Na wt% concentration (Table 1).

The increment of X-ray intensities for sodium (Na) in cryolite depends on the operating conditions and is accompanied by a strong migration of fluorine (F) from the beam excitation volume, leading to a decrease in F X-ray counting rates (figures 1 and 2). This phenomenon is in clear contrast with what has been reported to occur normally during EPMA analysis of aluminosilicates, feldspars and silicate glasses, where one usually observes a decrease in Na Kα intensities, as Na^+ ions move away from the bombarded area [7-10]. A comparison between F and Na X-rays intensities variation has been made for cryolite, and for the calibrating standards fluorite and albite (figure 3). As can be observed, there is always a decrease in the number of X-rays counts, when compared to their initial values, except for

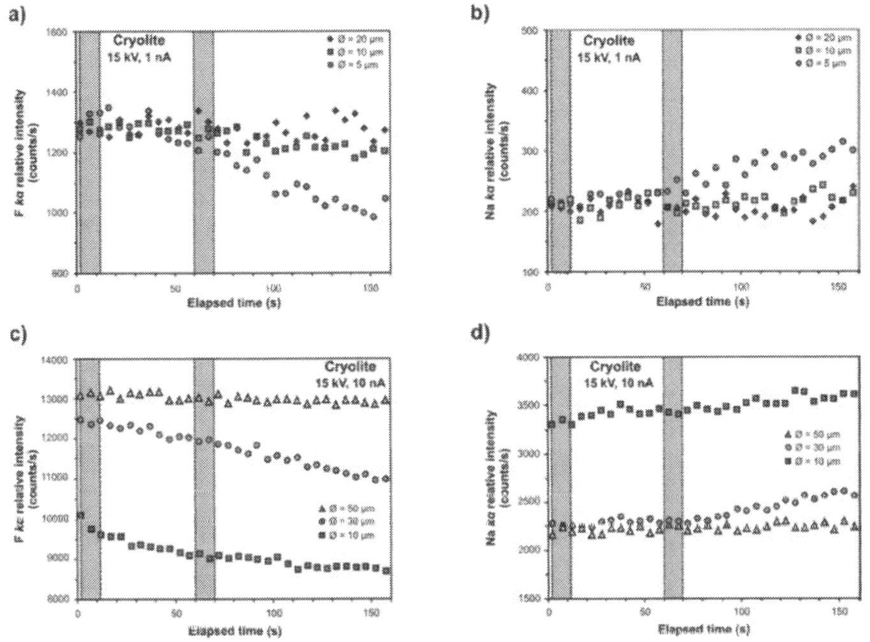

Figure 1. Plots showing the F (a, c) and Na (b, d) count rates in cryolite versus beam exposure time, with 15 kV accelerating voltage and a beam current of 1 nA (a, b) and 10 nA (c, d). Each point represents the average of the number of X-rays obtained in 3 seconds every 5.2 seconds, starting from t = 2.08 s. The shaded areas are the moments (10 s) of peak measurement, either for F or Na, during analyses (which last about 150 s each).

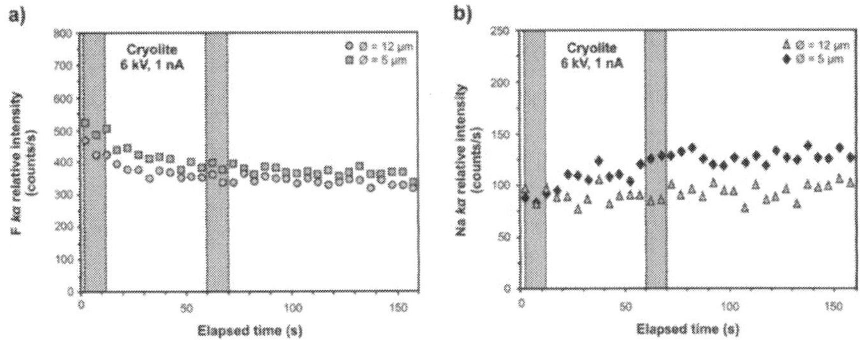

Figure 2. Plots showing the F (a) and Na (b) count rates in cryolite versus beam exposure time, with 6 kV accelerating voltage and a beam current of 1 nA. Each point represents the average of the number of X-rays obtained in 3 seconds every 5.2 seconds, starting from t = 2.08 s. The shaded areas are the moments (10 s) of peak measurement, either for F or Na, during analyses (which last about 150 s each).

Table 1. Electron microprobe analysis of cryolite (wt%), under different operating conditions.

Accelerating voltage	15 kV							
Sequence of analysis	F – Na – Al				Na – F – Al			
Probe current (nA)	0.5 nA	1 nA	5 nA	10 nA	**0.5 nA**	**1 nA**	**5 nA**	10 nA
Beam size (µm)	2 µm	5 µm	12 µm	20 µm	**2 µm**	**5 µm**	**12 µm**	20 µm
F	57.70	58.38	58.05	57.68	**54.18**	**55.02**	**53.85**	56.78
Na	37.25	34.47	34.54	32.86	**31.22**	**32.04**	**31.68**	32.09
Al	13.35	13.66	14.14	12.78	**14.03**	**13.31**	**13.94**	13.01
Total	108.30	106.51	106.73	103.32	**99.43**	**100.37**	**99.47**	101.88
n	6	2	4	6	4	6	9	12

Accelerating voltage	6 kV			
Sequence of analysis	F – Na – Al		Na – F – Al	
Probe current (nA)	1 nA	1 nA	1 nA	**1 nA**
Beam size (µm)	5 µm	12 µm	5 µm	**12 µm**
F	55.06	57.76	40.68	**54.67**
Na	43.66	32.36	40.42	**31.47**
Al	15.20	13.72	15.23	**13.70**
Total	113.92	103.84	96.33	**99.84**
n	5	5	4	5

Pure cryolite [a]	
F	54.29
Na	32.86
Al	12.85
Total	100.00

[a] Palache *et al.* 1951 [14]; n = number of analysis.
Values shown in bold are in good agreement with the ideal cryolite composition.

Na Kα counts on cryolite. For the same time interval (160 s), F Kα dropped about 16 % in cryolite and only 6 % in fluorite. The Na Kα count rates increased about 9 % in cryolite and decreased 6 % in albite (figure 3).

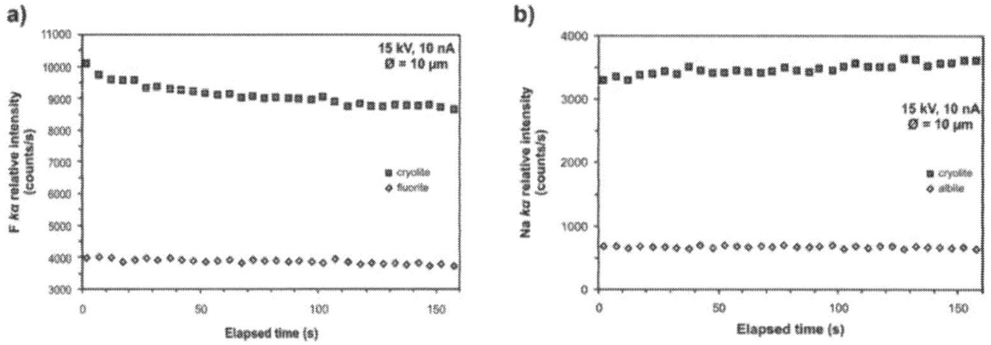

Figure 3. F (a) and Na (b) count rates variation with time in cryolite, fluorite (F) and albite (Na), at 15 kV, 10 nA and 10 µm beam diameter. Each point represents the average of the number of X-rays obtained in 3 seconds, every 5.2 seconds, starting from t = 2.08 s. Note that the decrease in F Kα intensities in cryolite is accompanied by an increase in Na Kα count rates during time, and neither of these situations occur in fluorite or albite at the same operating conditions.

Some secondary electron images (SEI) and electron backscattered images (COMP) were taken at the end of three WDS quantitative analysis on cryolite, all at 15 kV and 3 different probe currents: 0.5 nA, 1 nA and 10 nA. The electron beam damage induced on cryolite is evident, and, as it would be expected, higher incident beam currents induce higher radiation damage in the mineral (figure 4).

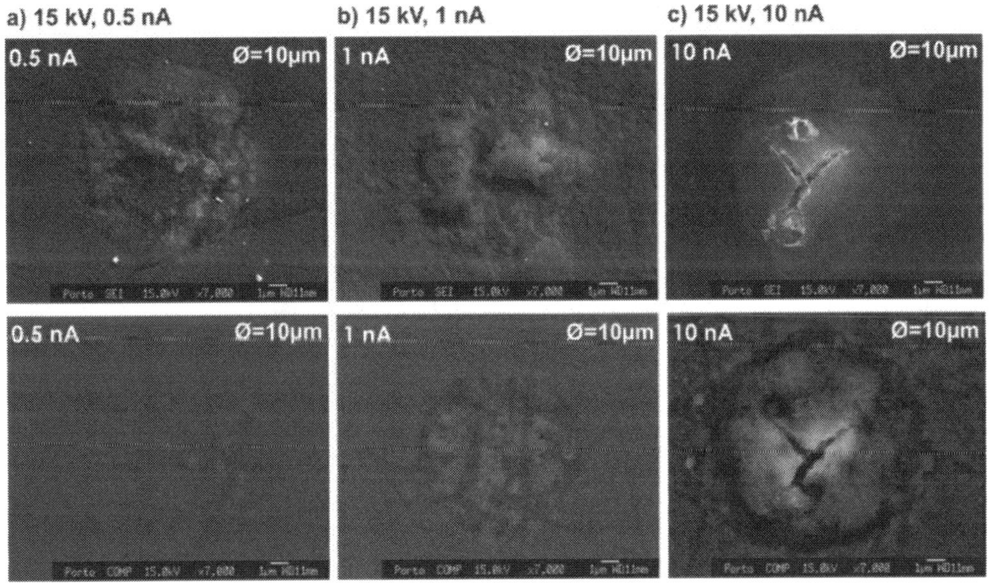

Figure 4. Secondary electron images (SEI) and backscattered electron images (COMP) of cryolite after EPMA analyses with 20 s counting times for each element. Only one WDS spectrometer was used to analyse the 3 elements (F, Na and Al).

In addition to the beam damage induced within the interaction volume in the mineral, one could also observe considerable fluctuations on sample current (probe current measured from the sample) with the elapsed time (figure 5). For 15 kV, 1 nA (nominal beam current) and Ø = 10 μm, the sample current measured decreases drastically after about 105 s for cryolite and remains relatively unchanged for albite and fluorite.

Some calculated and experimental data on thermal conductivity for NaF-AlF_3 molten mixtures have been published [11, 12]. However, the authors are unaware of the exact value of solid cryolite thermal conductivity, which precludes the application of the equation proposed by Castaing to calculate the temperature rise under electron bombardment [13]. Nevertheless, assuming that cryolite might have a low thermal conductivity value, one may expect a considerable local rise in temperature induced by the electron beam.

In order to investigate the influence of the temperature rise in the cryolite structure, several heating tests followed by XRD spectra acquisition were performed. The results obtained showed that heating a cryolite sample up to 300 °C does not change the structure of the mineral (figure 6). Therefore, the observed damage induced on cryolite may not be attributed to a temperature increase, but to the electron beam interaction with the structural lattice.

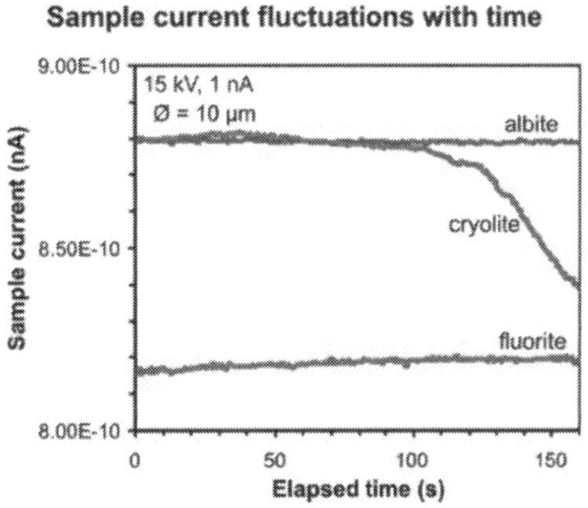

Figure 5. Fluctuations of sample current (nominally set at 1 nA) versus time, in albite, fluorite and cryolite, with 15 kV and a 10 µm beam diameter. Sample current decreases drastically after about 105 s exposure time in cryolite.

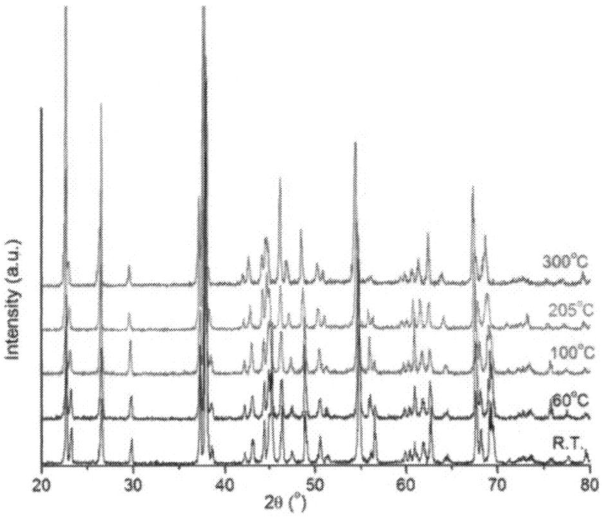

Figure 6. Some XRD spectra obtained during heating of a cryolite sample, showing that from room temperature (RT) to 300 °C there is no structural dissociation of cryolite.

Table 1 presents some results of WDS analysis on cryolite performed with only 1 spectrometer for 2 values of accelerating voltage: 15 kV and 6 kV, 2 sequences of analysis (F-Na-Al and Na-F-Al) and several incident beam currents and beam diameters. Some WDS analysis were also performed scanning the beam in a raster (15 kV, 1 nA) for the analysis sequence Na-F-Al, but the results were unacceptable for fluorine (F = 57.69 % average of 4 analysis), although acceptable for the determination of sodium (Na = 31.38 %).

4. Discussion

Variations in time of the apparent concentrations of sodium and/or fluorine upon electron bombardment of minerals and glasses have been described over the years [7-9, 15-18]. Sample damage under electron beam bombardment and an unexpected behaviour of the so-called sodium migration in alkali silicate glasses have also been discussed earlier [15, 16]. These authors had observed that, after an initial loss of Na within the first seconds, there is a reversal effect with increasing Na cps values, in particular when using small beam diameters and relatively high probe currents. A similar behaviour of Na migration during electron probe microanalysis was described for hydrated silicate minerals, such as zeolites [19, 20]. Even minerals like petzite, a silver-gold telluride made up of conductive heavy elements, can reveal significant compositional changes during electron microprobe analysis, which can be ascribed to the semi-conducting properties of the mineral [21].

The charging effects due to electron bombardment on metallic coated insulators have been studied previously [22]. The model calculates the non-uniform electric field distribution which influences the secondary electron yield value (δ) and the surface potential (V), causing the ion migration. The oxygen desorption from glasses, measured by mass spectrometry [7], can be explained by this model, as well as the local compositional changes in alkali silicate glasses [15].

In the specific case of cryolite, a superionic compound [23], the diffusion of fluorine and sodium under electron bombardment is clearly enhanced.

During an investigation on the effect of electron bombardment on the secondary electron emission from cryolite, at a low value of primary energy beam (~ 400eV), it was observed that, while the secondary electron emission coefficient (δ) was falling, fluorine atoms escaped from the surface, eventually producing a surface metallic layer of Al and Na [24]. As demonstrated by [25], electron beam induced dissociation occurs in cryolite. Fluorine loss from the electron bombarded cryolite was observed and attributed to the emission of fluorine atoms and, to a lesser extent, simultaneous diffusion of fluorine from the surrounding mineral structure into the area depleted by the electron bombardment. However, no desorbed F was observed when heating the sample at ≈ 200 °C [25]. This electron induced dissociation of cryolite can be inferred to have also occurred under the operating conditions used in this study. The imbalance between desorbed F atoms and adsorbed F- ions that occur at different operating conditions, may explain the difficulties in obtaining reliable analyses of F in this mineral.

5. Conclusions

Cryolite is highly sensitive to electron beam radiation even with a defocussed beam. The problem gets worse with the increase of beam current. Cryolite damage observed under the electron bombardment may occur sooner (higher currents), or later (lower currents). The high drop of sample current, as shown in figure 5 after about 105 s elapsed time, is an indication of a severe damage to the structure.

Heating tests followed by XRD spectra acquisition showed that heating a cryolite sample up to 300 °C does not change the structure of the mineral (figure 6). Therefore, the observed cryolite damage may not be solely attributed to the temperature increase, but mainly to the electron beam interaction with the structure lattice.

Despite these problems, it is possible to perform quantitative EPMA analysis of cryolite, using only one WDS spectrometer, providing that:

- short counting times are used;
- Na must be the first element in the analysis sequence;
- F is the second element in the analysis sequence;
- a low probe current is used;
- the beam must be defocused within a limited range.

Acknowledgments

This research is partially sponsored by FEDER funds through the programme COMPETE (Programa Operacional Factores de Competitividade) and by national funds through FCT (Fundação para a Ciência e a Tecnologia), under the project PEst-C/EME/UI0285/2013. This paper was substantially improved by the helpful comments of Prof. Romano Rinaldi and an anonymous reviewer.

References

[1] Bailey J 1980 *Bull. Geol. Soc. Denmark* **29** 1-45
[2] Bastos Neto A C, Pereira, V P, Ronchi L H, Lima, E F and Frantz J C 2009 *Can. Mineral.* **47** 1329-1357
[3] Costi H T, Dall'Agnol R, Pichavant M and Rämö O T 2009 *Can. Mineral.* **47** 1301-1327

[4] Dolejš D and Baker D R 2005 *Geochim. Cosmochim. Acta* **69** 5537-5556
[5] Dolejš D and Baker D R 2006 *Am. Mineral.* **91** 97-103
[6] Dolejš D and Baker D R 2007 *J. Petrol.* **48** 785-806
[7] Lineweaver J L 1962 *J. Appl. Phys.* **34** 1786-1791
[8] Nielssen H N and Sigurdsson H 1981 *Am. Mineral.* **66** 547-552
[9] Morgan VI G B and London D 1996 *Am. Mineral.* **81** 1176-1185
[10] Reed S J B 2005 *Electron microprobe analysis and scanning electron microscopy in geology.* (Cambridge: Cambridge University Press)
[11] Redkin A, Zaikov Y, Tkatcheva O and Filatov E 2006 Empirical evaluation of properties of molten electrolytes used for aluminum production. *210th ECS Meeting*, Abstract #1998
[12] Kjelstrup S, Qian J and Haarberg G M 2000 *Electrochim. Acta* **45** 2707-2717
[13] Castaing R 1951 Ph.D.-thesis, Université de Paris.
[14] Palache C, Berman H and Frondel C 1951 *The system of mineralogy of James Dwight Dana and Edward Salisbury Dana, 7th Edition, vol. II.* (New York: John Wiley & Sons) 110-113
[15] Borom M P and Hanneman R E 1967 *J. Appl. Phys.* **38** 2406-2408
[16] Ribbe P H and Smith J V 1966 *J. Geol.* **74** 217-233
[17] Stormer J C, Pierson M L and Tacker R C 1993 *Am. Mineral.* **78** 641-648
[18] Goldoff B, Webster J D and Harlov D E 2012 *Am. Mineral.* **97** 1103-1115
[19] Rinaldi R 1981 *La microsonda elettronica microscopia elettronica a scansione e microanalisi (vol. II).* (Armigliato A and Valdré U ; Eds.) (Bologna: Università di Bologna, Laboratorio di Microscopia Elettronica ; Centro Stampa "Lo Scarabeo") Cap. 3 pp 242-94
[20] Rinaldi R 1984 Mineralogy of natural zeolites: present status. in: *Proc. 6th International Zeolite Conference.* (Guildford, Surrey: Butterworth Publ. Co.) 570-583
[21] Rucklidge J and Stumpfl E F 1968 *Neues Jahrb. Mineral. Monatsh.* **1-2** 61-68
[22] Cazaux J 1985 *J. Appl. Phys.* **59** 1418-1430
[23] Hull S 2004 *Rep. Prog. Phys.* **67** 1233-1314
[24] Knapp A G 1979 *J. Appl. Phys.* **50** 5961-5965
[25] Knapp A G and Hughes J R 1981 *J. Appl. Phys.* **52** 438-442

IOP Conf. Series: Materials Science and Engineering **55** (2014) 012007 doi:10.1088/1757-899X/55/1/012007

Electron channelling contrast observations in deformed Mg alloys prepared with ion milling

S Kaboli[1], P T Pinard[2], J Su[1], S Yue[1] and R Gauvin[1]

[1] McGill University, Department of Mining and Materials Engineering, M.H. Wong Bldg., 3610 University Street, Montreal H3A 0C5, Quebec, Canada
[2] R.W.T.H. Aachen, Central Facility for Electron Microscopy (GFE), Ahornstrasse 55, DE-52074 Aachen, Germany

E-mail: shirin.kaboli@mail.mcgill.ca

Abstract. Electron channelling contrast imaging (ECCI) was used in the cold-field emission scanning electron microscope (CFE-SEM) to image the microstructure on deformed bulk specimen. Imaging was conducted with a pole-piece mounted silicon photodiode detector at 5 keV to collect backscattered electrons generated from a low-tilted (0 - 3 degrees) specimen. Broad ion beam milling surface preparation technique was used to remove surface layers and reveal near-surface deformation features. The uniaxial hot-compression tests were conducted on Mg-0.3 wt% Al-0.2 wt% Ca alloy. ECCI observations on deformed bulk specimen showed irregular and complex channelling contrast variations inside parent grains and low angle grain boundaries originated from parent grain boundaries. ECCI on an ion milled prepared surface provides non-destructive and rapid visualisation and characterisation of strain fields along with near-surface deformation substructures in CFE-SEM.

1. Introduction

Understanding hot deformation behaviour of Mg alloys is essential to improve their formability through the activation of multiple slip systems at elevated temperatures. Work hardening, softening and phase transformation processes may occur simultaneously during deformation at high temperatures. The interactions between the three processes may lead to a complex deformed microstructure. The deformed microstructure varies with different deformation variables such as temperature, strain rate and strain [1, 2]. Generally, microstructural evolution during hot deformation involves changes in the shape of grains, increase in grain boundary area and formation of substructures such as low angle grain boundaries inside original grains. At low strains, parent grain boundaries start to bulge out to form serrations. Migration of parent grain boundaries initiates the dynamically recrystallized (DRX) grains formation. At high strain, the microstructure reaches a steady state at which more grain boundaries are covered with DRX grains [1].

1.1. Characterisation technique

Visualisation and semi-quantitative analysis of deformed microstructures have been traditionally performed using colour metallography with polarized light microscopy [3-5]. The advantages of this technique include general microstructure observations at low magnifications, phase identification and semi-quantitative crystallographic data acquisition. The drawbacks of this technique include low

Content from this work may be used under the terms of the Creative Commons Attribution 3.0 licence. Any further distribution of this work must maintain attribution to the author(s) and the title of the work, journal citation and DOI.
Published under licence by IOP Publishing Ltd

resolution (~ 20 μm) and low crystallographic sensitivity (misorientation angle > 10°) [6]. Atomic scale microstructure observation and characterisation of deformed materials have been performed using electron diffraction-based techniques in transmission electron microscope (TEM) [7]. However, TEM studies require time consuming and destructive thin foil preparation [8, 9] and may result in non-representative examination area and statistically unreliable results [10, 11]. In addition, deformed materials can be studied using electron channelling contrast imaging (ECCI) in a scanning electron microscope (SEM). SEM studies provide nanoscale spatial resolution, large field of view and statistically reliable information on a bulk specimen [8-10]. Elastic or plastic strain fields can be visualized with ECCI, due to changes in crystal orientation [9]. An elastic strain field causes lateral shifts and rotations in crystal lattice while plastic strain field causes lattice distortion through introduction of dislocations into the crystal. As a result, the intensity of collected backscattered electrons (BSEs) is modulated by the orientation of lattice planes with respect to incident or exit electron trajectories. ECCI micrographs of deformed bulk specimens contain direct evidence of deformation state [12, 13]. Performing ECCI requires appropriate geometry for specimen/microscope, optics conditions and suitable specimen surface [14]. Channelling contrast is commonly detected using either pole piece-mounted Si photodiode detector at low-tilt specimen position or forescatter electron detectors mounted on electron backscattered diffraction camera at high-tilt specimen position [8].

1.2. Surface preparation

For any microstructural observations in a scanning electron microscope (SEM), appropriate specimen preparation is essential such that a mechanical damage free, oxide free, clean and reasonably flat surface is achieved. Suitable specimen surface for ECCI observations can be achieved using chemical or, more commonly, electrochemical polishing techniques after mechanical polishing steps [9]. However, successful electrochemical polishing involves proper selection of variables such as chemical composition and temperature of electrolyte, voltage and time which, sometimes, can be tedious particularly for polycrystalline and multiphase specimens with strained microstructure [13]. Broad ion beam milling is used as an alternative preparation technique to replace electrochemical polishing technique to prepare bulk specimens for ECCI observations [15].

Using appropriate milling conditions for each type of material, an oxide free, clean and reasonably flat surface can be achieved. In addition, rate of material removal from surface can be controlled using the milling parameters. A variety of sputter-induced topographical features have been reported which depend on both specimen conditions and ion beam parameters [16]. A wide range of topographical features was observed in grains with different crystallographic orientations. Such observations include cones, etch pits, bubbles, and periodic and wavelike ripples. Similar artefacts were reported on surfaces prepared with focussed ion beam milling. Formation of all topographical features was attributed to the different localized sputtering yield relative to the surrounding area. For the same ion beam milling parameters (i.e., energy, flow, and angle of incidence), the sputtering yield depends on specimen parameters (i.e., crystallographic orientation, purity, defects and pre-existed topography), and on the surrounding environment (i.e., presence of contamination and neighbouring material). The type and extent of topography development for an arbitrarily chosen specimen is not accurately predicted. Ion beam parameters can be optimized to control the extent of topography development. However, the optimal ion beam milling parameters depend on the specimen conditions and experiments are necessary to determine the best operating conditions for surface preparation [16].

2. **Materials and methods**

2.1. Casting and heat-treatment

Pure Mg with 99.9 % purity, pure Al with 99.9 % purity and Mg-30 wt% Ca master alloy were used to cast the Mg-0.2 wt% Al-0.3 wt% Ca alloy. Pure Mg was melted at 690 °C in a graphite crucible heated with a high-frequency induction furnace. Pure Al and Mg-30 wt% Ca master alloy were added to the molten Mg. The melt was held for 15 minutes to ensure Al and master alloy were melted. The melt was cast under SF_6 and CO_2 gas mixture atmosphere into a preheated copper mould. Homogenisation heat treatment was carried out on as-cast specimens at 500 °C for 8 hours to eliminate the micro-segregation such as precipitates.

2.2. Uniaxial hot-compression tests

Homogenized specimens were machined into cylindrical specimens of 11.4 mm in height and 7.6 mm in diameter. Two ends of the cylinder were covered with a BN paste to minimize the friction between the machine gauges and specimen during deformation. Uniaxial hot-compression tests were carried out using a computer controlled 100 kN servo-hydraulic materials testing system. The tests were performed at 350 °C, strain rate of 0.01 s^{-1} and 0.1, 0.3, 0.6 strains. Specimen was held at deformation temperature for 10 minutes prior to deformation to ensure temperature uniformity throughout the entire specimen. The test was conducted under Ar gas to prevent specimen oxidation. The specimens were quenched immediately after deformation to preserve the dynamically deformed microstructure.

2.3. Specimen preparation

The deformed specimens were cut into halves along the compression axis. Cold mounting was carried out with resin and epoxy. The mounts were ground down to # 1200 grit SiC paper. Subsequently, specimens were polished down to 0.05 μm particle size colloidal silica suspension until a scratch-free surface was observed using an optical microscope. In order to remove damaged surface layers caused by the cold-working effects of the grinding and polishing steps, specimen was ion milled at 2 keV ion beam energy and 0.425 s^{-1} specimen rotational speed, using a Hitachi IM-3000 flat ion milling system. For low magnification SE imaging, specimens were ion milled at 0° specimen tilt angle (ion beam perpendicular to specimen surface) for 10 minutes. Ion milling at normal incidence angle induces severe surface topography and is not normally used for bulk surface preparation. For ECCI observations, specimens were ion milled at 85° specimen tilt angle (ion beam almost parallel to specimen surface) for 30 minutes. Finally, prior to insertion into the microscope, a Hitachi ZONEsem sample surface cleaning system was used for 2 hours and 30 % ozone pressure (~ 227 Torr), to minimize surface contamination.

2.4. Characterisation

A Hitachi SU-8000 FE-SEM was used to study the microstructure after hot-compression tests. ECCI was performed at 0° tilt specimen position (sample surface perpendicular to the electron beam axis), using a photodiode BSE detector inserted below the pole piece (sample surface parallel to the detector). SEM was operated at beam energy of 5 keV and a working distance between 7 - 9 mm for ECCI observations. Acquisition time for each micrograph was 120 seconds. In addition, a SE detector was used with the same parameters. This detector is placed in upper part of the SEM column, and collects SEI and II signals which carry topographical information. All micrographs were obtained from the centre of the specimen to ensure uniform deformation across the examined area. In all micrographs, the compression axis was aligned along the vertical direction. The average grain size of as-received specimen was calculated using the line intercept method [17].

3. Results and discussion

3.1. Deformed microstructure

A uniaxial hot compression test was carried out on polycrystalline bulk Mg-Al-Ca alloy specimen. Figure 1 shows the low magnification SE micrographs of specimens deformed at low and high strains. The specimens were ion milled at 0° specimen tilt angle for SE imaging. Figures 1a and b show the microstructure deformed at 0.1 strain and 0.6 strain, respectively. At 0.1 strain, the microstructure consists of Mg parent grains with internal sub-grain boundaries and extensive serrations at original grain boundaries. At 0.6 strain, a duplex microstructure was observed consisting of elongated parent grains surrounded with equi-axed DRX grains of 10 ± 3 μm average grain size. Parent grains were elongated perpendicular to the compression axis to accommodate deformation during hot-compression. A light grey grain had a high SE yield and a dark grey grain had a low SE yield. The topographical contrast observed between grains was driven by ion-beam-induced surface topography. For a polycrystalline material, milling rate is a function of local crystallographic orientation [16]. Hence, misoriented regions had different milling rates which consequently, led to different SE yields observed at low magnifications.

Figure 1. Polycrystalline bulk Mg-Al-Ca alloy microstructure after ion milling at 0° specimen tilt angle. Secondary electron micrographs of deformed specimen at (a) 0.1 strain and (b) 0.6 strain. The compression axis was aligned along the vertical direction. The topographical contrast observed between grains was driven by ion-beam-induced surface topography.

For ECCI observations, the specimens were ion milled at 85° specimen tilt angle to remove surface topography and obtain a flat surface. Figure 2 shows low magnification BSE micrographs of specimens deformed at low and high strains. Figure 2a and b show the microstructure deformed at 0.1 strain and 0.6 strain, respectively. In both micrographs, irregular and complex channelling contrast variations were observed across the parent grains. A light grey region was oriented at a non-channelling position and had a high backscattering coefficient. A dark grey region was oriented at a channelling position and had a low backscattering coefficient [12]. In a polycrystalline material subjected to deformation tests, grains deform in a complex and inhomogeneous manner. Each grain is subjected to microscopic constrain from neighbouring grains in addition to the macroscopic applied stress and hence, shows a unique BSE contrast with irregular and regular contrast variations. Type and number of activated slip or twinning systems in individual grains control the microstructure evolution [1].

3.2. Deformation effects in ECCI micrographs

Among Mg parent grains with irregular and complex channelling contrast variations, certain grains had unique channelling contrast in the form of bright bands. Two examples of these grains in specimens deformed at 0.3 strain are shown in figure 3. Figure 3a shows straight bright bands traversing the width of the grains, indicated with black squares. The bands had bright central region surrounded with two lateral dark bands on each side. Figure 3b shows a dark grain containing wavy bright bands with six-fold symmetry originated from a pole, indicated with a black square. Also, the electron backscattered diffraction (EBSD) orientation maps acquired on areas shown in figures 3a

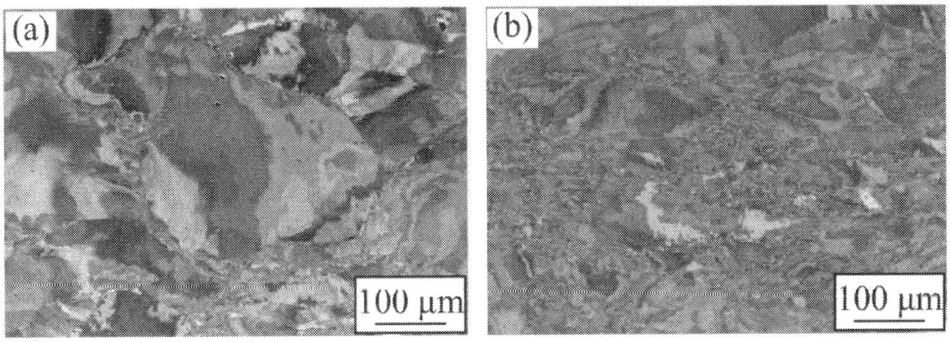

Figure 2. Polycrystalline bulk Mg-Al-Ca alloy microstructure after ion milling at 85° specimen tilt angle. Backscattered electron micrographs of deformed specimen at (a) 0.1 strain and (b) 0.6 strain. The compression axis was aligned along the vertical direction. Irregular and complex channelling contrast variations were observed across the parent grains.

and b did not show the same misorientation gradient as was observed in ECCI micrographs. The origin for the differences between the contrast variations in ECCI micrograph and colour variations in EBSD orientation maps is not yet well understood.

Figure 3. Polycrystalline bulk Mg-Al-Ca alloy microstructure after hot-compression test at 0.3 strain. (a) Backscattered electron micrograph of deformed microstructure with bright bands crossing the grains; and (b) backscattered electron micrograph of deformed microstructure containing a grain with wavy bright bands with six-fold symmetry originated from a pole.

In Figure 3b, the shape, contrast and geometry of the observed intersecting bands closely resembled a disturbed electron channelling pattern [12]. Similar channelling contrast observations in the form of irregular dark and bright bands running across the width of the grains were reported on deformed polycrystalline pure Au and Fe-3% Si alloy specimens [12]. Joy *et al.* [10] used "bend contour" terminology to describe such contrast variations based on the analogy to the bend contours observed on bent foils in TEM. Origin and appearance of observed bend contours in ECCI micrographs were described as follows. At high magnifications, the angular scan of the electron beam is small such that the beam can be considered normal to the surface at all positions. If the grain itself is bent during deformation tests, the angle between the lattice planes and electron beam changes during scanning the surface from point to point which in turn, changes the BSE intensity [12].

During hot-compression test, a regular change in orientation occurs across the grains due to even bending of the lattice planes. For grains subjected to intense deformation, long-range systematic misorientations in lattice planes result in complex shape and contrast of bend contours. The shape and contrast of bend contours is a function of beam-crystal orientation and incident beam energy and varies from grain to grain due to changes in grain orientation [18]. In a deformed polycrystalline material, grains can be bent about one or multiple axes. If a grain is bent about one axis cylindrically, one single straight band appears in the ECCI micrograph, as shown in figure 3a. If a grain is bent in two dimensions and grain normal is positioned near a low index zone axis, bend contours form a two dimensional zone axis pattern, as shown in figure 3b. In figure 3b, the grain was bent about the zone axis positioned at central pole of the intersecting contours. Hence, bend contours observed on ECCI micrographs can be used to reconstruct the three dimensional concave/convex curvature of the underlying deformed grain. In addition, real-space crystallographic analysis can be performed on bend contours observed in ECCI micrographs since each contour is related to one particular set of diffracting planes [12, 14, 19, 20].

3.3. Appearance of bend contours

In Figures 3b, the appearance of the observed bands has been distorted while the symmetry of the pattern remained unchanged. Plastic deformation causes lattice distortion through introduction of defects such as dislocations into the grains. The presence of dislocations increases the electron beam divergence angle inside the bulk specimen. Increase in divergence degrades the electron channelling pattern quality through decrease in contrast and increase in angular width of the channelling bands. Such quality degradation depends on orientation, local strain conditions and relative deformation temperatures (T/T_m), where T_m is the absolute melting point and T is the absolute deformation temperature. Deformation at low relative temperatures ($T/T_m < 0.3$) degrades the pattern quality, however, the pattern geometry remains unaffected and the bands remain straight. Deformation at higher relative temperatures ($T/T_m > 0.3$) distorts the channelling pattern geometry and bands become wavy. This is due to the fact that the orientation being scanned (angular and spatial scan) by the beam varies due to the presence of low angle boundaries [12]. In the present case of $T/T_m = 0.6$, the channelling bands and patterns observed in figures 3a and b were distorted, due to the presence of the low angle grain boundaries. Furthermore, the appearance of bend contours may not be uniform due to the non-uniform bending of the corresponding lattice planes during heterogeneous deformation [21]. For more quantitative analysis on the degree of misorientation inside grains, ECCI micrographs should be compared with the corresponding EBSD orientation maps.

4. Summary

Ion milling surface preparation technique was proved to be beneficial to study deformation of Mg alloys. Milling at low incidence angles revealed low angle grain boundaries and sub-grains with low angle misorientation. Milling at high incidence angle revealed substructures inside grains such as low angle grain boundaries. Furthermore, ECCI observations showed that plastic deformation in Mg grains can be mapped with channelling contrast on an ion milled prepared specimens in FE-SEM. The complicated bend contour patterns observed after ion milling directly maps out the region of severe deformation inside the specimen.

Acknowledgements

The authors would like to thank the General Motors (GM) for their financial support.

References

[1] Humphreys F J and Hatherly M 1995 *Recrystallization and related annealing phenomena.* (Oxford: Pergamon Press)
[2] Al-Samman T and Gottstein G 2008 *Mater. Sci. Engng. A* **490** 411
[3] Thomas M 2012 On the characterisation of subsurface deformation microstructures in aerostructural titanium alloys. (Sheffield: University of Sheffield)
[4] Urai J L 1983 *Tectonophysics* **96** 125-157
[5] Urai J L, Humphreys F J and Burrows S E 1980 *J. Mater. Sci.* **15** 1231-1240
[6] Trimby P W and Prior D J 1999 *Tectonophysics* **303** 71-81
[7] Williams D B and Carter C B 2009 *Transmission electron microscopy a textbook for materials science.* (New York: Springer)
[8] Kamaladasa R J and Picard Y N 2011 *Microsc: Sci., Tech., Educ. and App.* **4** 1583-1590
[9] Wilkinson A J and Hirsch P B 1997 *Micron* **28** 279-308
[10] Joy D C, Newbury D E and Davidson D L 1982 *J. App. Phys.* **53** 81-122
[11] Zauter R, *et al.* 1992 *Phil. Mag. A* **66** 425-436
[12] Newbury D E, *et al.* 1986 *Electron channeling contrast in SEM.* in: Advanced scanning electron microscopy and X-ray microanalysis. (New York: Plenum Press)
[13] Prior D J, *et al.* 1999 *Am. Mineral.* **84** 1741-1759
[14] Holt D B 1974 *Quantitative scanning electron microscopy.* (London: Academic Press)
[15] Ulirich C, *et al.* 2012 *Pract. Metallogr.* **49** 290-304
[16] Czanderna A W, Madey T E and Powell C J 1998 *Beam effects, surface topography, and depth profiling in surface analysis.* (New York: Springer)
[17] Vander Voort G F 1999 *Metallography: principles and practices.* (ASM International)
[18] Joy D C, Newbury D E and Hazzledine P M 1972 Anomalous crystallographic contrast on rolled and annealed specimens. in: *Proc. 5th annual scanning electron microscope symposium part I and part II, workshop on biological specimen preparation for scanning electron microscopy.* (Chicago: IIT Research Institute)
[19] Bolotov I E, Kolosov V Y and Kozhyn A V 1982 *Physica Status Solidi A* **72** 645-654
[20] Rodríguez-González B, Pastoriza-Santos I and Liz-Marzán L M 2006 *J. Phys. Chem. B* **110** 11796-11799
[21] Shannon M D and Eades J A 1979 *Phil. Mag. A* **40** 125-129

SEM, TEM and SLEEM (scanning low energy electron microscopy) of CB2 steel after creep testing

J Kasl[1], Š Mikmeková[2] and D Jandová[1]

[1] VZÚ - Research and Testing Institute Plzeň, Tylova 1581/46, CZ-30100 Plzeň, Czech Republic
[2] Academy of Sciences of the Czech Republic, Institute of Scientific Instruments of the ASCR, v.v.i., Brno, Czech Republic

E-mail: kasl@vzuplzen.cz

Abstract. The demand to produce electrical power with higher efficiency and with lower environmental pollution is leading to the use of new advanced materials in the production of power plant equipment. To understand the processes taking place in parts produced from these materials during their operation under severe conditions (such as high temperature, high stress, and environmental corrosion) requires detailed evaluation of their substructure. It is usually necessary to use transmission electron microscopy (TEM). However, this method is very exacting and time-consuming. So there is an effort to use new scanning electron microscopy techniques instead of TEM. One of them is scanning low energy electron microscopy (SLEEM). This paper deals with an assessment of the possibility to use SLEEM for describing the substructure of creep resistant steel CB2 after long-term creep testing. In the SLEEM images more information is contained about the microstructure of the material in comparison with standard scanning electron microscopy. Study of materials using slow and very slow electrons opens the way to better understanding their microstructures.

1. Introduction

Turbines, boilers and steam piping are among the most exposed parts of steam power plants. They operate under severe service conditions for several decades. Therefore, high mechanical strength, good corrosion/oxidation resistance and high structural stability of the materials used for their production are desired. Several new grades developed on the basis of modified 9Cr-1Mo steels in the last few decades are currently used for the production of various parts of high efficiency power plants. Their chemical composition is balanced with the aim of improving oxidation and creep properties. Alloying with chromium increases the oxidation resistance, however its high content (higher than 9 %) causes precipitation of Z-phase, dissolution of fine MX precipitate and a decrease in long-term creep strength. Limitation of nickel content improves creep strength. Addition of cobalt compensates the low Ni content and restrains the formation of δ-ferrite. Nitrogen together with vanadium and niobium can increase precipitation strengthening and boron can restrict coarsening of chromium carbides that pin grain and sub-grain boundaries and causes an increase in sub-structural strengthening. A balanced content of molybdenum and tungsten increases solid solution strengthening and long-term structure stability. Precipitation of secondary phases during service may have an adverse effect on the mechanical properties such as tensile and creep strength, fatigue, etc. Cast steel CB2 developed in Europe is one of the most promising alloys for production of steam turbines working in

Content from this work may be used under the terms of the Creative Commons Attribution 3.0 licence. Any further distribution of this work must maintain attribution to the author(s) and the title of the work, journal citation and DOI.
Published under licence by IOP Publishing Ltd

ultra-supercritical conditions [1]. The changes of substructure – precipitation and dissolution of precipitates, dislocation density – have to be observed using transmission electron microscopy (TEM). It is very labour intensive and time-consuming. Standard scanning electron microscopy (SEM) is inapplicable in many cases due its insufficient resolution for investigation of some structural features (e.g., fine precipitates). Thus, new methods of SEM are tested with the goal of improving observation capability.

Scanning low energy electron microscopy (SLEEM) is beneficial for the investigation of many different types of materials. This technique can be used for improving image parameters, such as atomic number and crystallographic contrast. The aim of this study is to analyze the microstructure of cast CB2 steel using ultra-high and standard high vacuum scanning electron microscopes equipped with cathode lens mode, which enables us to observe samples at arbitrary landing energy of primary electrons. In the SLEEM images more information is contained about the microstructure than with standard scanning electron microscopy (SEM). Study of materials using slow and very slow electrons open the way to better elucidation of their microstructures.

This paper deals with the possibility of using SLEEM for the study of advanced creep resistant steel COST CB2.

2. Experimental materials and methods

2.1. Experimental materials

Research and Testing Institute Plzeň Ltd. performed measurement of mechanical properties, long-term creep test to ruptures at a temperature of 650 °C, fractography analyses and a quantitative assessment of microstructure as a part of a collaborative effort in the framework of the COST 501 and 522 programmes. The tested samples were machined from the 5t pilot valve CB2P produced by PHB Stahlguss GmbH (Germany) with dimensions which corresponded to the real valve of the steam turbine. The heat treatment consisted of annealing at 1100 °C for 9 h, cooling with compressed air, annealing at 740 °C for 10 h and at 730 °C for 24 h. Mechanical and creep testing of the valve were carried out at three localities marked A, B and C showing the various wall thicknesses (130, 85 and 50 mm, respectively) [1]. Three of the creep tests are still running and the longest time to rupture at the time of publication is about 90,000 hours.

The microstructure of the virgin material and crept specimens (tested at 650 °C and stresses at 60 and 85 MPa) taken from locality B were investigated.

2.2. Experimental methods

The samples were studied at Research and Testing Institute Plzeň Ltd. using "standard" methods of SEM and TEM. Observation was performed using microscopes JEOL JSM 6490 LV, JEOL JEM 1200 EX, and a microanalyser INCA ENERGY 300. Carbon extraction replicas were prepared from electrolytic polished and etched metallographic samples. Thin foils were prepared using a jet electropolishing. Secondary phases were identified on the basis of EDS microanalysis and electron diffraction.

In addition evaluation of selected samples was performed at the Institute of Scientific Instruments of the Academy of Sciences of the Czech Republic. The microstructures of specimens were studied under standard vacuum in scanning electron microscope JEOL 6170F and in the ultra-high vacuum SEM (UHV SLEEM) of own design. UHV SLEEM consists of three separate vacuum chambers. One of them is the observation chamber equipped with a two-lens field emission electrostatic electron optical column (FEI Company) and CL system. The second chamber is intended for in-situ cleaning of the samples and incorporates an argon ion beam gun for surface cleaning. The third chamber is the loading chamber of the air lock. Both these microscopes are equipped with cathode lens (CL) mode which enables us to observe specimens at the arbitrary landing energy of primary electrons. The principle of the CL mode is shown in figure 1 and as you can see the cathode of the CL is formed by a negatively biased specimen and the anode of the CL is created by the detector based on the ground potential. The primary beam is retarded in the field of the CL to the final landing energy and the electrons emitted from the specimen are collimated and accelerated towards the detector [2]. The information about crystallographic orientation was obtained using electron backscattered diffraction (EBSD) in a Philips XL30 SEM.

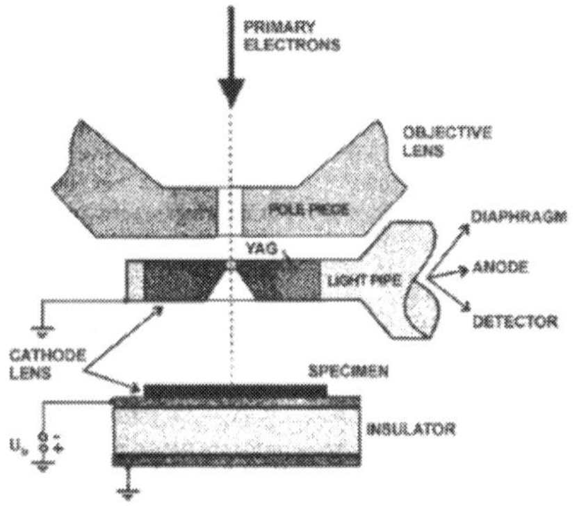

Figure 1. Schematic sketch of the cathode lens.

Observation of specimens using very slow electrons requires in-situ cleaning of the surface from native oxide layers. The native oxide is impenetrable for a low energy primary beam and its presence prevents observation of the real microstructure. This situation is demonstrated in figure 2, which shows the comparison between the in-situ cleaned and rough areas together with X-ray photoelectron spectroscopy (XPS). XPS analysis shows that the specimen surface is covered by Fe_3O_4, Fe_2O_3 and Cr_2O_3 oxides.

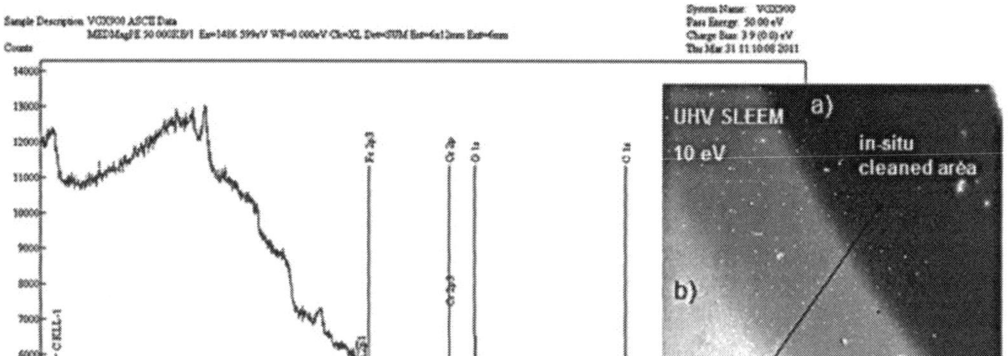

Figure 2. Surface of CB2 steel: a) in-situ cleaned area; b) as-inserted oxidized area, together with the XPS spectra of the oxidized area.

3. Results

3.1. Substructure assessment using standard SEM and TEM

The microstructure of the virgin CB2 sample corresponds to tempered martensite without δ-ferrite, with some evidence of dendritic segregation and shrinkage porosity. Aluminate inclusions and large BN precipitates up to 5 μm in size were observed in etched metallographic samples. They occurred close to the prior austenite grain boundaries, in regions where the boron content exceeded the limit of boron solubility in solid solution (figure 3). Thus the boron atoms bonded in primary nitrides cannot be active in the stabilisation of chromium carbides. A high density of relatively large chromium rich $M_{23}C_6$ carbides was found in the specimens before and after the creep tests. Their size did not exceed 0.5 μm (figure 3).

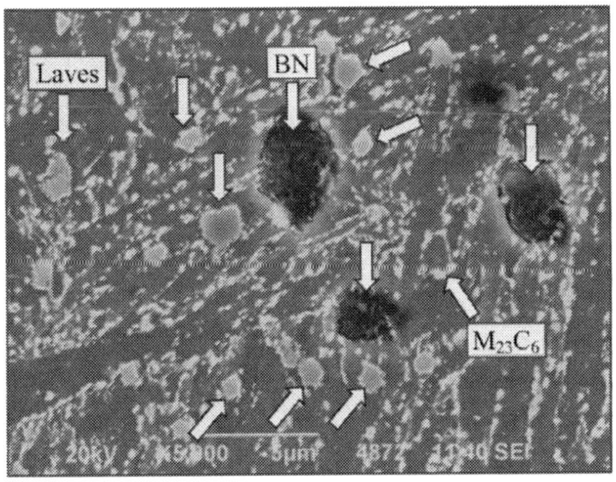

Figure 3. Metallographic sample of virgin specimen etched in Villela-Bain's reagent. SEM micrograph.

A lot of Laves phase particles were observed in the etched metallographic samples and also in the carbon extraction replicas (figure 4). These particles were concentrated around coarse BN precipitates or formed clusters with coarse NbN particles. Fine (V,Nb)N precipitates were observed in the replicas and were detected also in thin foils.

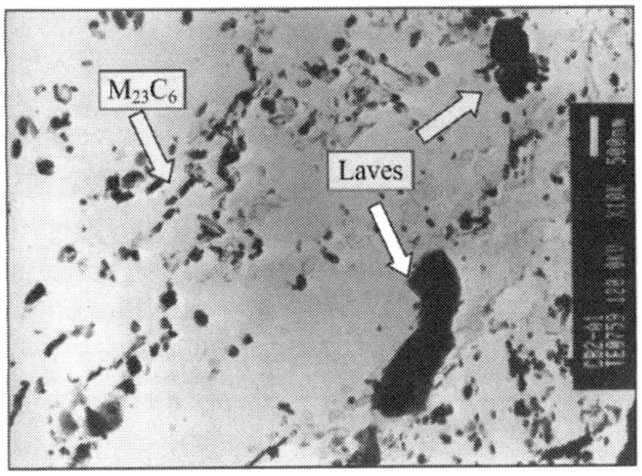

Figure 4. Carbon extraction replica of virgin specimen. TEM micrograph.

Various types of grain boundaries were evident in the structure of tempered martensite. High angle boundaries separated prior austenite grains, which consisted of blocks of martensite laths at different levels of recovery, which occurred during tempering and creep testing. Before creep tests blocks consisted of laths with relatively high dislocation density (in the order of 10^{14} m^{-2}) and laths divided by dislocation walls into sub-grains with significantly lower dislocation density (in the order of 10^{13} m^{-2}). Relatively coarse $M_{23}C_6$ carbides and Fe_2Mo Laves phase pin dislocations at the sub-grain boundaries, fine MX nitrides pin slip dislocations within the sub-grains (figure 5).

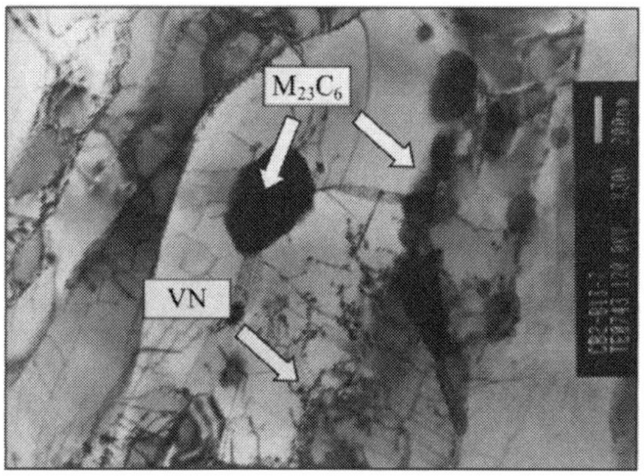

Figure 5. Thin foil of virgin specimen. TEM micrograph.

Coarsening of Laves phase occurred during creep tests and new Laves phase particles nucleated on the site of $M_{23}C_6$ carbides, NbN and BN nitrides. Significant growth of Laves phase occurred after creep exposure for 17,924 h and 13,962 h respectively. With regard to $M_{23}C_6$ carbides, no important changes were found in their size and distribution. Density of fine MX nitride was higher than in as received conditions.

After a creep test with the time to rupture of 67,857 h further coarsening of Laves phase was observed and a significant decrease in the density of fine precipitates. Some new plate-like particles containing V, Cr, Nb and N were observed in the carbon extraction replicas, which could correspond to undesirable the Z-phase.

3.2. Substructure assessment using advanced SEM

The atomic number and crystallographic contrast is caused mainly by backscattered (BSE) electrons (i.e., electrons that are leaving the sample with energy higher than 50 eV and by elastic scattering). The BSE leave the specimen surface at a wide range of angles relative to the optical axis. The electrons backscattered at low angles from the optical axis mainly carry information about chemical composition and electrons backscattered at high angles are sensitive to the density of atoms within the crystal lattice [3].

3.2.1. Tuning of material contrast. The low beam energy and thus the low penetration depth are necessary for observing small surface features, such as fine precipitates. A mean free path of inelastically scattered electrons decreases with decreasing landing energy of the primary electrons, which makes this method extremely surface sensitive. Figure 6 shows the UHV SLEEM images of the in-situ cleaned specimen acquired at conditions of the landing energies from 6017 eV (= primary beam energy) to 4 eV. The electrons were retarded by means of the negative bias applied on the specimen surface. The micrographs obtained from the same field of view demonstrate increasing crystallographic contrast with decreasing landing energy until 50 eV, where the minimum of the escape depth is reached [9]. At this landing energy the escape depth of electrons is the least and thereby detected electrons are the least affected by the arrangement of atoms in the matrix, which finally causes a significant deprivation of crystallographic contrast. On the other hand, the fine precipitates located on the surface are visible.

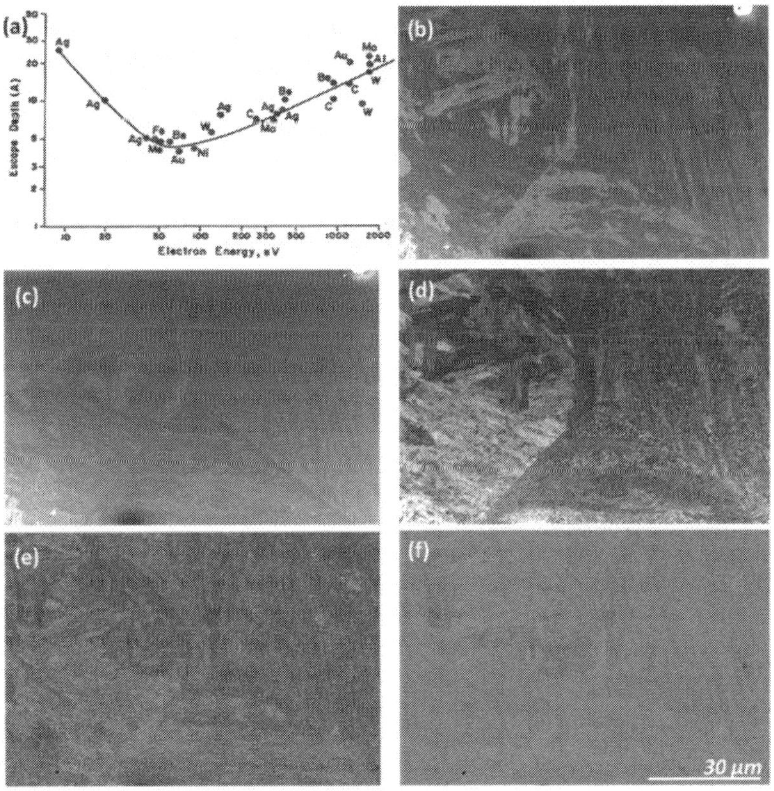

Figure 6. a) Dependence of the escape depth on the kinetic energy of electrons [9]; together with the UHV SLEEM images of the same area acquired at: b) 4 eV, c) 50 eV, d) 500 eV, e) 3000 eV, and f) 6017 eV landing energy of electrons. All images were taken at the same image contrast.

Figure 7 shows images of a specimen cleaned in-situ obtained under ultra-high vacuum conditions (figures 7a and b) in the UHV SLEEM microscope, together with the image of the same area in a JEOL 6170F SEM (standard high vacuum, without in-situ cleaning). Figure 7 demonstrates that the small precipitates of Laves phase near shrinkage porosity are not visible in standard SEM and can be successfully imaged only in the CL mode. Figure 7d and the attached table show the results of ED microanalyses of Laves phase particles and matrix. Thus, using the CL mode it is possible to visualize areas with very small difference in chemical composition and this technique provides a powerful tool for examining multi-component materials.

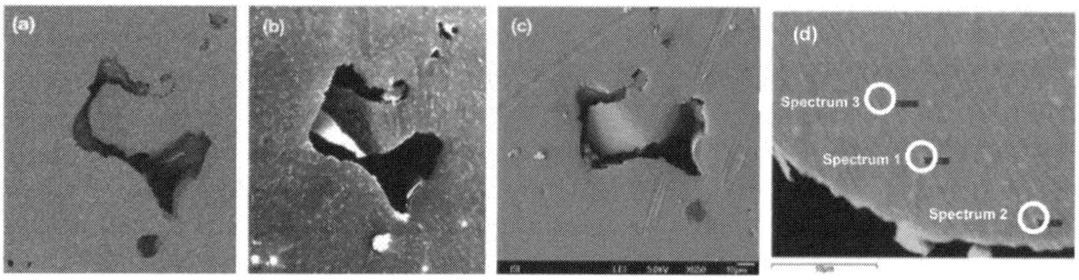

Element	B	V	Cr	Mn	Fe	Mo	W	
Spectrum 1	7.1	3.4	12.2		26.1	50.5	0.8	Laves phase
Spectrum 2	10.2	3.6	12.5		23.5	50.2		Laves phase
Spectrum 3			11.0	1.4	87.7			Matrix

Figure 7. The effect of beam energy and cleanness of the specimen surface on material contrast. UHV SLEEM images obtained at a) 6017 eV, and b) at 10 eV (primary beam energy = 6017 eV). c) shows standard SEM image obtained at 5 keV landing energy of electrons; and d) attached table shows positions and results of EDS microanalysis in wt%.

3.2.2. Sensitivity to crystallographic orientation. Backscattering of electrons is the source of one of the most important image signals in SEM. In the CL mode of SEM a combination of secondary electrons with (slow) backscattered electrons (including electrons backscattered at high angles from the optical axis) is acquired, providing micrographs with high crystallographic contrast even under conditions available in a standard SEM. Figure 8 shows the UHV SLEEM images obtained from CB2 steel at 4 keV (a) and 1 keV (b) landing energy of electrons, together with the corresponding crystallographic orientation map obtained from the EBSD measurement (c) and with simulations of the BSE trajectories (d, e). Acquisition of the high angle electrons enables the contrast between martensite laths to be observed with extremely high contrast.

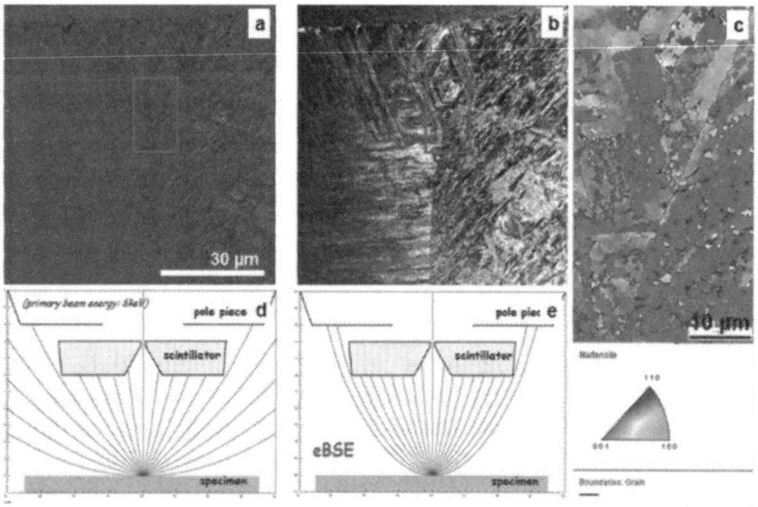

Figure 8. UHV SLEEM images obtained at a) 4 keV and b) 1 keV, together with c) EBSD map of the same area; and d, e) corresponding simulations of BSE trajectories in the electron optical design (EOD) software [6].

Evolution of the contrast between the individual martensitic laths is displayed in figure 9. The primary beam energy was fixed at 6019 eV and the landing energy was governed by means of electrostatic field formed by the cathode lens system. Series of the UHV SLEEM images shows the increasing contrast of martensitic laths with the increasing immersion ration (primary beam energy / landing energy ratio). Such high sensitivity to the structure details using the UHV SLEEM cannot be achieved using the standard SEM BSE image (figure 9z). At electron energies in the order of keV, the crystal orientation contrast is week due to the detection of the BSE scattered close to the optical axis, which carry the major information about the chemical composition. At lower energies, the electrons are collimated towards the optical axis and detection of the high angle BSE causes increasing crystallographic contrast [4, 5, 7, 8].

Figure 10 shows the graph of the image brightness variation across the line X-Y marked in figure 9o. The comparability of the images is guaranteed using of the same settings of image contrast during their acquisition. From this graph is evident that the best contrast between the laths of different orientation is obtained at 500 eV landing energy of primary electrons.

4. Conclusion

Scanning low energy electron microscopy is a promising new method for evaluating the microstructure of advanced creep resistant steels used in the power industry.

It is possible to assess local crystal structure or phase composition at high resolution in relation to the incident electron energy.

The main advantages of scanning low energy electron microscopy are the many opportunities for using the method not only in fundamental research but also in practice. This method offers the possibility of studying materials with an extremely high sensitivity to the crystallography and chemical composition. Knowledge of the crystallography and chemical composition of materials is important for a better understanding of their microstructures and provides a key for the development of materials with unique mechanical and physical properties.

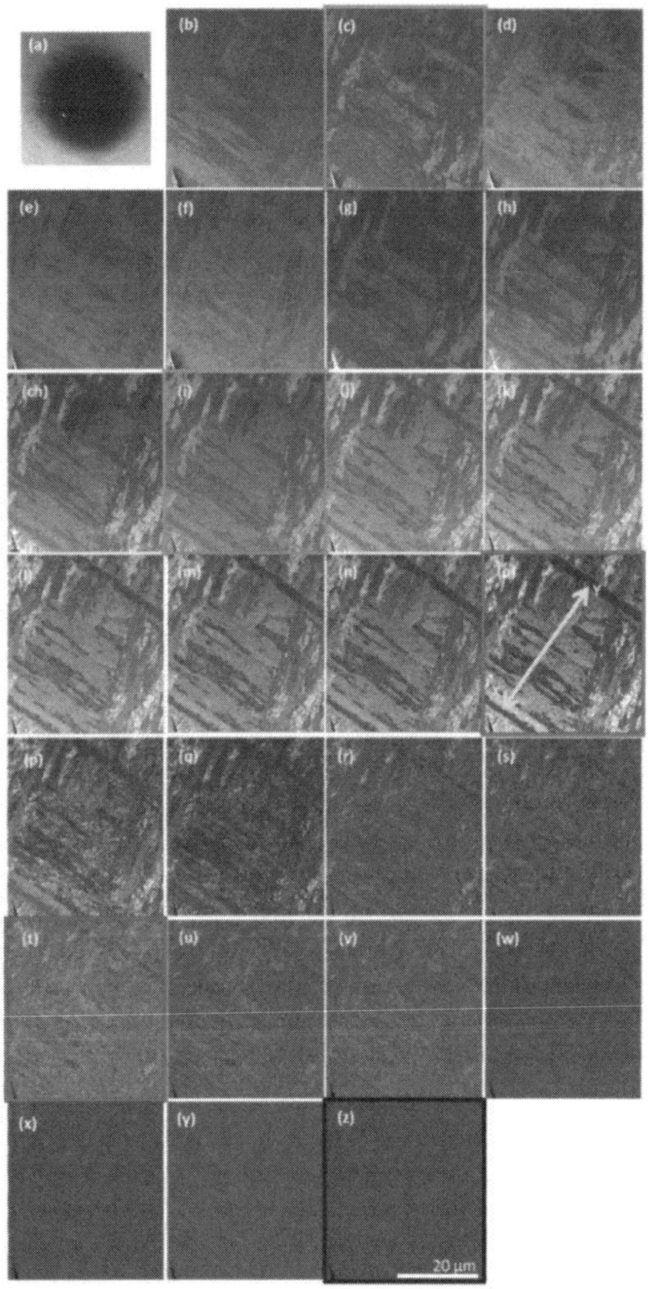

Figure 9. The same area acquired at landing energies of: a) 0 eV (i.e., total reflection); b) 2 eV; c) 8 eV; d) 11 eV; e) 16 eV; f) 26 eV; g) 60 eV; h) 100 eV; ch) 150 eV; i) 200 eV; j) 250 eV; k) 300 eV; l) 350 eV; m) 400 eV; n) 450 eV; o) 500 eV; p) 1000 eV; q) 1500 eV; r) 2000 eV; s) 2500 eV; t) 3000 eV; u) 3500 eV; v) 4000 eV; w) 4500 eV; x), 5000 eV; y) 5500 eV; and z) 6019 eV.

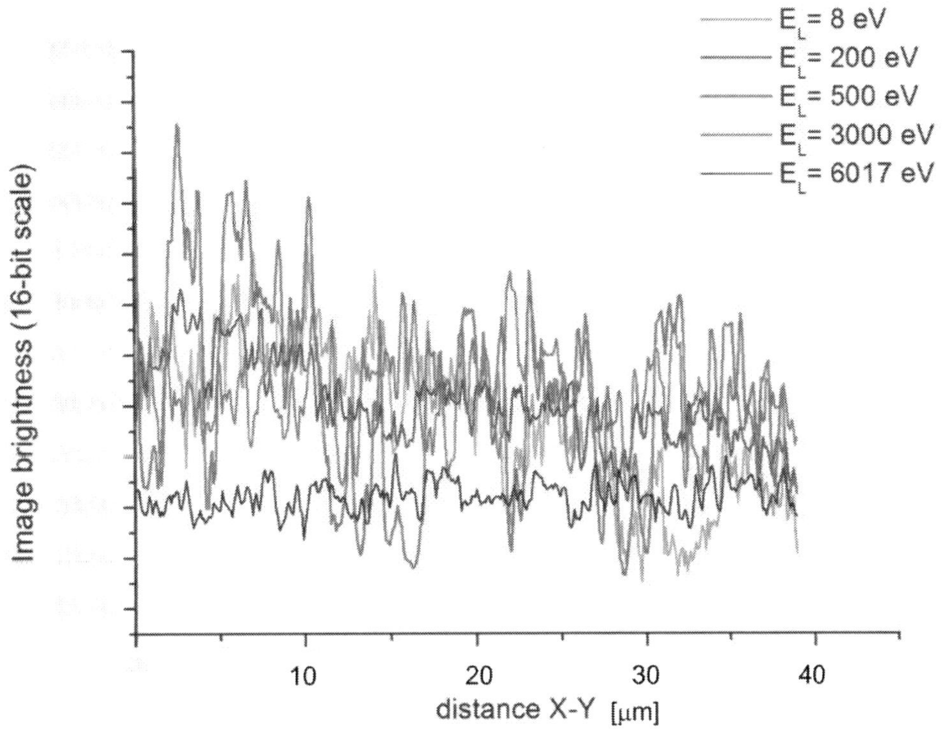

Figure 10. A graph of the intensity variation across the martensitic structure along the line X-Y (in figure 9o).

Acknowledgements
We are very grateful for the financial support of our research from the project no. TE01020118 (Competence centre: Electron microscopy) from the Technology Agency of the Czech Republic.

References

[1] Staubli M, Hanus R, Weber T, Mayer K-H and Kern T U 2006 in: Proc. 8th Liege Conf. Materials for Advanced Power Engineering. (Lecomte-Beckers J, Carton M, Schubert F and Ennis P J; Eds.). (Jülich, Germany: Forchungszentrum Jülich GmbH) 855-870
[2] Müllerová I and Frank L 2003 *Adv. Imag. Electr. Phys.* **128** 309-443
[3] Reimer L 1998 *Scanning electron microscopy.* (Berlin: Springer Verlag)
[4] Mikmeková Š, *et al.* 2010 *Mat. Trans.* 51 292-296
[5] Mikmeková Š, *et al.* 2011 *Key Engng. Mat.* **465** 338-341
[6] Lencová B and Zlámal J 2007 *Microsc. Microanal.* **13** 2-3
[7] Bauer E 1998 *Surf. Rev. Lett.* **5** 1275-1286
[8] Heiner J 2008 *Microsc. Microanal.* **14** 1226-1227
[9] www.globalsino.com

Characterisation of sub-micrometre features with the FE-EPMA

P McSwiggen[1,2]

[1] McSwiggen & Associates, 2855 Anthony Lane South, Suite B1, US-55418-2883 St. Anthony, Minnesota, U.S.A.
[2] JEOL USA, Inc., 11 Dearborn Road, US-01960 Peabody, Massachusetts, U.S.A.

E-mail: pmcs@mcswiggen.com

Abstract. The goal of this work is to compare two strategies for doing sub-micrometre analyses using the Fe-Ni binary system, as an example. The first approach involves reducing the overvoltage to 1 - 3 kV over the critical ionisation energy of the K X-ray lines. Using such a small overvoltage greatly restricts the size of the analytical volume. Upon entering the sample, the beam electrons quickly lose the additional energy required to excite the X-rays of interest. As a result, the K X-ray line for Fe and Ni will only be produced very near the surface. The second strategy is to use the L-lines for Fe and Ni, and drop the accelerating voltage to a level that will produce the smallest overall interaction volume of the beam electrons. Each strategy has its advantages and disadvantages that depend on the ultimate goal of the analysis and the elements involved. Both methods produce small analytical volumes. However, the L-lines from the transition elements are more problematic because of uncertainties with their mass absorption coefficients. Therefore, using the L-lines in low-kV analyses are more challenging to get good quantitative results. The advantage of using L-lines is that they travel a shorter distance within the sample, and therefore secondary fluorescence becomes less of an issue. The best results will come from a combination of these strategies. Using a multiple kV approach allows the user to select the optimum conditions for each element.

1. Introduction

The field emission electron column has dramatically changed electron imaging. The higher electron density and smaller electron energy spread allows for an electron beam spot size that is nearly an order of magnitude smaller than that produced by the traditional thermionic tungsten filament (figure 1). This smaller spot size results in images that have far improved spatial resolution particularly at high beam currents (figure 2). Earlier cold field emission electron guns were not suitable for the electron microprobe because of their poor beam stability (~ 10 %/hour) and their low beam currents (~ 1×10^{-9} A). With the thermal, field emission electron guns (TFE), both beam stability (< 0.1 %/hour) and high beam currents (~ 1×10^{-6} A) can be achieved. This now makes the field emission gun ideal for the electron microprobe.

Some have previously argued that the field emission gun does not add to the analytical range of the electron microprobe because it's the scatter of electrons within the sample that controls the analytical area, not the beam diameter. The electron scatter is in turn controlled by the energy of the electron beam and the density/atomic number of the sample being analysed. However, with the thermal field

Content from this work may be used under the terms of the Creative Commons Attribution 3.0 licence. Any further distribution of this work must maintain attribution to the author(s) and the title of the work, journal citation and DOI.
Published under licence by IOP Publishing Ltd

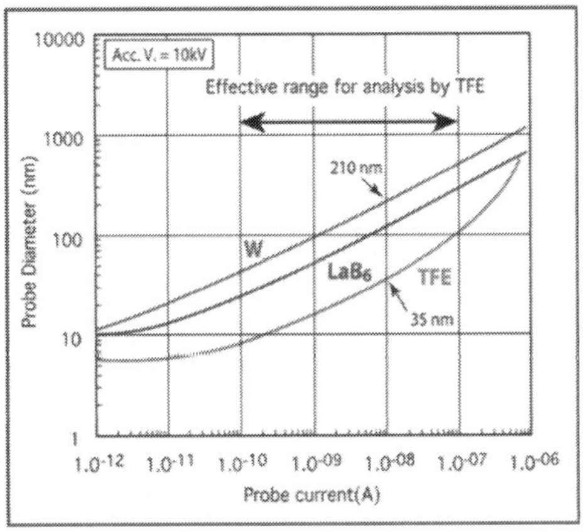

Figure 1. Comparison of the relative beam diameters produce by a tungsten filament (W), LaB_6, and a thermal field emission gun (TFE).

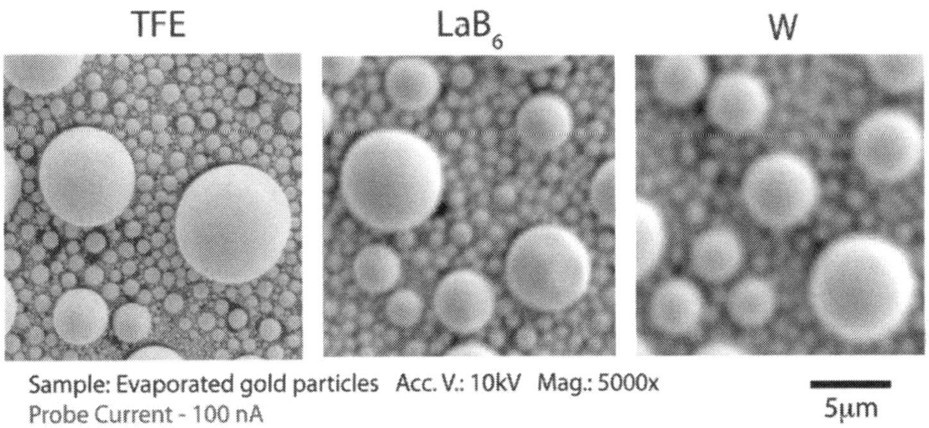

Figure 2. Differences in imaging between a thermionic W filament gun (W), a LaB_6 gun, and a thermal field emission gun (TFE). Sample is evaporated gold particles on carbon (imaging by JEOL, Ltd.).

emission (TFE) electron gun, it is possible to work at much lower accelerating voltages and still maintain a very small spot size. This greatly reduces the analytical area.

It should be pointed out that high-resolution imaging can also greatly improve the quality of microanalyses, even if one is not working at low-accelerating voltages. High spatial resolution imaging not only assists in determining what to analyze, but helps identify where not to analyse. Figure 3 is such an example. Shown are small crystallites in a volcanic glass. If the objective is to analyse the glass, it is critical to be able to identify areas where the crystallites exist in order to avoid them. Therefore, even if the analytical volume is much larger than the size of the crystallites, finding larger areas clear of them becomes paramount when attempting to analyse the glass.

Two general strategies have been suggested for analysing very small volumes [1]: (1) low voltage analyses, and (2) low overvoltage analyses. Both have advantages and disadvantages. This work will attempt to summarize the many issues that should be considered when attempting to decide which approach is best for a given analytical situation. These include: (1) the size of the analytical volumes, (2) minimum detection limits, (3) quality of the matrix corrections, (4) secondary fluorescence, and (5) effects of surface contamination, oxide layers, and carbon coatings.

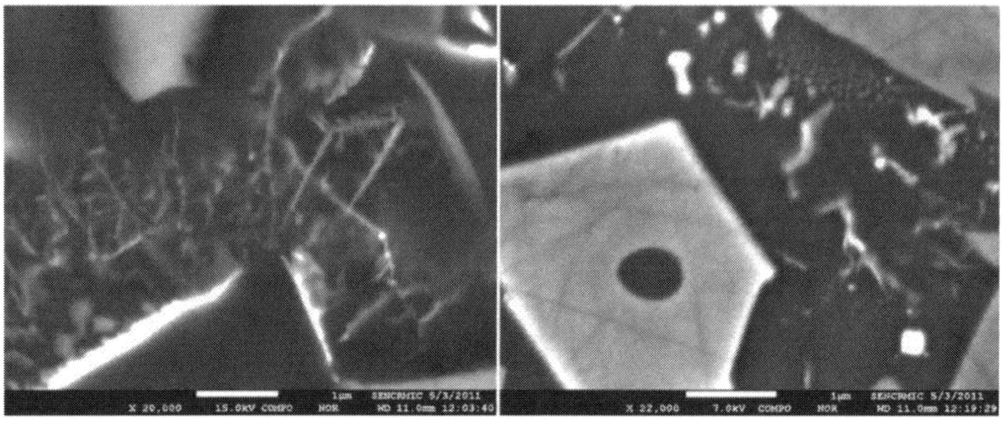

Figure 3. Backscattered electron images of sub-micrometre dendritic crystallites in a volcanic glass collected on a JEOL JXA-8530F field emission electron microprobe. The crystallites are too small to analyse, but high-resolution imaging means that they can be avoided when preforming analyses of the glass.

2. Low voltage analyses (low-kV)

As one decreases the accelerating voltage, and therefore the energy of the beam electrons, the spread of the electrons within the sample is also reduced. The electrons have less energy and therefore can travel only a short distance before they have lost all their energy. As a result, one can greatly reduce the volume being analysed by dropping the accelerating voltage from the typical microprobe analytical range of 15 - 25 kV to a low voltage range of 5 - 10 kV (figure 4). Often, this will require switching to a lower energy X-ray line. However, there is a limit as to how low the accelerating voltage can be decreased and still improve the spatial resolution of the analyses. As the accelerating voltage is decreased, the interaction volume decreases, but the diameter of the electron beam increases (figure 5). Therefore, for each material, there is an optimum accelerating voltage, where the size of the area of electron scattering within the sample is equal to the diameter of the beam. A lower accelerating voltage will increase the analytical area because the beam diameter is bigger, and a higher accelerating voltage will increase the analytical areas because of the larger scatter volume. The optimum accelerating voltage depends on the material, but typically falls in the 5 to 8 kV range.

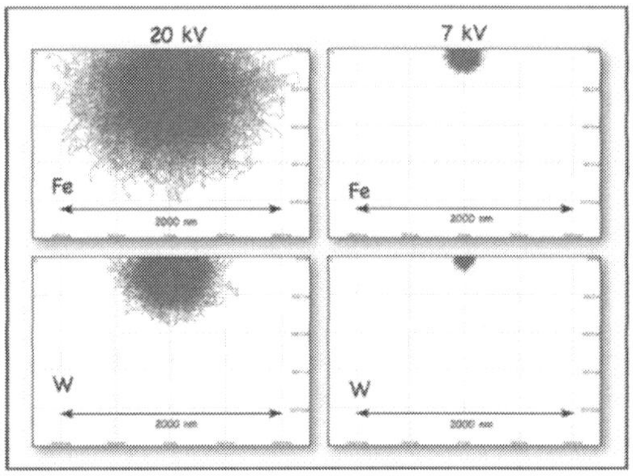

Figure 4. Monte Carlo modelling shows the improved spatial resolution with both increased atomic number and a decrease in the accelerating voltage. The diameter of the interaction volume for iron (Fe) drops from 2000 nm at 20 kV to about 300 nm at 7 kV. For tungsten (W), the diameter drops from about 1000 nm at 20 kV to about 160 nm at 7 kV. (Models generated by CASINO).

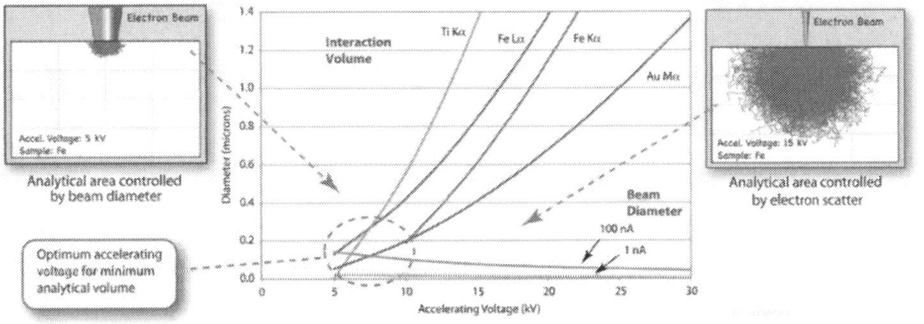

Figure 5. The analytical area is controlled by both the beam diameter and the scatter within the sample. For every material there is an optimum accelerating voltage that will produce the minimum analytical area. Commonly, this is in the 5 to 8 kV range. At a higher accelerating voltage, scattering within the sample generates a larger analytical area; at a lower accelerating voltage the expanding beam diameter makes for a larger analytical area. Beam diameters were measured on JEOL JXA-8530F field emission electron microprobe. Size of X-ray generation volume calculated from the Castaing equation [2].

3. Low overvoltage analyses (low-U)

A second method that can be used to generate a small analytical volume is to minimize the overvoltage (U) (U = accelerating voltage / critical ionisation energy of X-ray) [1]. Typically an overvoltage of around 2 is recommended for routine analyses. This gives each beam electron enough energy to generate at least one characteristic X-ray, even if it has lost considerable energy due to other types of inelastic interactions within the sample. However, this high overvoltage results in considerable scattering of the beam electrons within the sample, making a large volume from which the X-rays will be generated.

A low overvoltage analysis reduces the accelerating voltage to the minimum required to produce the characteristic X-ray of interest. An accelerating voltage of just 1 - 3 kV over the critical ionisation energy is used. This dramatically reduces the number of X-rays generated, but also greatly reduces the volume from which the X-rays will be produced. The total electron interaction volume can still be very large, but the beam electrons quickly lose the required energy to create the characteristic X-rays of interest. As a result, all of the X-rays of interest are generated from a small volume near the surface (figure 6).

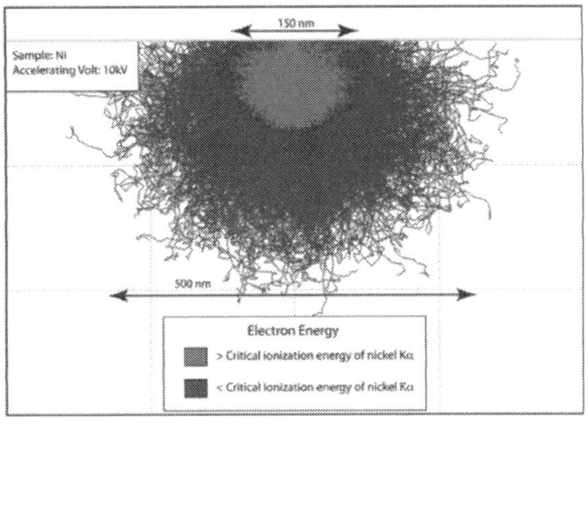

Figure 6. Shown are Monte Carlo simulations of the size of the total electron interaction volume within Ni metal (blue) for a 10 kV accelerating voltage. Compared to that is the volume in which the electrons have enough energy (> 8.332 keV) to produce Ni-Kα X-rays (analytical volume in red). As the 10 keV beam electrons scatter within the sample, they quickly lose sufficient energy to produce the Ni-Kα X-rays, resulting in a small analytical volume.

For the low-U analysis, 10 kV was selected for the accelerating voltage. This is only 1.7 kV higher than the critical ionisation energy required to produce the Ni Kα X-ray line (Table 1). A lower accelerating voltage would reduce the number of Ni Kα X-rays generated to a level less than acceptable. The alternative is to use the L-lines for Ni and Fe. The critical ionisation energy for these X-ray lines would, in theory, allow for an accelerating voltage as low as 1 kV to be used. However, considering the issues discussed in figure 5, and others to be discussed later, the low-kV analyses for this work will be limited to 7 kV.

Table 1. Critical ionisation energy of Fe and Ni X-ray lines [3].

	Kα (keV)	Lα (keV)
Fe	7.111	0.707
Ni	8.332	0.854

Both the low voltage method and the low overvoltage method for analysing sub-micron areas have advantages and disadvantages. The remainder of the paper will evaluate one binary system (Fe-Ni) in order to compare the differences between the two approaches. The purpose is not to determine which method is better overall. That will depend on the material being analysed and the objective of the analyses. For a given material, one method might work better for map analysis, while the other for quantitative analysis. One method might be better for the detection limits in trace element analyses, the other for accuracy in major element analyses. The objective is to highlight some of the issues that should be considered when attempting to achieve the smallest analytical volumes.

4. Analytical volumes

Monte Carlo modelling can be used to determine the size of the volume in which X-rays are generated by simply putting a threshold on the electrons, truncating the path once they have dropped below the critical ionisation energy of the X-ray in question. A comparison of the X-ray generation volumes for the low-U analysis method (using Fe-Kα, Ni-Kα, and 10 kV) and the low-kV analysis method (Fe-Lα, Ni-Lα, and 7 kV) is shown in figure 7.

For both Fe and Ni, the volume in which the L-lines can be generated at 7 kV in a metal containing 50 wt% Fe and 50 wt% Ni (figures 7a and b), is about 200 nm in diameter. For the K-lines at 10 kV, the Fe volume is also about 200 nm in diameter (figure 7c), whereas the Ni volume is closer to about 150 nm (figure 7d). The depths from which the X-rays can escape are generally smaller than the diameter of the X-ray generation volume. The $\phi(\rho z)$-plots in figure 8 show the depths at which the various X-rays are generated, and from which the X-rays can escape. For the K-lines, the generated and escaped profiles are nearly identical. For the L-lines, there is a significant difference.

Due to the low energy of the L-lines, there is a large discrepancy between the number of X-rays generated in the sample and the number that ultimately escape from the surface (figures 8a and b). The sample absorbs many of the generated X-rays before they reach the surface. This discrepancy must be corrected for by the matrix correction routines. These large corrections are one of the reasons for the errors when using the lower energy L-lines in quantitative analyses. Because of the higher energy of the K-lines, most of the generated Kα X-rays for Fe and Ni will escape; therefore, the matrix corrections are much smaller.

The analytical depth for these X-ray lines in a metal that is 50:50 Fe:Ni can be defined using the 99 % escaped depths show in figure 8. This is the depth above which 99 % of the escaped X-rays came. For Fe-Lα and Ni-Lα this is between about 120 to 150 nm. For Fe-Kα this depth is about 156 nm, and for Ni-Kα it is about 100 nm. Therefore, there is not a huge difference in the analytical

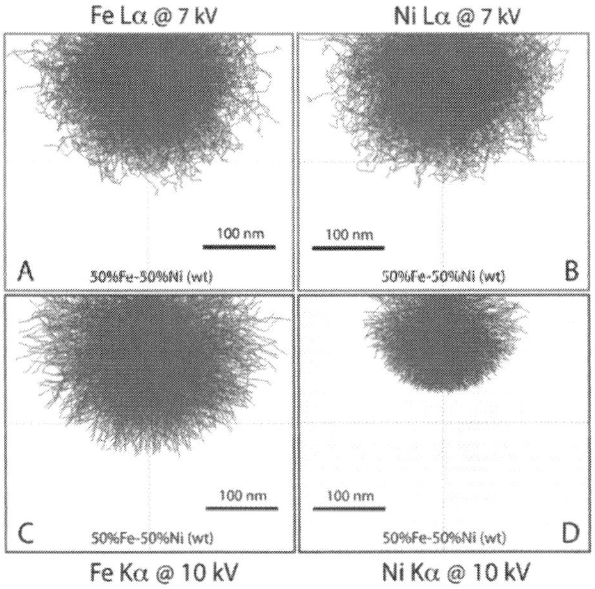

Figure 7. Monte Carlo modelling of the analytical volume for Fe-Lα and Ni-Lα X-ray lines at 7 kV, and Fe-Kα and Ni-Kα X-ray lines at 10 kV for a metal containing 50 wt% Fe and 50 wt% Ni. These volumes were determined from the electron interaction volumes truncated at the critical ionisation energy for the X-ray line in question. (Models generated using CASINO).

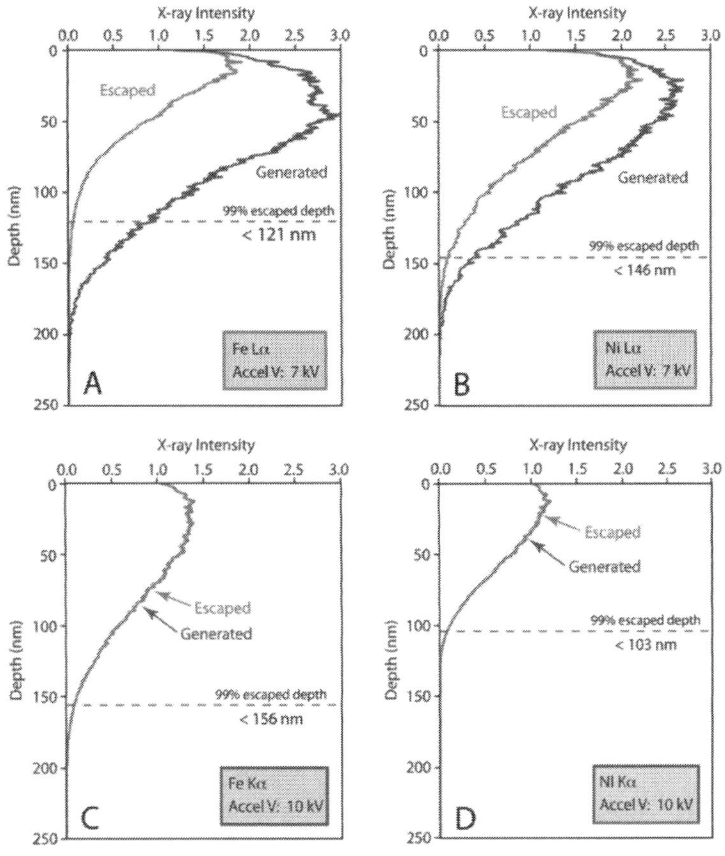

Figure 8. $\phi(\rho z)$-plots for Fe-Lα, Ni-Lα, Fe-Kα and Ni-Kα in a material consisting of 50 wt% Fe and 50 wt% Ni. The dashed lines on each plot show the depth above which 99 % of the escaped X-rays will come.

156 nm, and for Ni-Kα it is about 100 nm. Therefore, there is not a huge difference in the analytical area between the low-kV and the low-U analytical method for this binary system. Both should allow for analyses of volumes that are in the range of about 200 nm in diameter, and about 150 nm in depth. This is similar to the empirical observations of McSwiggen *et al.* [4], where they report consistent quantitative analyses of Ag_3Sn grains down to about 250 nm.

5. Secondary fluorescence

X-rays can be generated in a sample from more than just the beam electrons. In many samples, secondary fluorescence can also generate characteristic X-rays. X-rays produced by the primary electrons can fluoresce X-rays in another phase some distance from the primary electron interaction volume. These secondary X-rays are then mistakenly assumed to be coming from the phase being analysed [5, 6]. Whether a secondary X-ray is fluoresced depends mostly on the energy of the primary X-ray relative to the critical ionisation energy of the secondary X-ray. The probability of fluorescence is at its greatest when the energies are the same. As the difference between the two energies increases, the probability of fluorescence decreases. If the primary X-ray has an energy greater than the critical ionisation energy of the secondary X-ray of more than about 3 keV, then there will be no significant fluorescence [3]. If the primary X-ray has less energy than the critical ionisation energy of the secondary X-ray, then it is not possible to create the secondary fluorescence.

To determine the degree to which secondary fluorescence may be a problem in the Fe-Ni system, the NIST programme DTSA II was used to model the effect [7]. It was assumed that an area of pure nickel was in contact with a pure iron metal phase. Spectra were generated for spots within a pure Ni phase at various distances from the interface with Fe. The weight percent of Fe was determined using both the Fe-Kα line and the Fe-Lα line. The results are shown in figure 9. When the Fe-Kα line is used, one can see that the effect of secondary Fe fluorescence is very significant (figure 9a). Even 10 to 20 μm from the interface, about 0.5 wt% Fe could be measured. This would be a major problem if trying to measure minor or trace concentrations of Fe in a nickel phase that was surrounded by an iron phase. However, figure 9b shows that if the Fe-Lα line is used, no Fe was detected beyond about 0.2 μm from the interface when a 7 kV accelerating voltage was used. Therefore, a sub-micrometre grain of nickel, surrounded by an iron phase, could still be analysed for trace amounts of Fe. The reason there is little fluorescence of the Fe-Lα line is that the energy of the Ni-Kα line is too high to fluoresce a Fe-Lα line, and the energy of the Ni-Lα is too low to allow it to travel very far within the sample.

The Fe-Ni system is particularly sensitive to secondary fluorescence when using the K-lines. Other systems would have to be evaluated individually to determine the importance of this process when analysing sub-micrometre grains.

6. Detection limits

Both methods for analysing small volumes, whether using a low kV or using a low overvoltage will result in a drop in X-ray counts relative to those produced using the typical microprobe analytical range of 15 - 25 kV. As the accelerating voltage is reduced, each beam electron is capable of producing fewer X-rays. This in turn reduces the detection limits for the element in question. Figure 10 shows the calculated detection limits for Ni-Kα and Ni-Lα in Fe metal as a function of accelerating voltage. These detection limits were calculated using eq. 1, while keeping the counting time and beam current constant at 20 seconds and 100 nA, respectively. The absolute values shown in figure 10 are not significant. These will vary with the sensitivity of the spectrometer being used, the counting time, and the beam current. However, if one keeps these constant, the detection limit can degrade significantly as one attempts to minimize the analytical area by dropping the accelerating voltage. For the Ni-Kα line, the detection limits become asymptotic as the accelerating voltage approaches the critical ionisation energy (8.332 keV) (figure 10a). For the Ni-Lα line, the best detection limit is achieved at an accelerating voltage of around 10 to 15 kV (figure 10b). The

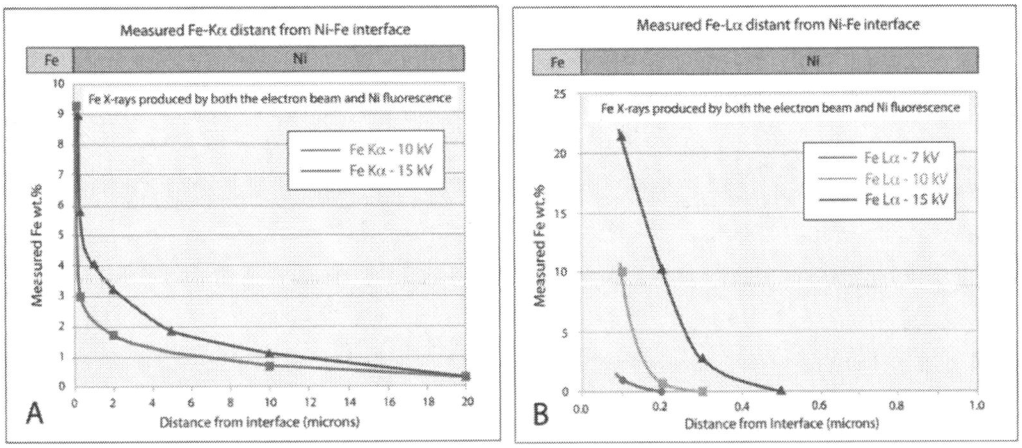

Figure 9. Plots showing the measured Fe resulting from secondary fluorescence when analysing Ni metal near an interface of Fe, using Fe-Kα (a) and Fe-Lα (b) X-ray lines. Even 10 to 20 μm from the surface, 0.5 wt% Fe can be measured if the Fe-Kα lines are used and a 10 kV accelerating voltage. If the Fe-Lα line is used, no measureable Fe can be detected past 0.2 μm from the interface. Modelling was done using the NIST programme DTSA-II [7].

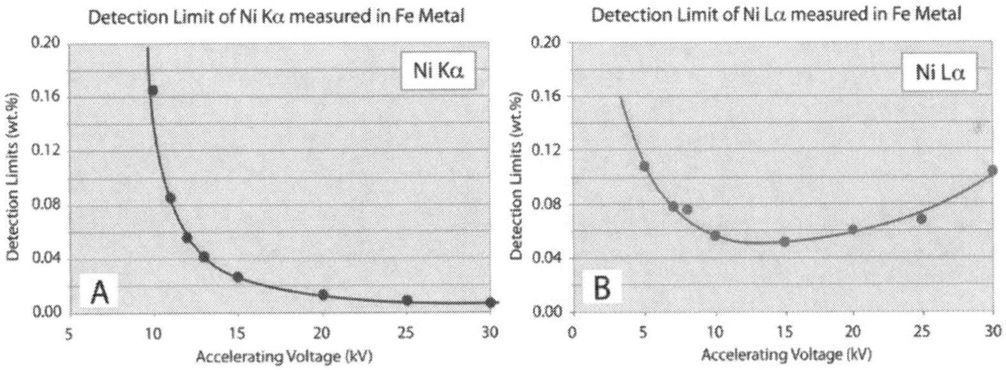

Figure 10. Detection limits for Ni-Kα and Ni-Lα as a function of accelerating voltage. Values were calculated from the total measured counts and Equation 1.

detection limit worsens when the accelerating voltage is either increased or decreased. Decreasing the accelerating voltage produces fewer X-rays, while increasing the voltage will produce more X-rays. However, the X-rays are produced deeper in the sample, thereby getting absorbed before they reach the surface. Therefore the maximum measured Ni-Lα X-rays occur at the intermediate accelerating voltage.

$$MDL = \frac{3\sqrt{I_B}}{I_P} * 100 \qquad (1)$$

where I_B = average counts of the background at the Ni peak position on Fe metal, and I_P = net counts of the Ni peak on pure Ni.

It is clear that there is a trade-off between trying to obtain the smallest analytical area and the best detection limits. Attempting to minimize the analytical area will result in a worse detection limit. Comparing a 10 kV Ni-Kα analysis with a 7 kV Ni-Lα analysis shows that the Ni-Lα analysis has the better detection limit for a comparable analytical area. However, for Fe the reverse is true. This is due to the low count rate for Fe-Lα on the TAP crystal. Again, every situation is different. Spectrometer performances vary, as do intensities from one element to another. Therefore, one cannot generalize to other element systems without preforming the measures on the specific electron microprobe that will be used.

7. Quality of quantitative analyses

One problem with low-kV analyses is that for many of the transition elements the K X-ray lines can no longer be used, if a 7 kV accelerating voltage is used, and the L-lines can be problematic in some cases. For example, Fe has an absorption edge between the Lα- and Lβ-lines. This absorption edge causes a shift in the combined peak position on a layered synthetic crystal (2d = 6 nm) with changes in Fe abundance. On the TAP crystal, Lα- and Lβ-peaks are separate, but their mass absorption coefficients are dramatically different and poorly known because of the position of this absorption edge [1]. For the low energy L-lines, this is a particular problem because of the massive absorption that they undergo within the sample. Figure 11 shows a plot of the measured Ni-Lα *k*-ratios at 5 different accelerating voltages for a set of standards along the Fe-Ni binary. At 20 kV, the measured *k*-ratio for a 50 wt% Ni standard is about 0.25. About half of the expected X-rays are being absorbed by the sample. At successively lower accelerating voltages, the depth the beam electrons can penetrate diminishes, and therefore, the volume the X-rays are generated in becomes smaller and closer to the surface. Since the path to the surface is shorter, fewer of the X-rays get absorbed. At a 2 kV accelerating voltage, the Ni-Lα *k*-ratio is equal to the weight fraction of the Ni in the binary sample, and no matrix corrections are required to obtain the correct values. In this case, simply using a sufficiently low accelerating voltage can overcome the uncertainty in the mass absorption coefficient.

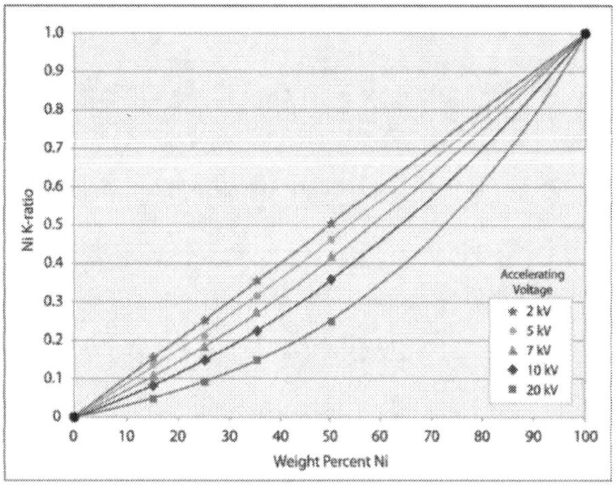

Figure 11. Plot of Ni-Lα *k*-ratios against the Ni wt% as a function of accelerating voltage for sample along the Fe-Ni binary.

Above a 2 kV accelerating voltage, the matrix corrections perform poorly (figure 12a) for the L-lines. For the Ni Lα-line, the matrix correction dramatically over-corrects for the absorption. For the Fe Lα-line, the matrix corrections are acceptable between 15 and 25 kV, but they do not properly correct at lower accelerating voltages. The K-lines perform much more reliably over the range of accelerating voltages from 10 to 25 kV for both Fe-Kα and Ni-Kα (figure 12b).

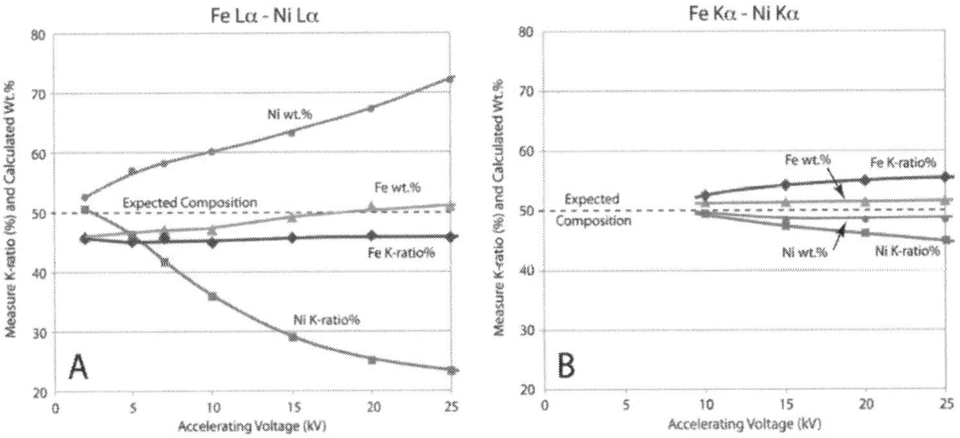

Figure 12. Measured *k*-ratios and matrix corrected wt% as a function of accelerating voltage for Fe-Lα and Ni-Lα (a) and for Fe-Kα and Ni-Kα (b) for a metal containing 50 wt% Fe and 50 wt% Ni. Pure end-member compositions were used as standards.

For quantitative analyses of sub-micrometre grains, the low overvoltage method will work better in the Fe-Ni system because the K-lines perform better in the matrix corrections. However, analyses using the L-lines can be improved by using standards that are closer to the "unknown" in composition. Table 2 compares the results of three metals along the Fe-Ni binary. Consistently, the 10 kV Kα analyses are better than those of 7 kV Lα analyses using 100 % metals for standards. However, the Ni-Lα results are greatly improved when a 50 % Ni standard is used.

Table 2. Quantitative analyses of three materials along the Fe-Ni binary using different X-ray lines, accelerating voltages, and standards.

weight% Ni				
Material	7 kV Lα 100 % Ni Std	7 kV Lα 50 % Ni Std	10 kV Kα 100 % Ni Std	Expected
Ni35	43.1	36.3	35.6	35.3
Ni25	31.0	26.0	24.8	25.2
Ni15	19.1	15.9	16.1	15.2

weight% Fe				
Material	7 kV Lα 100 % Fe Std	7 kV Lα 50 % Fe Std	10 kV Kα 100 % Fe Std	Expected
Ni35	62.8	66.4	64.4	64.7
Ni25	72.6	76.8	75.2	74.8
Ni15	84.1	89.0	83.9	84.8

If normal standards are not available, internal standards may greatly benefit low-kV analyses. In these situations, larger grains within the sample are analysed at a higher kV to determine their composition. These grains are then used as a standard for the low-kV analyses. These internal standards have the advantage of (1) potentially being more similar in composition, (2) having the same thickness of carbon coating, and (3) having the same surface oxidation or contamination as the unknown.

8. Surface coatings

Quantitative analyses at low kV can be very sensitive to surface coatings. Figure 13 shows the effect of various carbon coat thicknesses on the intensity of the Fe-Kα and Fe-Lα X-ray lines when compared to an uncoated Fe standard. Plotted are both the calculated curves using the equation of Reed (1997), assuming a density of 1.8 g cm^{-3} for carbon, and measured values (red dots) from an iron standard with an approximately 15 nm carbon film. These results show close agreement between the measured and calculated results. One can interpolate the amount of error that will result when different thicknesses of carbon are used on the standard versus the unknown.

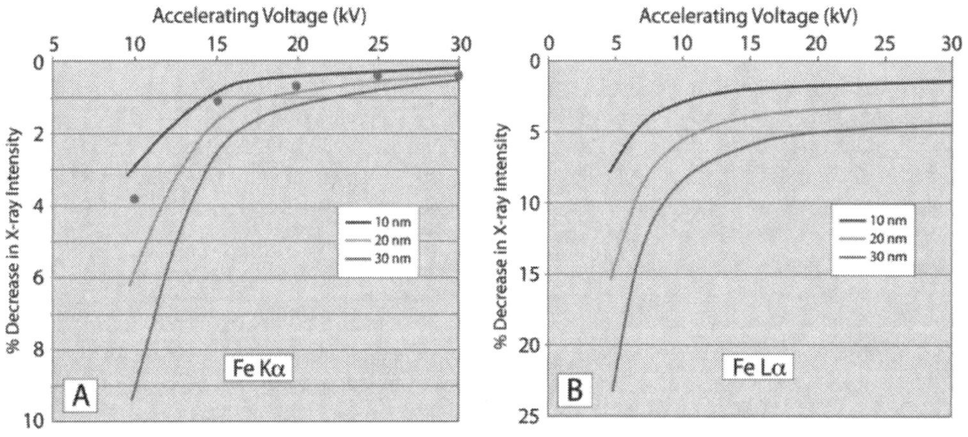

Figure 13. Effect of carbon coating on X-ray intensity. Plotted are the calculated curves showing the decrease in X-ray Fe counts resulting from carbon films that are 10 nm, 20 nm and 30 nm thick, with a density of 1.8 g cm^{-3} and using the formula from Reed [8]. The red dots are measured values from an iron metal standard which was expected to have about 15 nm of carbon on its surface.

At high accelerating voltages, when using the Fe Kα-line, the difference in counts between a coated and uncoated sample is in the 0.5 wt% range. However, at 10 kV, a standard with a 10 nm coating and an unknown with a 30 nm coating will result in a difference of up to 6 wt% in Fe Kα-counts. For the Fe Lα-line (figure 13b), the impact of the carbon coating is much greater. At 7 kV there will be approximately a 15 wt% difference in X-ray counts between iron samples coated with 10 nm of carbon versus 30 nm.

Similar impacts can be expected as a result of oxide layers on sulphide minerals, sputtered metal coatings, or residue from polishing. To minimize these effects, make sure the surface of both the standard and the unknown are recently polished and clean, and that any deposited coatings are applied simultaneously to the standards and unknowns to ensure consistency.

9. Conclusions

Using either the L-lines and a low accelerating voltage, or the K-lines and a very low overvoltage will produce very small analytical volumes in the Fe-Ni binary. However, both methods have their advantages and disadvantages. If the objective is to produce the best quality, quantitative analyses of the major elements, then the low overvoltage method is preferable, because of the better-understood behaviour of the K X-rays lines. However, for quantitative analyses involving minor or trace elements, then one should consider using the L-line, because they are less likely to be produced by secondary fluorescence, and a better spatial resolution can be obtained. For many systems, though, using multiple accelerating voltages to perform a single analysis might be the best approach. It would allow the user to optimize their analytical conditions for each element.

This strategy allows the user to measure the *k*-ratio of each element at the accelerating voltage that is best for that particular X-ray line. The *k*-ratios are then converted to weight percent using an off-line matrix correction programme that allows for multiple accelerating voltages. In a mineral system involving SiO_2, Al_2O_3, CaO, MgO and FeO, for example, the FeO could be analysed at 10 kV, using the Kα X-ray line for better quantitative accuracy, and the other elements could be analysed at 6 or 7 kV. For a typical silicate, the analytical volume for all of the elements would be around 0.5 μm under these conditions, and good quantitative results could also be obtained.

Regardless of the methods used, care should be taken when carbon coating the samples to ensure that identical thickness are applied to both the standards and the unknowns, and the surfaces should be cleaned of any oxide layers or polishing residue.

References

[1] Armstrong J T 2011 Low voltage and low overvoltage X-ray nanoanalysis with field emission electron microprobes and SEMs: problems in quantitation for first-row transition elements. in: *2011 AGU Fall Meeting, San Francisco, Calif., 5-9 December*; abstract V31C-2538

[2] Castaing R 1960 Electron probe microanalysis. in: *Advances in electronics and electron physics*. (Marton L L and Marton C; Eds.) (New York: Academic Press), vol. 13, 317-386

[3] Goldstein J I, Newbury D E, Echlin P, Joy D C, Romig Jr. A D, Lyman C E, Fiori C and Lifshin E 1992 *Scanning electron microscopy and X-ray microanalysis: a text for biologists, material scientists, and geologists*. (Plenum Press), 820 pp.

[4] McSwiggen P, Mori N, Takakura M and Nielsen C 2012 Improving analytical spatial resolution with the JEOL field emission electron microprobe. *Microsc. Microanal.* **17**, Suppl. 2 (abstract)

[5] Fournelle J H 2007 Problems in trace element EPMA: modeling secondary fluorescence with PENEPMA. *Eos Trans. AGU* **88**(52) Fall Meet. Suppl., abstract V51A-0329

[6] Fournelle J H, Gosses J, Kelly C, Staffier K, Waters J and Webber C 2005 Secondary fluorescence corrections for EPMA: PENELOPE MC simulations. *Microsc. Microanal.* **11**, Suppl. 2 (abstract) 1282-1283

[7] Ritchie N W M 2009 *Microsc. Microanal.* **15** 454-468

[8] Reed S J B 1997 *Electron microprobe analysis*. (Cambridge: Cambridge University Press) 326 pp.

Phase analysis in duplex stainless steel: comparison of EBSD and quantitative metallography methods

J Michalska and B Chmiela

Silesian University of Technology, Institute of Materials Science, ul. Krasinskiego 8, PL-40019 Katowice, Poland

E-mail: joanna.k.michalska@polsl.pl

Abstract. The purpose of the research was to work out the qualitative and quantitative analysis of phases in DSS in as-received state and after thermal aging. For quantitative purposes, SEM observations, EDS analyses and electron backscattered diffraction (EBSD) methods were employed. Qualitative analysis of phases was performed by two methods: EBSD and classical quantitative metallography. A juxtaposition of different etchants for the revealing of microstructure and brief review of sample preparation methods for EBSD studies were presented. Different ways of sample preparation were tested and based on these results a detailed methodology of DSS phase analysis was developed including: surface finishing, selective etching methods and image acquisition. The advantages and disadvantages of applied methods were pointed out and compared the accuracy of the analysis phase performed by both methods.

1. Introduction

Duplex stainless steels (DSS) represent an important and expanding class of stainless steels with an attractive combination of properties related to the fact that their microstructure allows a beneficial mixture of austenite (γ) and ferrite (α). However, the ferritic matrix in duplex stainless steels undergoes a variety of decomposition processes when aged in the temperature range 650 - 900 °C. Undesirable phases such as intermetallic phases, carbides and nitrides may exist in the steel if the manufacturing processes are not carefully controlled. High levels of elements stabilising ferrite, such as chromium, molybdenum and silicon, can promote the formation of sigma-phase (σ). σ-phase is a hard and brittle intermetallic phase which is generally formed between 600 and 950 °C with rapid formation kinetics [1-3]. Additional phases found in duplex stainless steels can include chi (χ), R, and α' [4, 5]. χ-phase belongs to the topologically close-packed (TCP) phases. It frequently occurs in steels, as a ternary compound containing Fe, Cr and Mo, according to the composition $Fe_{36}Cr_{12}Mo_{10}$ [6]. The nucleation sites for σ- and χ-phases are grain boundaries, incoherent twin boundaries and dislocations. It is well established that the precipitation of these phases leads to a reduction of the creep ductility. It has also a reverse effect on the toughness and corrosion properties [7, 8]. A substantial depletion of solid solution strengtheners (like Cr, Mo, C and N) mainly due to a copious precipitation of the σ- and χ-phases beyond one hundred hours results in a strong decrease of the yield strength [9]. Therefore, it is very useful to be able to identify correctly the quantity of σ- and χ-phases in the DSS.

Content from this work may be used under the terms of the Creative Commons Attribution 3.0 licence. Any further distribution of this work must maintain attribution to the author(s) and the title of the work, journal citation and DOI.
Published under licence by IOP Publishing Ltd

There are a lot of contradictions in the literature regarding the correct identification of σ- and χ-phases, mainly due to the similar chemical composition of these phases. Electron backscatter diffraction (EBSD) can be used to distinguish the various phases within the DSS [10-13] and may solve these problems. The ability to rapidly analyse large areas of samples and easier specimen preparation when compared to transmission electron microscopy (TEM) is a major advantage promoting its use. However, selection of sample preparation method for EBSD studies is a crucial point and may lead to significant impediment.

In the present work, EBSD phase analysis was employed to detect and evaluate the quantity of basis balance phases and intermetallic precipitations in 2205 DSS. The results of EBSD analyses were compared with quantitative metallography methods. A brief review of sample preparation methods for EBSD analysis was presented.

2. Experimental procedure

2.1. Material

The material used in this investigation was 2205 DSS (plate of 25 mm in thickness and rod of 25 mm in diameter) with its chemical composition shown in Table 1. All the specimens were water-quenched for 30 min at 1100 °C in order to homogenize the structure. Afterwards, isothermal heat treatments were performed: 30 min at 900 °C, 2 h at 750 °C, and 5 h at 750 °C.

Table 1. Chemical composition of 2205 DSS.

	Content of chemical elements (wt%)									
	C	Cr	Ni	Mo	Mn	Si	P	S	N	Fe
plate	0.021	22.70	5.95	3.20	0.88	0.39	0.020	0.001	0.16	balance
rod	0.017	22.82	5.45	3.12	1.57	0.30	0.019	0.001	0.17	balance

2.2. SEM observations and EDS analysis

Microstructural observations were carried out using a Hitachi S-3400 scanning electron microscope (SEM). The individual phases were analysed with a Thermo Noran energy-dispersive X-ray spectrometer system (EDS) with Six System software coupled to the SEM.

2.3. Quantitative metallography

For the metallographic research, specimens were cut out from the as-rolled steel plate in the longitudinal (L) direction of rolling and in the transverse (T) direction from the rod. Grinding was performed using water-cooled silicon carbide papers of 120-, 280-, 500- and 1200-grit size. After being ground to a 1200-grit finish, the samples were polished to produce flat, scratch-free surfaces. Coarse polishing was performed in two steps, using 6 μm and 3 μm diamond pastes; for the fine polishing step 1 μm diamond paste was used. Specimens were cleaned properly after each grinding and polishing step. Electrolytic etching was performed just after examination of the as-polished specimen. To maximize the contrast among different phases, 1 N KOH aqueous solution was used. In the electrolytic system, stainless steel was used as the cathode material.

The phase area fraction A was measured for each sample by Metllo [14]. As many as 10 fields were measured at 500x magnification. The procedure of the measurements is illustrated in figure 1.

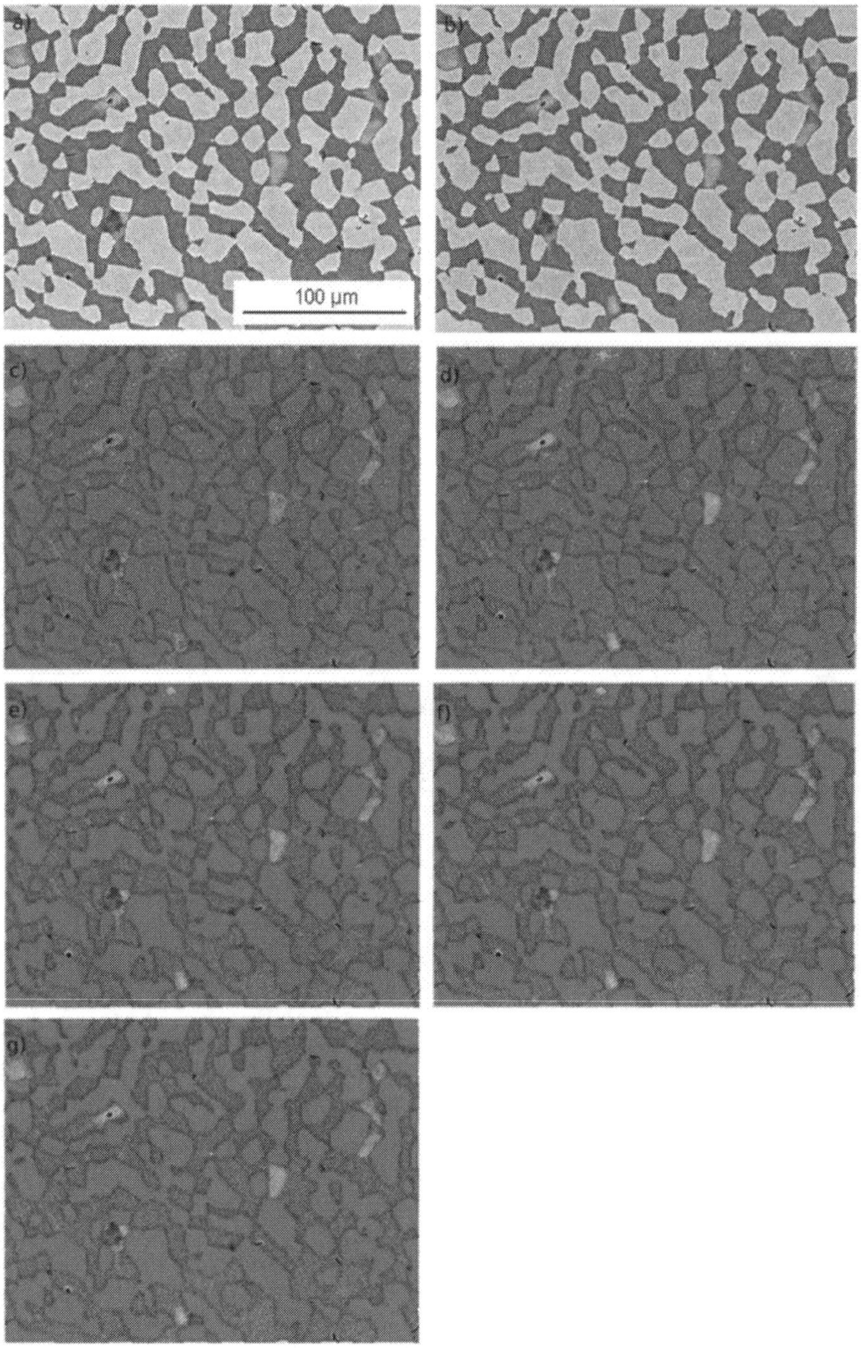

Figure 1. The measurement procedure in Metllo software for the quantitative evaluation of phases in 2205 DSS: a) uploading of the image; b) putting the median filter 3x3 in aim to smooth of the grey image; c) manual binarisation of the chosen phase in the image; d) selection of the objects (> 200 pixels); e) closing (with a step: 3); f) manual correction - filling of large holes; and g) measurement of the detected surface.

2.4. EBSD analysis

Specimens were cut from the DSS, mounted in conductive bakelite and were prepared using standard metallographic procedures, followed by prescribed method. Specimens were ground and polished using first a diamond suspension with grit sizes of 9, 3, 1, and 0.25 μm, then a colloidal silica suspension of 0.05 μm grit size. The last stage of specimen preparation consisted of vibratory polishing using a colloidal silica suspension with 0.05 μm grit size for 20 hours to obtain the lowest possible surface roughness. Several methods of surface finishing for EBSD analysis were studied.

The crystallographic orientation were characterized with a scanning electron microscope (SEM, Hitachi S-3400N) equipped with an electron backscatter diffraction (EBSD) detector INCA HKL Nordlys II (Channel 5). The primary SEM electron beam for imaging and EDS analyses was 15 keV. EBSD analyses were done with a 20 keV primary electron beam and a 0.3 μm step size. Because the indexation was better than 90 %, raw maps were submitted to the standard cleaning procedure as per the manufacturer's user instructions (Oxford Instruments HKL Technology, 2006). The cleaning procedure consisted of first removing wild spikes (isolated points that were incorrectly indexed) and then removing zero solutions (low-level extrapolation). Finally, we applied the 3×3 orientation-averaging filter (Kuwahara filter).

In Table 2 the list of phases and their crystallographic information is shown, which were supposed to exist in the 2205 DSS alloy.

Table 2. List of phases investigated in the DSS alloy using EBSD.

Phase	Crystal structure	Lattice parameters, Å
ferrite	body centred cubic (bcc)	a = 2.87
austenite	face centred cubic (fcc)	a = 3.66
σ	tetragonal	a = 8.80, c = 4.56
χ	body centred cubic (bcc)	a = 8.92

The orientation relationships between the phases are clearly seen in the inverse pole figures shown in figure 6. The austenite and ferrite phases were found to have the Kurdjumov-Sachs relationship, i.e., (110)Ferrite||(111)Austenite. The σ- and χ-phases can be seen to have an orientation relationship with the ferrite phase, (001) σ||(110)ferrite, (110)χ||(110)ferrite.

3. Results and discussion

3.1. SEM observations and EDS analyses

Electrolytic etching (3 V, 2 - 15 s) in 1 N KOH solution was used to reveal the microstructure of the steel. The respective microstructures of DSS in the as-received state and after heat treatment are shown in figure 2. As can be seen in these micrographs, the austenite phase (light grey phase) was embedded in the semi-continuous ferrite matrix (dark phase). Precipitation of intermetallic phases were coloured bright and were found at the expense of the ferrite with the increase in time of heat treatment. The chemical composition of the phases given by energy-dispersive X-ray spectrometry (EDS) was applied to preliminary identification. The results of EDS analyses of phases depicted in figure 2 are presented in Table 3. The EDS analyses confirm the correct detection of main balance phases: austenite and ferrite. In case of intermetallic precipitates, the analysis did not reflect any constant relation in the chemical composition. Some of precipitates are characterized by the higher content of Mo and Cr, while most of precipitates possess similar composition to the ferrite. These divergences in chemical composition of intermetallic phases raise doubts regarding the proper detection of phases in thermally aged DSS.

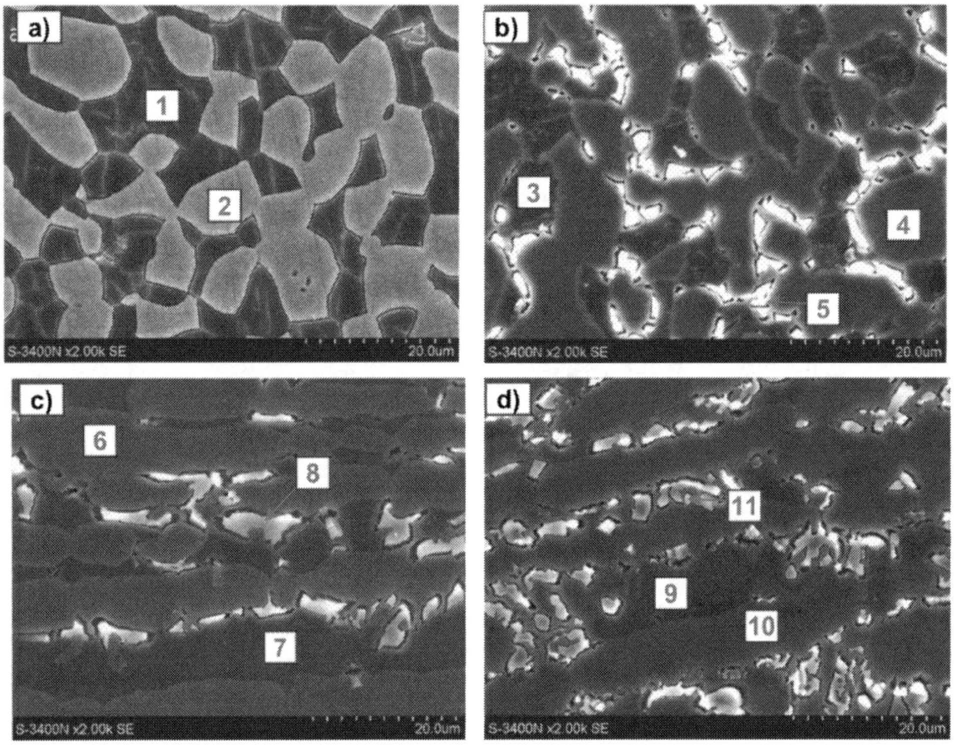

Figure 2. Microstructure of 2205 DSS: a) as-received (rod); b) heat treated at 900 °C for 30 minutes (rod); c) heat treated at 750 °C for 2 hours (plate); d) heat treated at 750 °C for 5 hours (plate).

Table 3. EDS analysis results of phases in 2205 DSS (average values for 5 analyses per each phase).

		Content of chemical elements (wt%)				
no.	**Phase**	**Cr**	**Ni**	**Mo**	**Mn**	**Fe**
1	ferrite	24.4	4.7	3.7	2.0	65.2
2	austenite	21.2	7.0	1.6	2.1	68.0
3	ferrite	25.5	2.6	1.9	1.7	68.3
4	austenite	21.8	6.4	1.9	1.9	67.8
5	precipitate	28.9	2.9	6.4	1.8	60.0
6	ferrite	25.4	3.8	3.1	1.9	65.7
7	austenite	21.5	5.9	2.0	1.9	67.9
8	precipitate	22.2	4.3	3.9	1.6	66.4
9	ferrite	24.4	3.3	4.1	1.5	68.4
10	austenite	21.6	6.4	2.1	1.8	68.2
11	precipitate	21.8	4.2	2.4	1.9	68.4

3.2. Quantitative analysis of phases

The quantitative analysis results of phases in DSS in the as-received state and after ageing are shown in Table 4. The results were obtained on the basis of measurements performed on SEM images (magnification 1000x, 10 images per one sample condition) by means of Metllo software. With the measurement methodology applied, the relative error of the volume fraction evaluation of phases and structural components in the steel did not exceed 5 %.

Table 4. Quantitative phase evaluation results for precipitates in 2205 DSS.

Sample	Phase	A, %
as-received	ferrite	53.5
	austenite	46.5
900 °C / 30 min	ferrite	26.0
	austenite	58.5
	precipitate	15.5
750 °C / 2 h	ferrite	23.9
	austenite	57.9
	precipitate	18.2
750 °C / 5 h	ferrite	8.5
	austenite	69.7
	precipitate	21.8

3.3. EBSD analysis - selection of preparation method

The preparation methods tested for EBSD purposes in this study are listed in Table 5. All the methods were tested on as-received DSS.

Table 5. Preparation methods for EBSD sample preparation for DSS.

no.	Procedure	Indexation	Quality of Kikuchi lines
1	**Vibratory polishing** (Al_2O_3 suspension 0.05 μm), time: 4 h	~ 85 %	good
2	**Vibratory polishing** (Al_2O_3 suspension 0.05 μm), time: 8 h	82 - 86 %	good
3	**Vibratory polishing** (Al_2O_3 suspension 0.05 μm), time: 4 h **Electrolytic polishing** (70 % ethanol, 20 % $HClO_4$, 10 % ethyleneglycol monobuthyl ether), 20 °C, 28 V, time: 10 s	84 - 87 %	good
4	**Vibratory polishing** (OP-S* 0.04 μm, pH = 10), time: 15 mins	50 - 60 %	poor
5	**Vibratory polishing** (OP-S 0.04 μm, pH = 10), time: 4 h	90 - 95 %	very good
6	**Vibratory polishing** (OP-S 0.04 μm, pH = 10), time: 8 h	90 - 95 %	very good
7	**Vibratory polishing** (OP-S 0.04 μm, pH = 10), time: 3x8 h (3 days with intervals)	90 - 92 %	very good
8	**Vibratory polishing** (OP-S 0.04 μm, pH = 10), time: 20 h (constant polishing)	94 - 96 %	very good
9	**Electrolytic polishing** (10 % $HClO_4$, 90 % ethylene glycol monobuthyl ether), 6 °C, 30 V, time: 30 mins	< 50 %	-

*colloid silica suspension

Among the tested methods, the best results of sample preparation were obtained using vibratory polishing. Alumina suspension did not work well for a final polishing; the best for EBSD pattern quality is colloidal silica. Constant long term vibratory polishing (no. 8) assures good quality of sample finishing and reliable OIM data (figure 3a). The solution polishes and slightly etches the specimen, removing most of the surface deformation layer. It was also reported that a key role in this method plays time of polishing. Shorter periods of polishing or using intervals during polishing cause the adverse effect of "ferrite shredding" (figure 3b). This effect was also observed for other methods tested in this study.

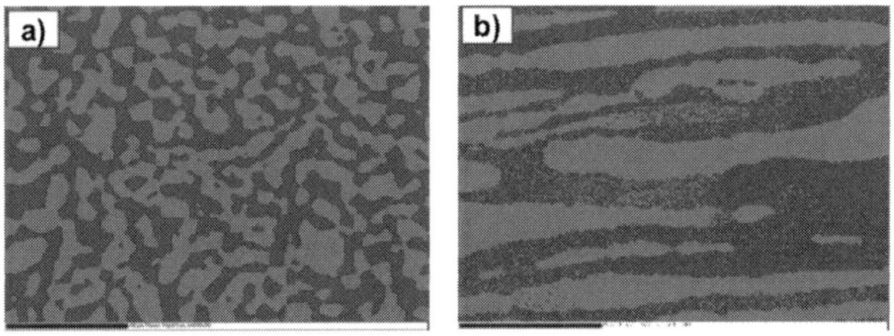

Figure 3. EBSD phase maps for 2205 DSS in the as-received state: a) well prepared sample (method no. 8; rod); b) poorly prepared sample (method no. 7; plate).

3.4. EBSD phase analysis

For the chosen preparation method (no. 8), samples in the as-received state and after heat treatment were studied for the phase quantification. Figure 4 shows the EBSD phase maps for all the samples tested in this study.

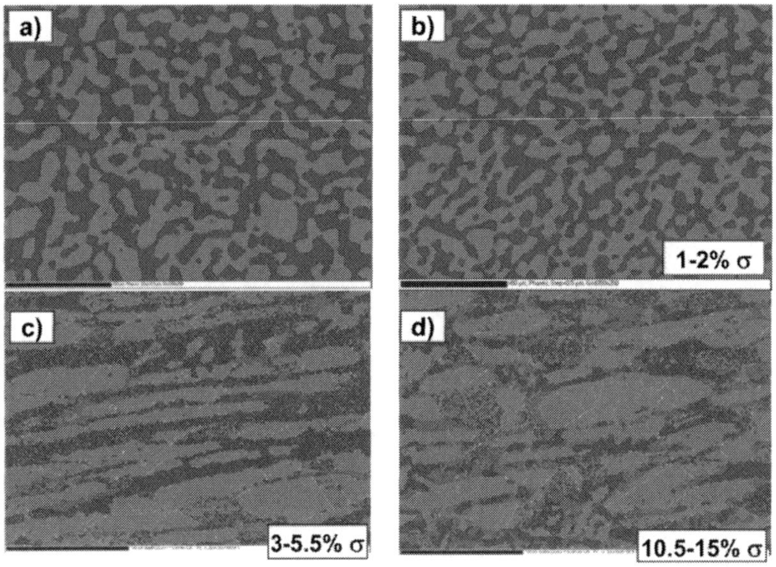

Figure 4. EBSD phase maps for 2205 DSS: a) as-received (rod); b) heated at 900 °C for 30 minutes (rod); c) heated at 750 °C for 2 hours (plate); d) heated at 750 °C for 5 hours (plate). Ferrite (blue), austenite (red), sigma (green).

The maps clearly show the distribution of the different phases in DSS. In the case of heat treated sample, the existence of the σ-phase was confirmed, while there was no evidence of χ-phase. The σ-phase was formed at the expense of the ferrite phase with the increase in time of the heat treatment. Unfortunately, in samples with higher content of the σ-phase re-calling of adverse effect of "ferrite shredding" on the phase map was observed. This phenomenon causes incorrect analysis of ferrite area fraction in favour of the austenite in DSS. No effect on σ-phase identification was observed.

Identification of σ-phase in DSS by EBSD method is poorly presented in the literature. Only few authors showed results concerning the quantification of σ-phase. However, the preparation method for EBSD measurements is usually has been neglected. Souza *et al.* [15] mentioned the difficulties with a proper quantitative analysis of σ-phase in DSS. The number of wrongly indexed points strongly increased with the work hardening produced before aging. This was the main limitation of the quantification by EBSD – cold rolled and aged (800 °C / 1 h) samples were very difficult to obtain good pattern. They also proved that the increase of the amount of σ-phase in DSS changes the austenite texture [15]. Probably, it could also influence the ferrite texture what may result in wrong indexation of ferrite. Up to the present, it has not been unequivocally proved and requires further investigations.

The area fractions measured from EBSD results are shown in Table 6. The area fractions of the σ-phase increased from 1 % to 15 % and the ferrite phase decreased from 41 % to 19.1 %, as the heat treatment time increased. Comparison of these results unequivocally depicts that identification of σ- and χ-phases should be performed on the basis of crystallographic data only.

Table 6. Area fraction of phases measured by EBSD in DSS.

Sample	Phase	A, %
as-received	Ferrite	45.3
	austenite	54.7
900 °C / 30 min	Ferrite	41.0
	austenite	57.5
	Sigma	1.5
750 °C / 2 h	Ferrite	22.8
	austenite	73.0
	Sigma	4.2
750 °C / 5 h	Ferrite	19.1
	austenite	67.8
	Sigma	13.1

4. Conclusions

EBSD has been shown to be a useful tool for phase identification in DSS, which is particularly evident in the analysis of the σ and χ intermetallic phases. The following conclusions can be drawn:

1. Standard metallographic polishing followed by a final colloidal silica polish, results in strain free surfaces suitable for EBSD studies assuring high level of indexation and good quality of Kikuchi lines.
2. This study clearly shows that evolution of EBSD patterns is dependent not only on surface polishing but also on the phases present in the sample. Higher content of sigma phase (> 2 %) in DSS disrupts correct identification of phases by the effect of "ferrite shredding". New improved method should be developed for thermally treated DSS.

3. Significant differences in the phase area fraction (A) values for the σ-phase between quantitative metallography and EBSD methods show that identification of this intermetallic phase should not be performed on a basis of microstructural observation only. Identification of the σ-phase based on the analysis of crystallographic orientation is absolutely necessary!

Acknowledgements
Scientific work was supported by the Polish National Science Centre under research project no. N507 230040 (2011-2013).

References
[1] Kim J S and Kwon H S 1999 *Corrosion* **55** 512
[2] Horvarth W, Prantl W, Stuwe H P and Werner E 1995 *Mater. Charact.* **34** 277
[3] Chen T H and Yang J R 2001 *Mater. Sci. Engng. A* **311** 28
[4] DeBold T A 1989 *J. Metals* March 12
[5] Michalska J and Sozańska M 2006 *Mater. Charact.* **56** 355
[6] Lee Y H, Kim K T, Kim K Y and Lee Y D 1998 *Mater. Sci. Technol.* **14** 757
[7] Wilms M E, Gadgil V J, Krougman J M and Kolster B H 1991 *Mater. High. Temp.* **9** 160
[8] Bernhardsson S 1991 in: *Proc. Int. Conf. Duplex Stainless Steels 1991* **1** 187
[9] Lopez N, Cid M and Puiggali M 1999 *Corrosion Sci.* **41** 1615
[10] Perret J, Boehm-Courjault E, Cantoni M, Mischler S, Beaudouin D, Chitty W and Vernot J-P 2010 *Wear* **269** 383
[11] Wroński S, Tarasiuk J, Bacroix B, Baczmański A and Braham C 2012 *Mater. Charact.* **73** 52
[12] Bozzolo N, Jacomet S and Logé R E 2012 *Mater. Charact.* **70** 28
[13] Marinelli M C, Krupp U, Kübbeler M, Hereñu S and Alvarez-Armas I 2013 *Engng. Fracture Mechan.* **110** 421
[14] Szala J *Metllo, authorized program.* (Katowice: Silesian University of Technology, Institute of Materials Science) version 12.01a
[15] Souza C M, Abreu H F G, Tavares S S M and Rebello J M A 2008 *Mater. Charact.* **59** 1301

Recent developments in two fundamental aspects of electron backscatter diffraction

K P Mingard[1], A P Day[2] and P N Quested[1]

[1] National Physical Laboratory, Hampton Road, Teddington TW11 0LW, Great Britain

[2] Aunt Daisy Scientific, Claremont House, Lydney GL15 5DX, Great Britain

E-mail: ken.mingard@npl.co.uk

Abstract. Two very different aspects of electron backscatter diffraction (EBSD) are considered in this paper. Firstly, the use of the technique for the measurement of grain size is discussed with particular reference to the development of international standards to help ensure reproducible and repeatable measurements. In particular the lessons learnt for both calibration of the complete SEM-EBSD system and in choice of the correct data acquisition and processing parameters from an international round robin are summarized. Secondly, extending the capability of EBSD through development of new detectors is discussed. New shadow casting methods provide a means to achieve better accuracy in definition of sample-pattern geometry, while increased detail can be obtained by larger cameras and ultimately direct electron detection.

1. Introduction

A full review of the entire state of the art in electron backscatter diffraction (EBSD) would be impossible in a short article so it is intended here to look at a two particular aspects of EBSD, one being an application of the technique in common use and the other a development of the technique through possible improvements to detector technology.

One of the most common and important applications of EBSD is the measurement of grain size and phase, particularly for industrial quality control and materials characterisation. This is because of its capability to not only group pixels of one particular orientation, but also to unambiguously define grain boundaries by the misorientation across a boundary. Both enable software routines to automatically calculate sizes based on area or linear intercept. However, this automation brings with it the usual risk of accepting any number produced without understanding the factors which affect the accuracy of the result. The first section will review a recent study [1] of the repeatability and reproducibility of this fundamental measurement and ways of improving it through, for example, international standardisation [2, 3].

Advances in EBSD detector technology have been evolutionary rather than revolutionary in the past decade. Acquisition rates are one of the most commonly quoted parameters for a detector, and the maximum rates have increased perhaps by a factor of 10 over 10 years, with quoted maxima of between 600 and 700 patterns per second. At such rates the drive for faster detectors is perhaps lessening and other factors are beginning to receive attention such as accuracy and detail possible in acquisition of electron backscatter patterns (EBSPs). This includes the possibility of moving from current phosphor/CCD cameras to direct electron detection.

Content from this work may be used under the terms of the Creative Commons Attribution 3.0 licence. Any further distribution of this work must maintain attribution to the author(s) and the title of the work, journal citation and DOI.
Published under licence by IOP Publishing Ltd

2. Measurement of grain size by EBSD

Studies comparing the results of grain size measurements by optical and EBSD methods have been carried out over the past 15 years [4-10], but reported comparisons of measurements by different laboratories have been few [11, 12] and are old enough that they do not reflect improvements in EBSD detectors software and scanning electron microscopes of the past 10 years. In some of these works differences were observed in grain size between optical and EBSD and besides the greater spatial resolution of EBSD technique other factors were noted, such as limitations of optical methods in detecting boundaries below a certain misorientation and the variable effectiveness of etching in different samples affecting optical measurements.

More recently, an interlaboratory comparison was carried out in support of a new international standard, ISO 13067 "Microbeam analysis - Electron backscatter diffraction - Measurement of grain size and Ddstribution" [2] to provide uncertainty data for grain size measurement by EBSD. Commercially pure (CP) titanium metal bar with an average grain size of approximately 30 µm was chosen to produce uniform and repeatable samples since, in addition to being suitable for the EBSD technique, this size would allow comparison with optical microscopy methods of grain size measurement such as ASTM E112 [13]. The following is a summary of the full report of this round robin [1]. The report also includes details of the studies undertaken prior to the round robin to confirm the consistency of the grain size in the central region of all the samples used. In this paper, the phrase "mean grain size" refers to the most commonly used number average mean: there is not room in this paper to discuss the advantages and disadvantages of using various methods of weighted averages based for example on grain size area, nor the advantages of using the full grain size distribution rather than a single value, but these issues are discussed further in reference [1].

Figure 1 summarizes the data for the mean equivalent circle diameter grain size, D_{circle}, as returned from 12 laboratories, each denoted by a letter. Each laboratory measured two samples but in some cases the laboratories measured the two samples with two different microscopes and EBSD systems or in one case with two magnifications; in these cases two sets of two values are shown. It should also be noted that laboratories A and F measured the same sets of samples; similarly the samples measured by C were then measured by H, and those measured by E were measured by L. The mean value of these raw results before removal of any outliers or corrections for errors was 33.7 µm.

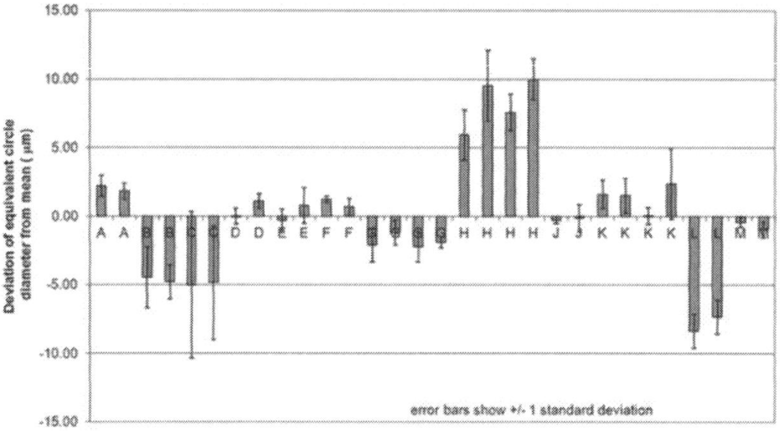

Figure 1. Mean values of D_{circle} for each sample as initially reported by the participants (before any corrections). Values are plotted as the difference from the mean of 33.7 µm. Error bars show ±1 standard deviation from 5 maps contributing to the mean.

Laboratories were informed of their results and were allowed to recheck their results where they appeared to have produced outliers. In two cases errors (discussed later) were identified which enabled some recalculations, after which all the results were analysed statistically following the procedures outlined in, for example, ISO 5725-2 [14, 15]. This analysis resulted in two sets of results being classed statistically as outliers (for reasons also described later) and after excluding these values, the final results of the round robin are summarized in Table 1.

Table 1. Summary of results after statistical analysis

		Equivalent circle diameter D_{circle}	Linear intercept D_{lin}
Overall mean (μm)		33.5	30.1
Repeatability standard deviation (μm)	s_r	0.88	0.74
Reproducibility standard deviation (μm)	s_R	1.98	1.15
Repeatability limit (μm)	r	2.46	2.07
Reproducibility limit (μm)	R	5.55	3.22
Number of measurements after removal of outliers		12	11

The repeatability and reproducibility limits are the values less than or equal to which the absolute difference between two test results obtained under repeatability/reproducibility conditions may be expected to be with a probability of 95 %.

2.1. Analysis of the variability of grain size measurements

Before exclusion of any outliers the mean value of D_{circle} was 33.7 μm but with a spread (± 2σ) of ±8.6 μm. This wide spread was attributed largely to significantly different values reported by just 4 of the laboratories, B, C, H and L, which if completely excluded changed the mean value slightly (34.0 μm), but reduced the spread significantly to ±2.9 μm (± 2σ). Examining the causes of variation for the 4 laboratories B, C, H and L helped identify some of the principal causes for much of the variation.

2.1.1. Magnification. Calibration of magnification was found to be a cause of variation for two laboratories (H and L). In one case this could be corrected, after tracing the problem to linking calibration of the SEM to the EBSD software. This type of software communication can easily occur without the user being aware of it as magnification is all too often taken as an absolute value for the SEM when it should refer only to the size at which the scanned image is displayed. It cannot be emphasized enough that what should be checked is the equivalence of scale bar markings or total image widths on the SEM and EBSD systems, not the magnification!

2.1.2. Step size and sample size. All but one laboratory used a step size of between 2 and 3 μm, or 6 to 9 % of the mean, agreeing well with the empirical rule established by Humphreys [5], that step sizes approximately 10 % of the mean grain size should produce good maps of most microstructures. The one laboratory that used a finer step size of 1 μm obtained a much smaller mean size 15 % below the mean. While this step size allowed resolution of small grains that might have been missed by coarser step sizes, the more important factor was that, for a given set of acquisition conditions and mapped area, acquisition would take 4 or 9 times longer than with, respectively, 2 or 3 μm step sizes, so instead much smaller areas were mapped. This not only reduced the total number of grains but also eliminated many of the larger grains from the analysis because such grains are more likely to touch the map edges and thus be discarded because they are incompletely analysed.

It is also worth noting that analysing the results from many of the laboratories by plotting running averages showed that, at least for this material, a minimum of between 200 and 300 grains is sufficient to obtain a stable average that will change little as further grains are measured. This is probably applicable to most materials with an equi-axed grain structure and a similar size distribution, but with this quantity of data a cumulative size distribution curve itself could be plotted to give a more comprehensive description of the grain structure.

2.1.3. Specification of minimum grain size. It recommended in the standard ISO 13067 that grains >10 pixels in area will give valid results since EBSD validates each pixel in a grain by measuring its orientation. Pixellation errors, which alter the true grain size at small sizes, are approximately 5 % at the 10 pixel area level. This cut off limit also agrees approximately with conventional image analysis where the minimum size of an object is 9 (3x3) pixels for a square grid or 7 for a hexagonal grid to avoid erosion of step size 1 deleting an object completely. Thus any grain comprised of <10 pixels should be ignored in any calculation because it is unlikely to be well defined in size: this size is called the cut off limit.

In the round robin however a variety of cut off limits were used. Most were stated in terms of pixels although a couple were in terms of a minimum size in micrometres. The cut offs selected by the participants included > 1, 4, 5, 6 or 8 pixels, but most chose 10 pixels. Because of the variation in step sizes chosen, the actual sizes of the cut off limits varied from 2 µm to 10.75 µm, equivalent to 3.46 μm^2 to 91 μm^2. Figure 2 plots the mean grain size against the cut off limit used (where this was known). The cut off is plotted in micrometres rather than pixels to take into account the step size used as well as the number of pixels and suggests that the mean size initially increases before beginning to plateau as the cut off is increased.

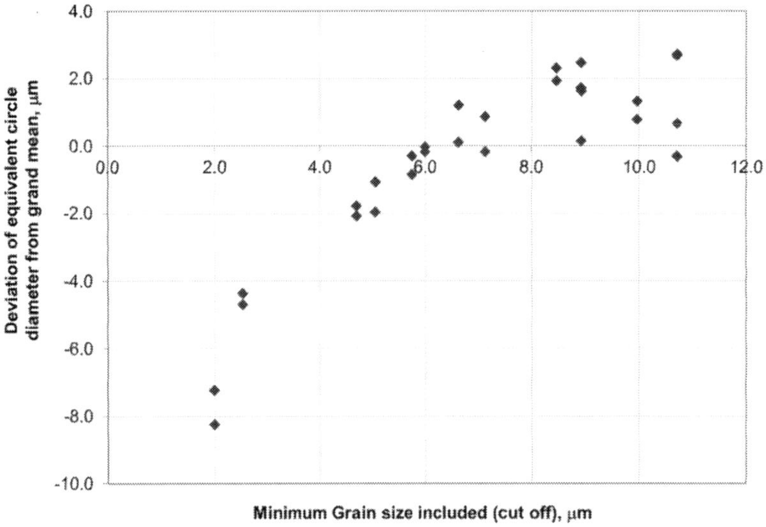

Figure 2. Mean values of D_{circle} for each sample (calculated as the difference from the overall mean value) plotted against the cut off minimum grain size (expressed in µm, not pixels) used by each participant in the calculation of their sample mean size.

The data supplied by individual laboratories was reanalysed to give a clearer idea of the effects of varying the cut off limit. This showed that if a single cut off value, specified in terms of micrometres rather than pixels, of 8 μm was used for all laboratories (with the exception of the outliers), the data spread in figure 3 was reduced from ≈ 5 μm to 2 μm. It should be noted that good agreement would be achieved at any given cut off above about 8 μm, as long as a single consistent cut off is applied to all the data.

2.1.4. "Clean-up" of map data. A difficult area to specify in the standard ISO 13067 was the amount of data "cleaning", or replacing poorly indexed pixels by more appropriate orientations that might be allowed, since many factors can affect how much and where the poor indexing occurs. Poor data cleaning (or noise reduction) did account for one of the initial outliers (C) in figure 1, and the full report [1] gives more detail than there is room for here about clean-up issues, but it may be noted that the maps in figure 3 show the extremes of the percentages indexed in the raw maps supplied (98.8 and 75 %) and both these maps with suitable clean up produced grain size values very close to the overall mean. One factor in this was the choice of a suitable cut off since this can remove most or all of the poorly indexed pixels. However, if too large a cut off is chosen before data cleaning, subsequent reindexing of the removed pixels may lead to overestimation of the remaining grains.

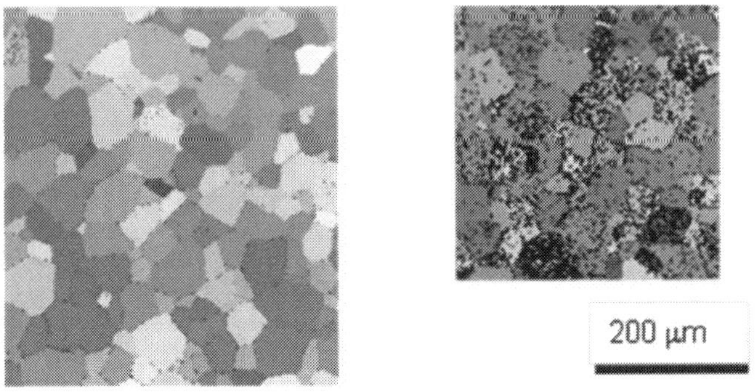

Figure 3. Raw EBSD orientation maps before noise reduction showing two extreme levels of good indexing (left – 98.8 %, right, 75.0 %) which gave very similar mean grain size measurements (34.2 μm and 34.7 μm respectively).

2.1.5. Other factors. A range of other factors can influence the mean grain size value. These include errors in acquisition, two of which can combine to affect the magnification calibration in the vertical direction: firstly, inaccurate measurement of the tilt angle of the specimen, and secondly, drift of the specimen/imaged area which usually occurs in the same direction. Other factors include definitions used such as the chosen value of angular misorientation used to define a grain boundary. In some combinations the effect of these factors can cancel each other out, but if combined to, say, all increase the mean they can add 15 - 20 % to the quoted value.

2.2. Linear intercept and optical grain size measurements

The results discussed in the preceding sections were all based on the use of circle equivalent diameter grain sizes D_{circle}, but linear intercept measurements D_{lin} are also commonly used, particularly when comparing with optical methods of measurement or where grains are not equi-axed. The participants in the round robin were all asked to report D_{lin} as well and these are summarized in Table 1 next to the D_{circle} values, from which it can be seen that:

i) Comparing the overall mean values, the D_{circle} value is about 9 % greater than D_{lin}, which is close to the ratio suggested in the appendix to ISO 13067 [2].

ii) The repeatability of the two measurements is very similar (r_{circle} = 2.54 is 20 % greater than r_{lin} = 2.13), but the reproducibility of D_{lin} is much better than that of D_{circle} (R_{circle} = 5.79 is 70 % greater than R_{lin} = 3.36).

It was also noted that only one laboratory reported linear intercept measurements in both X and Y directions and the same laboratory was the only one to explicitly report using a cut off for minimum linear intercept size in the same way as a cut off value was reported for the equivalent circle diameter. One possible reason for the increased reproducibility range for the D_{circle} measurements is the unassessed error of measurement in the Y direction caused by drift and/or tilt errors which would only be revealed by comparison of D_{lin} measurements in both directions after using this same method to check that the material was truly equi-axed.

Linear intercept measurements were also made on optical images taken from one of the samples. This produced a mean value of 34.6 μm taken from 431 measurements. The good agreement with the EBSD measurements could be seen from plots of the size distribution (figure 4) to be in part a consequence of choosing a suitable cut off value for the minimum grain size chosen for EBSD such that it had a similar effective resolution to that of the optical microscope.

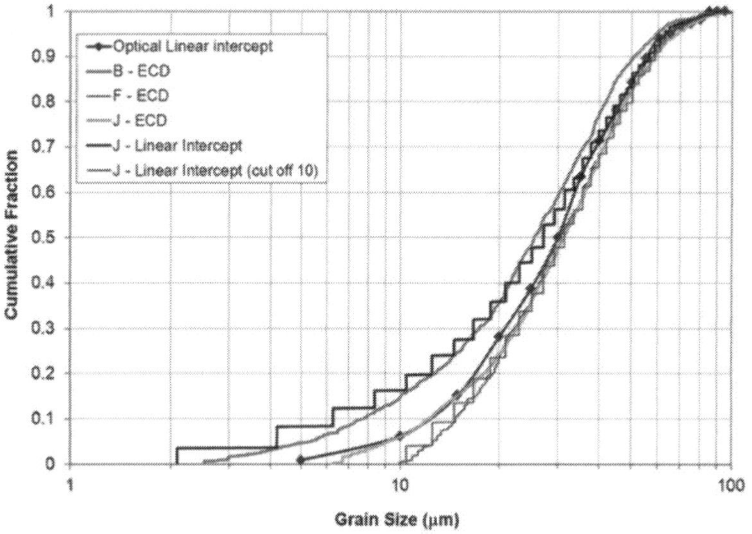

Figure 4. Distribution of optically measured linear intercepts compared with EBSD measurements by linear intercept and equivalent circle diameter from a selection of participating laboratories B, F and J.

2.3. Summary

The use of EBSD for measurement of grain size is now a widely used method in industry and academia and the round robin study described has shown reproducible measurements are possible, but only with careful attention to detail. Even obvious issues such as magnification calibration need to be checked in orthogonal directions and use should be made of both circle equivalent and linear intercept measurements to look for errors caused by, for example, tilt or drift. Use of agreed standards such as ISO 13067 will help with this, and for specific materials attention must be paid to agreeing definitions of grain boundary angles and particularly a minimum grain size (cut off value) used in the calculation. Further studies, extending perhaps to finer grain sized materials, would be useful to establish if greater reproducibility is possible by application of the findings from the current work.

3. EBSD cameras

It was noted in the introduction that acquisition rates have increased by a factor of about 10 in the past decade. However, increased speeds usually require the compromise of a large reduction in the information available from the EBSPs. For much day to day mapping of microstructures this is perfectly acceptable, but there are some applications where high detail and accuracy in EBSP acquisition is of more importance than speed. Examples of such applications include discrimination between phases with similar structures or small differences in lattice parameter such as tetragonal lead zirconium titanate piezoelectric materials with differences between a and b lattice parameters of < 3 % [16, 17].

A significant application which requires high accuracy and detail in acquired EBSPs is the measurement of strain: a paper by Wilkinson [18] this series will describe this method in more detail, but it is appropriate here to describe some of the implications this has for detector design and accuracy of EBSP acquisition. A key requirement for the measurement of elastic strain is the ability to compare (by cross-correlation) an EBSP from the strained material to an EBSP produced from unstrained material of the same orientation [19]. This can be difficult, if not impossible in, for example, small grained materials in which unstrained positions in each grain may not be identifiable. A way of overcoming this problem is to use a simulated strain free EBSP. This has two main requirements: obviously a method for producing a detailed simulation is needed, but also an accurate knowledge of the geometry of the EBSD system, relating the position of the beam on the specimen generating the EBSP precisely to the position of the camera phosphor and any distortions in the optical path between the phosphor and the CCD camera.

Considering pattern simulation first, there are two possible methods with very different requirements for computational time. Kinematical simulations can produce simple patterns (as shown in figure 5) relatively quickly but studies by Britton [20] have shown that their use can result in significant errors in calculations of strain because of the large differences in intensities, especially at the zone axes of the patterns. Dynamical simulation of EBSPs, pioneered by Winkelmann [21], produces much greater detail, as can be seen in figure 5b. With appropriate filtering of the cross-correlation between the real and dynamical simulations in figures 5a and b, realistic calculations of strain fields can be achieved.

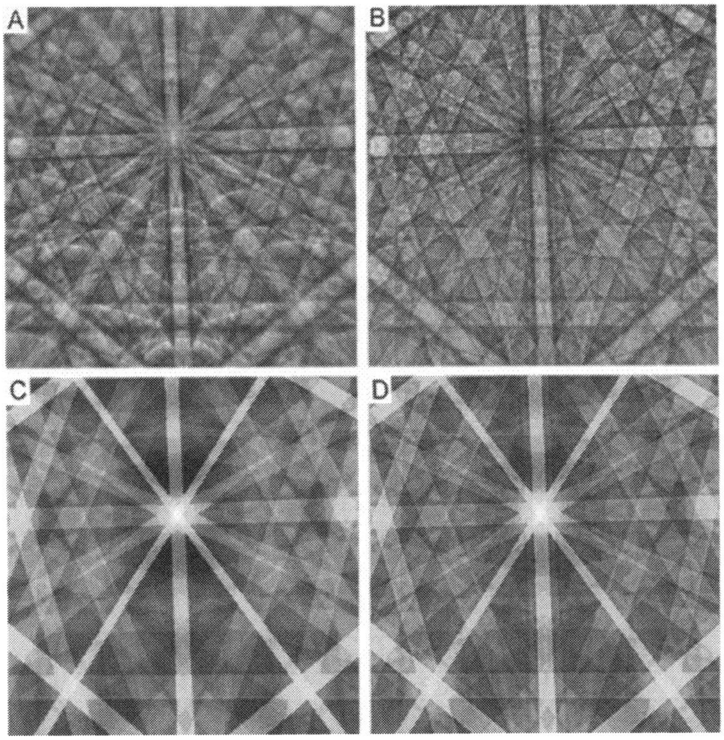

Figure 5. Cropped 20 kV EBSPs from Ge: a) experimental; b) dynamic simulation of pattern with minimum plane spacing of 0.5 Å and no inelastic scattering anistropy; c) Bragg simulation with 204 reflectors and no higher orders; d) improved Bragg simulation with simple band profile and higher order reflectors (after Britton *et al.* [19]).

3.1. Detector development

Given the detail possible from dynamical simulations, it is appropriate to consider if greater detail could be captured with improvements to EBSD detectors. Commercial detectors designed for higher quality pattern acquisition use CCD cameras with 12-bit acquisition and ranging from 1000 to 1600 pixels in their maximum dimension (although in most day to day operations they would be run with binning of pixels to reduce acquisition time and images generally stored after background correction in 8-bit grey scale). An experimental system with a 5 MPixel 16-bit CCD camera with 2448x2050 pixels and a 50 % larger phosphor screen has been demonstrated [22] and figure 6 compares the output of this camera with that of a commercial camera producing 1340x1024 pixel image when positioned 30 % closer. With only static background correction and no averaging, the greater detail, particularly in the black deficiency lines is obvious, as is the greater field of view. The ring and disc shadows visible in the 4 corners of the image will be discussed in the following section.

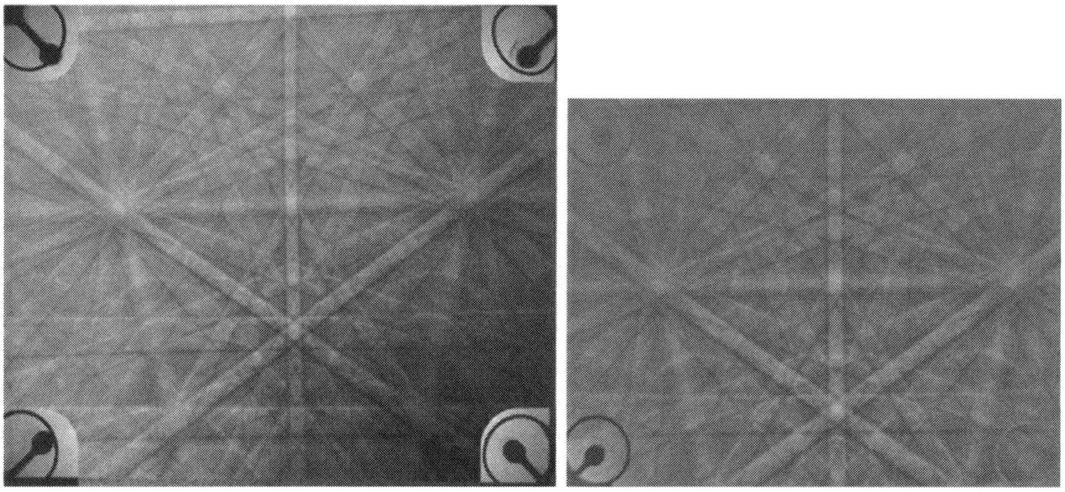

Figure 6. a) 2448 x 2050 pixel EBSP single frame image acquired from a <001> oriented silicon specimen, shown at the same magnification as b) a 1340x1024 pixel image (averaged over 10 frames) from a commercial camera positioned 30 % closer to the specimen.

An alternative method being investigated is to eliminate the phosphor screen and CCD camera altogether and detect electrons directly; previous work has looked at direct electron detection for transmission [23, 24] and low-energy electron microscopes[25] but is now looking at energies suitable for scanning electron microscopy [26, 27]. The conversion by the phosphor screen of the electrons to light and then reconversion of this light back to electrons by the CCD is obviously inefficient. Detecting the electrons directly increases the signal available and can in principle run faster with greater definition in the pattern as well. It also reduces errors arising from misalignment of the optic system and, being planar, removes distortions due to the lens, both of which cause problems for comparison of real and simulated patterns [28, 29]. Figure 7 shows an EBSD pattern from a GaN thin film acquired with a direct electron detection Timepix chip positioned approximately 3 cm in front of the sample [26]. The EBSD pattern shown in figure 7 was acquired at an acceleration voltage of 30 kV, at a beam current of order 0.5 nA, a beam spot size of around 2 nm and an acquisition time of 1 second. While the field of view is limited, the increase in detail available with this camera is noticeable.

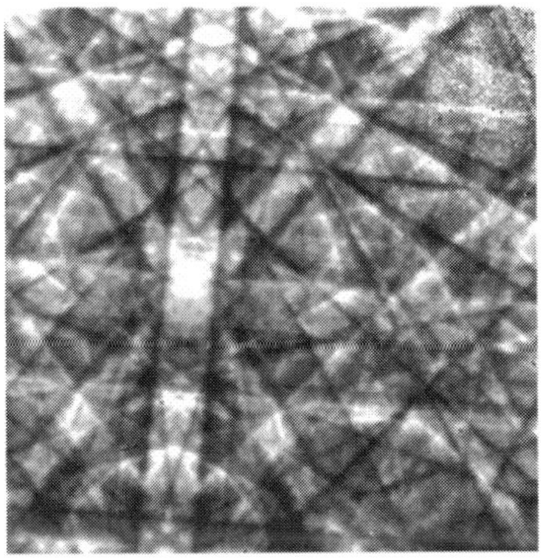

Figure 7. Direct electron detection EBSP of GaN acquired with a Timepix chip at 30 kV (after Vespucci *et al.* [26]).

3.2. EBSD camera geometry and pattern centre measurement

It has been shown [29] that, while not obvious in day to day use, detector geometry issues such as optic axis misalignment, spherical distortion by the EBSD system lens and vignetting can be significant and need to be minimized to reduce the effects that simulation must take into account. The 5 MPixel camera used to produce the pattern shown in figure 6 was designed to minimize these effects, for example selecting the camera lens to produce minimum spherical distortion even with the large apertures needed to increase light levels. Using the methods described in [29], these issues were accurately characterized since they would remain generally constant in operation. The other major design factor was to increase the rigidity of the system while allowing for fine adjustments to centre the system at precisely the required working distance before fixing in place. Even with this rigidity, the specimen-camera geometry, described by the pattern centre (PC) parameter, still needs to be measured accurately for each new specimen, and to do this a shadow casting technique was used [29]. With the geometry shown in figure 8, the solid discs projecting from a brass grid mounted in front of the camera phosphor produce shadows on the phosphor. Having assembled the front small grid in alignment with a larger grid mounted in the plane of the phosphor, the whole system of grids, specimen and camera can be aligned so that the common axis of the two grids marks the pattern centre when the circular disk artefacts are aligned with the rings. Even when the system is out of alignment, the true pattern centre can be calculated from the displacement of the disks relative to the circles.

Figure 9a shows an EBSP from a silicon specimen acquired with the 5 MPixel camera in which superimposed on the disc shadows are simulations of these discs produced using the calculated pattern centre parameters. To demonstrate the tracking of the movement of the pattern centre, the sample was displaced in the Z-direction by +0.30 mm, in X by 0.52 mm and Y by 0.24 mm. Figures 9b and c show enlargements of the disc shadows and simulations before and after the movement. At the centre between the four disc shadows the position of the pattern centre calculated from the disc positions is shown by the large "X" relative to the perfect alignment with the camera axis denoted by the smaller "+". It can be seen the "X" tracks the movement of the vertical {220} band from the EBSP with the change in PC caused by the sample displacement, indicating the calculations must be at least close to an accurate value. It may be noted that there is scope for refinement of this calculation as the ellipse fitting used to simulate the disc shadows could still be improved.

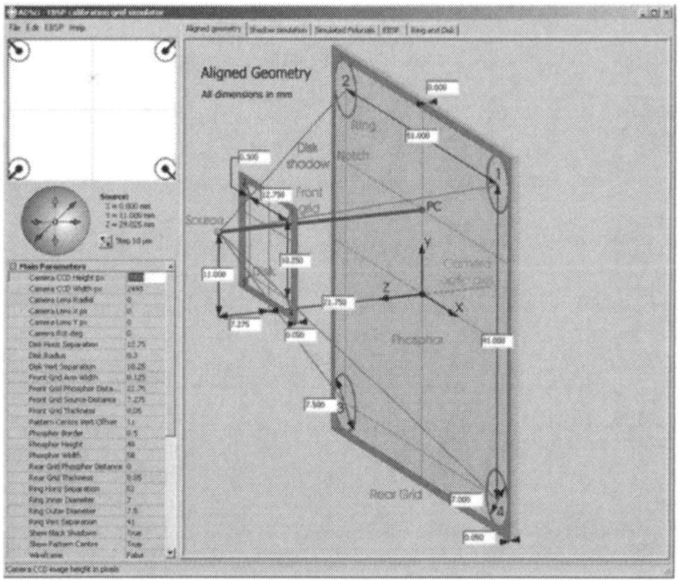

Figure 8. Geometry of shadow casting system

a)

Figure 9. Calculation of pattern centre shown by simulation of disc shadows. a) Full pattern with disc shadows seen in corners; b and c) Show enlargements of the disc shadows and simulations before and after movement of the sample by +0.30 mm in Z, 0.52 mm in X and 0.24 mm in Y. Position of the pattern centre indicated by green X, tracking the movement of the dark vertical {220} band.

b)

c)

4. Conclusions

EBSD technology is now sufficiently advanced that grain size measurement by this method for development and quality purposes in industry is becoming important. However care needs to be taken in both calibration of the complete SEM-EBSD system and in choice of the correct data acquisition and processing parameters. Development of international standards provides guidance to minimize errors.

Extending the capability of EBSD to enable more widespread use of, for example, the local measurement of strain requires improvements in the accuracy and detail available in EBSP acquisition. New shadow casting methods provide one way to achieve better accuracy in definition of sample-pattern geometry while increased detail can be obtained by larger cameras and ultimately direct electron detection.

Acknowledgements

Funding for organizing the work at the National Physical Laboratory was provided by the UK Government Department for Business, Innovation and Skills. The willing participants from five continents in the grain size round robin are gratefully acknowledged as is the assistance of Graham Sims and Stuart Saunders in setting up the VAMAS TWA 37 to oversee the work.

References

[1] Mingard K P, Quested P N and Peck M S 2012 *NPL Report MAT56*; ISSN 1754-2979 (London: National Physical Laboratory)
[2] ISO 13067: *Microbeam analysis - Electron backscatter diffraction - Measurement of average grain size.* (Geneva: International Standards Organisation)
[3] ASTM E2627-10: *Standard practice for determining average grain size using electron backscatter diffraction (EBSD in) fully recrystallized polycrystalline materials.* (West Conshohocken, PA: ASTM International)
[4] Gao N, Wang S C; Ubhi H S and Starink A 2005 *J. Mat. Sci. Lett.* **40** 4971-4974
[5] Humphreys F 1998 *J. Microsc.* **195** 170-185
[6] Day A P and Quested T E 1999 *J. Microsc.* **195** 186-196
[7] Bowles A, Dargusch M, Davidson C, Griffiths J and Nogita K 2004 *Mater. Trans.* **45** 3114-3119
[8] Piazolo S, Sursaeva V G and Prior D J 2005 in: *Proc. 14th Int. Conf. Textures of Materials - ICOTOM 14.* (Leuven, Belgium)
[9] Lins J F C, Sandim H R Z, Ribeiro R B and Pinto A L 2003 *Acta Microsc.* **12** 121-124
[10] Wright S 2010 *Prakt. Metallogr.* **47** 1
[11] Randle V 2003 *Pilot interlaboratory exercise for comparative EBSD measurements.* (Swansea: University of Wales)
[12] Ryde L 2004 *Report no.: IM-2004-121*; ISRN: SIMR/R- -04/080- -SE ISSN: 1403-848X
[13] ASTM E112-96: *Standard test methods for determining average grain size.* (West Conshohocken, PA: ASTM International)
[14] ASTM International Standard E691-09: *Standard practice for conducting an interlaboratory study to determine the precision of a test method.* (West Conshohocken, PA: ASTM International)
[15] BS ISO 5725-2: 1994: *Accuracy trueness and precision of measurement methods and results - Part 2: Basic method for the determination of repeatability and reproducibility of the standard measurement method.* (London: British Standards Institute)
[16] Farooq M U, Villarrutia R, MacLaren I, Kungl H, Hoffmann M J, Fundenberger J-J and Bouzy E 2008 *J. Microsc.* **230** 445-454
[17] Burnett T L, Comyn T P, Merson E and Mingard K P 2008 *IEEE Trans. Ultrasonics Ferroelectrics And Frequency Control* **55** 957-962

IOP Conf. Series: Materials Science and Engineering **55** (2014) 012011 doi:10.1088/1757-899X/55/1/012011

[18] Wilkinson A J, Britton T B, Jiang J and Karamched P S 2014 *IOP Conf. Ser.: Mat. Sci. Engng.* **55** 012020

[19] Wilkinson A J, Meaden G and Dingley D J 2006 *Ultramicroscopy* **106** 307-313

[20] Britton T B, Maurice C, Fortunier R, Driver J H, Day A P, Meaden G, Dingley D J, Mingard K and Wilkinson A J 2010 *Ultramicroscopy* **110** 1443-1453

[21] Winkelmann A, Trager-Cowan C, Sweeney F, Day A P and Parbrook P 2007 *Ultramicroscopy* **107** 414-421

[22] Mingard K, Day A, Quested P and England G 2011 in: *RMS EBSD Conference*. (Dusseldorf)

[23] Roberts P T E, Chapman J N and MacLeod A M 1982 *Ultramicroscopy* **8** 385-396

[24] Moldovan B J G, Nomerotski A and Kirkland A 2009 *Nucl. Inst. Meth. Phys. Res. A* **604** 108-110

[25] Sikharulidze I, van Gastel R, Schramm S, Abrahams J P, Poelsema B, Tromp R M and van der Molen S J 2011 *Nucl. Inst. Meth. Phys. Res. A* **633** Suppl. 1, S239-S242

[26] Vespucci S, Kumar N, Edwards P R, Maneuski D, O'Shea V, Day A, Mingard K and Trager-Cowan C 2013 in: *RMS EBSD Conference*. (Oxford)

[27] Wilkinson A, Molodovan G, Britton, T, Bewick, A, Clough R and Kirkland A 2013 *Phys. Rev. Lett.* **111** 065506

[28] Villert S, Maurice C, Wyon C and Fortunier R 2009 *J. Microsc.* **233** 290-301

[29] Mingard K, Day A, Maurice C and Quested P 2011 *Ultramicroscopy* **111** 320-329

Microstructure of nanocomposite carbon-, MoS_2- and MoO_3-based solid-lubricant coatings

T Moskalewicz[1], B Wendler[2] and A Czyrska-Filemonowicz[1]

[1] AGH University of Science and Technology, International Centre of Electron Microscopy for Materials Science and Faculty of Metals Engineering and Industrial Computer Science, Al. Mickiewicza 30, Pl-30059, Krakow, Poland
[2] Technical University of Lodz, Institute of Materials Science and Engineering, ul. Stefanowskiego 1, PL-90924 Lodz, Poland

E-mail: tmoskale@agh.edu.pl

Abstract. Three types of coatings (Type 1: nc-TiC/a-C where, a-C means amorphous hydrogen-free carbon; Type 2: MoS_2(Ti,W); and Type 3: MoO_3 with Ag addition) were deposited by magnetron sputtering onto oxygen hardened Ti-6Al-4V titanium alloy and model γ-TiAl alloy substrates in order to improve their tribological properties. The paper describes the coatings micro/nanostructure characterized by scanning and transmission electron microscopy as well as X-ray diffractometry. The Type 1 coatings were composed of titanium carbides nanocrystals embedded in a-C matrix. The MoS_2(Ti,W) coatings (Type 2) consisted mainly of MoS_2 nanoclusters embedded in an amorphous matrix with some Ti α and W nanocrystals. The as-deposited MoO_3-Ag coating (Type 3) consisted of nanocrystalline Ag embedded in an amorphous matrix containing Mo and O. The coating's microstructure was changed after annealing into a mixture of different silver molybdates, such as Ag_2MoO_4, $Ag_6Mo_{10}O_{33}$ and $Ag_2Mo_2O_7$. It was established that the applied coatings improved tribological properties of both substrates. Type 1 coatings are suitable for use at a temperature lower than 200 °C, Type 2 ones at a temperature lower than 350 °C, and Type 3 at an elevated temperature 450 °C - 550 °C.

1. Introduction

Nanocomposite coatings are novel, very important systems composed of two or more different nanocrystalline, or nanocrystalline and amorphous, phases [1, 2]. They are often used for improvement of tribological properties of metallic materials, decrease the coefficient of friction (CoF) and increase the wear resistance, under different operating conditions such as temperature and humidity. Thus, nanocomposite coatings enable to increase reliability, improve performance with less fuel consumption and reduce or eliminate greases as well as reduce environmental pollution.

Nanocomposite carbon-based coatings enable synthesis of numerous new materials with unique properties. Coatings contain a dispersion of nanocrystalline carbides in a solid low friction matrix, like hydrogen-free amorphous carbon a-C or hydrogenated amorphous carbon a-C:H, exhibit a reasonably high hardness as well as a relatively high wear resistance and low value of CoF against metallic and ceramic materials [1-3]. The performance of coatings is strongly influenced by their micro/nanostructure [2-6]. In the case of carbon-based coatings the amorphous matrix provides a low CoF, while nanocrystalline carbides increase the wear resistance of the coated materials [2-4].

Content from this work may be used under the terms of the Creative Commons Attribution 3.0 licence. Any further distribution of this work must maintain attribution to the author(s) and the title of the work, journal citation and DOI.
Published under licence by IOP Publishing Ltd

Molybdenum disulphide (MoS_2) is often used as a coating material with excellent self-lubricating properties [7, 8]. These properties result from a layered structure of the hexagonal elementary cell of this sulphide, in which the axial ratio c/a (where c and a are the parameters of the cell) amounts about 3.9. The bonding between the particular Mo and S atoms within the S-Mo-S layers of the crystal structure is strong covalent, while that between S-Mo-S layers is considerably weaker, van der Waals one [8]. These weak bondings are responsible for the low friction and good lubricating properties of the MoS_2-phase. The MoS_2-based coatings with admixture of other transition elements (such as Ti or Cr or W or Ta) exhibit higher hardness, better wear resistance, very low friction, high load bearing capacity, greater durability and oxidation resistance as well as lower sensitivity to water vapour than pure MoS_2 ones [8, 9]. Therefore, transition elements are often used as admixture to MoS_2-based coatings.

Oxides and their combinations are very interesting materials for lubrication at elevated temperature due to their high chemical stability [10]. However, the adhesion of pure oxides to the metallic substrate is poor due to their high brittleness. Therefore, in our work the Ag admixture was incorporated to the MoO_3-based coating.

The micro/nanostructure of the low-friction coatings, i.e., their chemical- and phase composition as well as the crystallite size, is very important because it influences tribological and micro-mechanical properties of the coatings and, as a result, the tribological and micro-mechanical properties of the system coating/substrate. The aim of the present study was to characteriZe the nano/microstructure of three types of relatively thick low-friction coatings: Type 1 - nc-TiC/a-C, Type 2 - MoS_2(Ti,W), and Type 3 - MoO_3 with Ag addition deposited on metallic substrates by magnetron sputtering.

2. Materials and methods

The Type 1 and Type 2 coatings were deposited on two-phase ($\alpha+\beta$) titanium alloy Ti-6Al-4V, while Type 3 coating was deposited on model γ-TiAl alloy. In addition, prior to the coatings deposition, the surface of the specimens made of Ti-6Al-4V alloy was hardened with interstitial oxygen atoms as a result of the diffusion process at 900 °C in the $Ar+O_2$ atmosphere in the presence of the glow discharge plasma. A detailed description of the microstructure of the oxygen hardened Ti-6Al-4V alloy was given in Ref. [11].

In order to increase the adhesion of the coating to the substrate at first a thin intermediate layer was deposited onto the substrate materials by magnetron sputtering. This intermediate layer was composed of Ti and C atoms in case of Type 1 coatings, of metallic Ti_α-phase in case of Type 2 ones and of Ag in case of Type 3 one. In the next step, the proper nanocomposite coating was deposited by magnetron sputtering as well. The MoO_3-Ag coated alloy was additionally annealed at 450 °C during 3 hours. A detailed description of the coatings deposition procedure was given in Refs. [11-15].

The investigations of the coatings nano/microstructure were carried out by scanning and transmission electron microscopy (SEM, TEM), as well as by X-ray diffractometry (XRD) and grazing incidence X-ray diffractometry (GIXRD). The SEM investigation was performed by NEON® 40EsB of Zeiss. TEM and HRTEM investigation was carried out by JEOL JEM-2010 ARP (200 kV), ARM 1250 (1250 kV), and Titan Cubed G2 60-300 (300 kV) FEI microscopes on cross-section lamellas. The lamellas were prepared by focussed ion beam (FIB) NEON CrossBeam 40EsB of Zeiss followed by a short ion-beam thinning using precision ion polishing system (PIPS) of Gatan. The XRD and GIXRD patterns were recorded using Siemens D500 Kristalloflex diffractometer with use of Cu Kα radiation on plane-view specimens.

Phase identification was performed by means of selected area electron diffraction (SAED) and fast Fourier transformation (FFT) as well as XRD and GIXRD techniques. The SAED and FFT patterns were interpreted with the help of a Java electron microscopy software (JEMS) [16]. The phase identification was supplemented by chemical microanalyses with use of energy-dispersive X-ray spectroscopy (STEM-EDS). The STEM images of a size of 512x512 pixels were acquired utilizing approx. 3 nm probe. Image analyses were performed using 'digital micrograph' software of Gatan.

The particles' size and spatial distribution were determined based on several different HRTEM micrographs supported by 'AnalySIS 3.1' and 'digital micrograph' software.

3. Results and discussion

3.1. nc-TiC/a-C coating

SEM and TEM investigation revealed that the coating was dense and had an uniform thickness of 3.3 ± 0.2 μm (figure 1). Due to a high volume fraction of the amorphous carbon, the electron diffraction patterns taken from the coating were diffused and difficult for unequivocal interpretation. However, the rings corresponding to {200} and {220} TiC (face-centred cubic; fcc) crystallographic planes might be distinguished and are visible on diffraction pattern intensity profile (figure 1, diffraction pattern no A, right up corner).

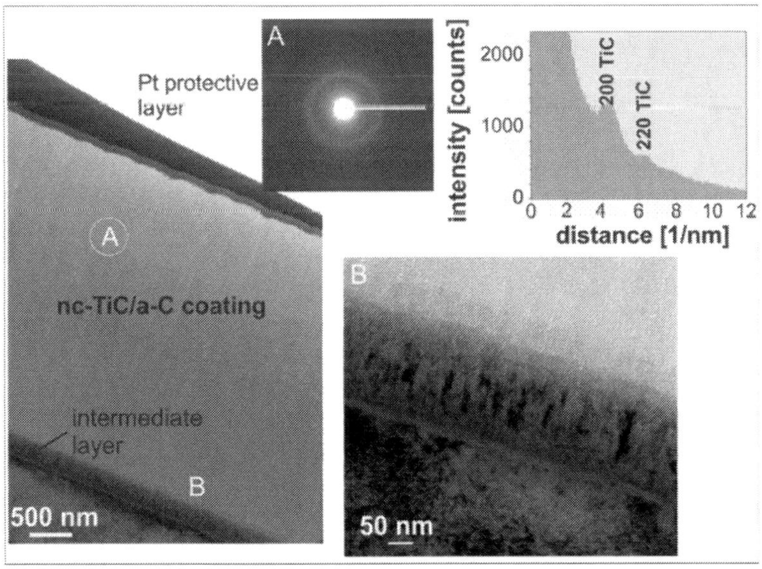

Figure 1. Microstructure of the nc-TiC/a-C coating on oxygen hardened Ti-6Al-4V alloy as well as SAED pattern taken from the area marked as A in the figure. An intensity profile along a line marked on the SAED pattern A is given as well. B) Magnified details of the intermediate layer; TEM-FIB lamella.

Identification of the carbides occurring in the coatings was performed by means of the XRD (GIXRD) as well as of the HRTEM techniques. Based on the XRD and GIXRD pattern analyses, nanocrystallities of two phases, TiC (fcc) and $TiC_{0.62}$ (fcc), were identified in the coating. The size and spatial distribution of the carbide nanocrystallites within the coating were estimated using HRTEM. It was found that the coating is composed of the nanocrystallites (equivalent circle diameter, ECD = 5 ± 4 nm) embedded in an amorphous carbon matrix (figure 2). The calculated diffraction patterns were obtained by performing FFT. Based on the FFT images, the particles present in the coating were identified as the nanocrystallities of the TiC (fcc) phase. As an example, a FFT image taken from the coating area containing several TiC particles, marked as a square is shown in figure 2. The titanium carbide nanocrystals were separated by an amorphous carbon matrix. The nanocrystallities were uniformly distributed within the coating.

In order to increase the coating adhesion to the underlying substrate a 200 ± 10 nm thick intermediate layer was deposited between the coating and the alloy (figures 1 and 3). Analyses of the electron diffraction patterns and FFT patterns (figure 3) taken from the intermediate layer revealed the occurrence of Ti α (hexagonal close-packed; hcp) and $TiC_{0.62}$ (fcc) phases. ECD of nanocrystallites,

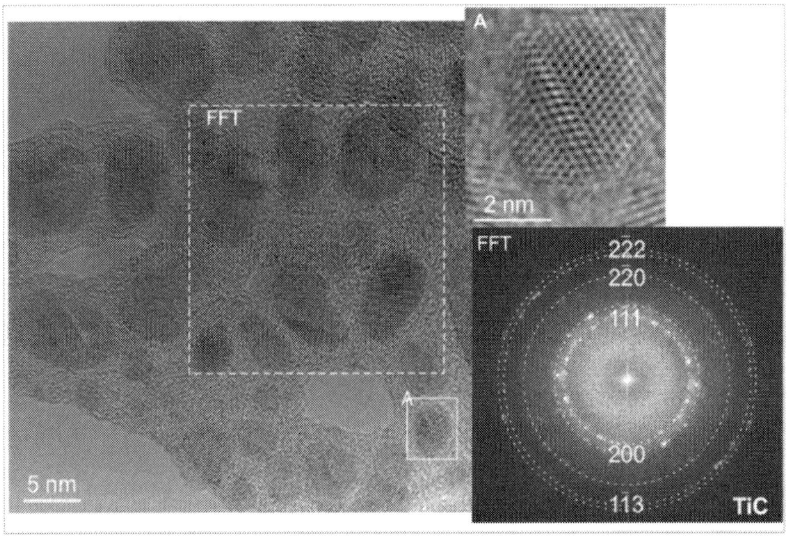

Figure 2. HRTEM micrograph of the nc-TiC/a-C coating on oxygen hardened Ti-6Al-4V alloy and a corresponding FFT pattern calculated from the area marked as a square in the micrograph and its identification.
A) Magnified detail of figure 2. FIB lamella.

evaluated from dark-field TEM and HRTEM images, was estimated in the range 10 ± 1 nm to 70 ± 1 nm. The Ti α and $TiC_{0.62}$ (fcc) nanocrystals were found in the coating/intermediate layer interface as well.

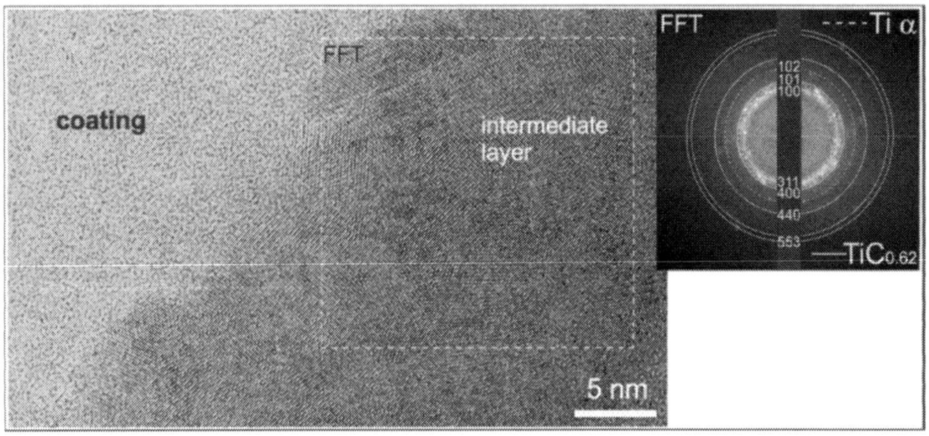

Figure 3. HRTEM micrograph of the intermediate layer between the nc-TiC/a-C coating and the oxygen hardened Ti-6Al-4V alloy and a corresponding FFT pattern calculated from the square area marked with the broken line in the micrograph and its identification; FIB lamella.

3.2. Microstructure of MoS_2(Ti,W) coating

The coating thickness evaluated from SEM and TEM images equals 3.1 ± 0.2 μm (figure 4). SEM-EDS microanalysis revealed the following mean chemical composition of the coating (in at%): 27 ± 1 Mo, 56 ± 1.2 S, 15.6 ± 0.5 Ti, and 1.4 ± 0.2 W.

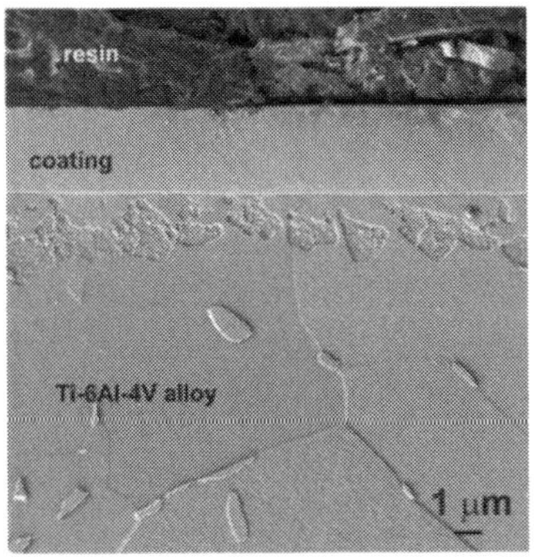

Figure 4. Microstructure of the oxygen hardened Ti-6Al-4V alloy with MoS_2(Ti,W) coating; SEM secondary electron image.

Analysis of the coating microstructure was performed using TEM by electron diffraction, STEM-EDS and HRTEM methods. The electron diffraction patterns and the FFT ones obtained from the coating were diffuse. Nevertheless, a strong incomplete {002} diffraction ring as well as two weaker ones {100} and {103} ones belong to the MoS_2-phase are clearly visible, as shown in figure 5 (FFT pattern calculated from figure 5a).

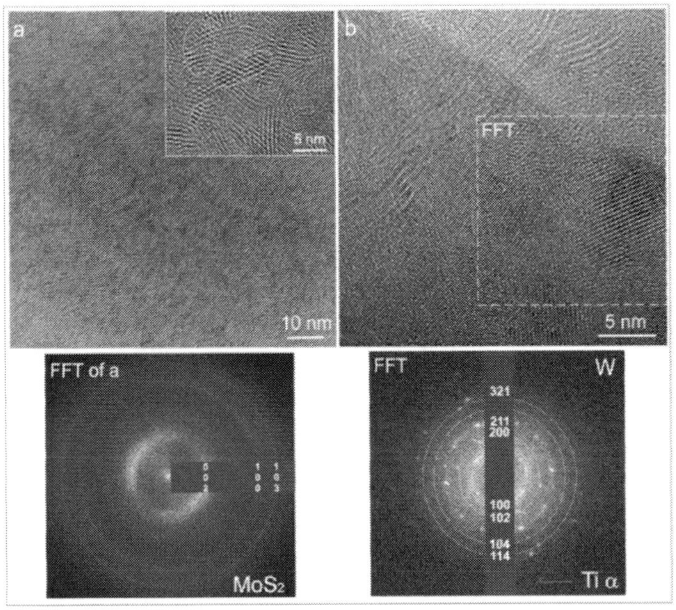

Figure 5. HRTEM micrographs of the MoS_2(Ti,W) coating on oxygen hardened Ti-6Al-4V alloy and corresponding FFT patterns calculated from the area of figure 5a and area marked as square in figure 5b and their identification. Magnified details of the figure 5a are given in right up corner, typical nanoclusters are marked as ovals; FIB lamella.

The HRTEM investigation revealed the occurrence of fine nanoclusters of ECD equalling 6 ± 3 nm (marked as ovals in figure 5a) embedded in an amorphous matrix (figure 5a). The HRTEM images show numerous short black lines with interplanar spacing close to {002} planes of MoS_2-phase. The planes of the nanoclusters were in general differently oriented to the surface of the coating. Nevertheless, in the area near to the coatings surface, the {002} planes were approx. parallel to the coating surface. Such orientation of nanoclusters with {002} planes parallel to the substrate surface is conductive to favourable tribological properties of these coatings.

Sporadically, inhomogeneously distributed nanocrystallities of the ECD equalling 7 ± 3 nm were observed in the coating as well. The interplanar spacings measured for particular nanocrystallities and analyses of the FFT patterns calculated from the nanocrystals revealed presence of Ti α (hcp) and W (body-centred cubic; bcc) phases. Typical FFT pattern calculated from the coating area marked as a square and indexed as Tiα and W is shown in figure 5b.

An intermediate titanium layer, 30 ± 5 nm thick, was deposited in order to increase the coating adhesion to the substrate. The layer was well adhering to the surface of the hardened titanium alloy substrate. Analysis of the SAED- and FFT-patterns confirmed that the intermediate layer was composed of Tiα (hcp) nanocrystallities. Nevertheless, an amorphous phase containing Mo and S was also found in the areas of the interlayer close to the coating.

3.3. Microstructure of MoO_3-Ag coating

The coating thickness was measured from TEM images as 1.9 ± 0.1 μm (figure 6). The as-deposited coating was dense; however, the adhesion of the coating to the underlying substrate was relatively poor.

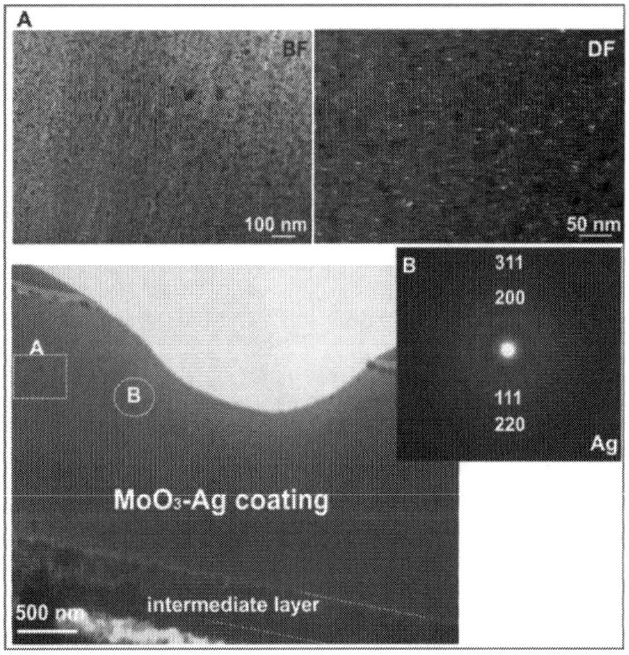

Figure 6. Microstructure of the MoO_3-Ag coating and of the interlayer on γ-TiAl alloy directly after deposition. A) Magnified details of the coating microstructure in bright field (BF) and dark field (DF) images as well as the SAED pattern taken from the coating and it's identification; TEM-FIB lamella.

Characterisation of the coating microstructure was performed by TEM with use of electron diffraction and EDS chemical microanalysis methods. It was found that the as deposited coating microstructure consisted of Ag (fcc) nanocrystallities (figure 6, SAED pattern marked as B) embedded in an amorphous matrix (a diffuse ring characteristic of an amorphous phase occurs in the SAED pattern). ECD of the Ag nanocrystallities was estimated from dark-field TEM images as up to 10 ± 1 nm.

A graded intermediate layer (250 ± 20 nm thick) was present between the coating and the substrate alloy (figure 7). This layer was composed mainly of Ag nanocrystallities of ECD up to 30 ± 1 nm. In addition, a diffuse ring was found in the SAED pattern taken from the intermediate layer (SAED pattern marked as 1 in figure 7). It proves that an amorphous phase occurs in the interlayer after deposition as well.

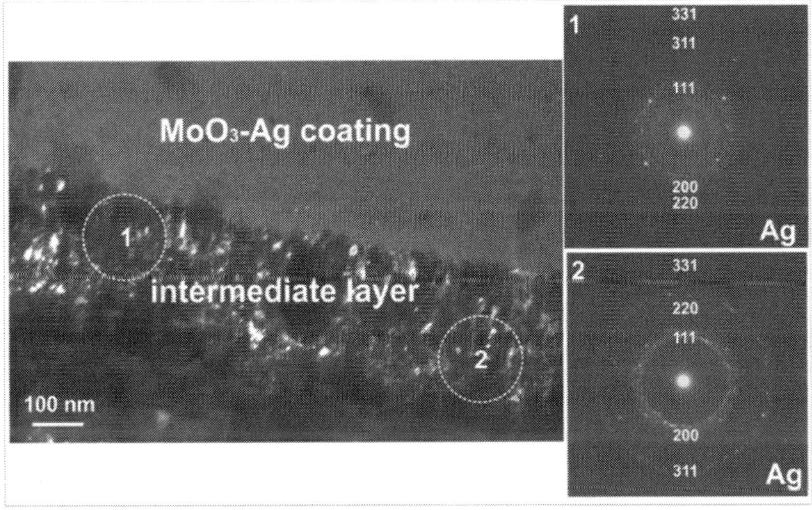

Figure 7. Microstructure of the MoO_3-Ag coating and intermediate layer as deposited on the γ-TiAl alloy (dark-field image taken from the 111 ring of Ag) as well as SAED patterns taken from areas of intermediate layer marked as 1,2 and their identification; TEM-FIB lamella.

STEM-EDS line analysis and chemical element distribution maps confirmed a presence of Mo, O and Ag in the coating as well as Ag (mainly) and Mo and O (sporadically) in the intermediate layer. The XRD and GIXRD investigation revealed also a presence of crystalline Ag and amorphous phase in the coating.

The coated specimen was next annealed at 450 °C in order to produce silver molybdates, which are characteristic of a layered graphite-like structure conductive to favourable tribological properties at elevated temperature. The coating microstructure was significantly changed after annealing. XRD investigation revealed a presence of Ag_2MoO_4 (fcc), $Ag_2Mo_2O_7$ (monoclinic primitive; mp), and $Ag_6Mo_{10}O_{33}$ (mp) phases in the coating. The TEM investigation was performed for wear track after ball-on disc tribotest conducted at 450 °C for 20000 rotations against Al_2O_3 ball (1 mm diameter under load of 1 N) on FIB lamella cut from an appropriate coating area. It was found that 820 ± 10 nm thick coating was not destroyed. Mainly Ag_2MoO_4 (fcc) grains were present in the $Ag_6Mo_{10}O_{33}$ (mp) matrix of the investigated area of the coating (figure 8). The grain size of Ag_2MoO_4 was measured as in the range from several nm to 250 ± 5 nm.

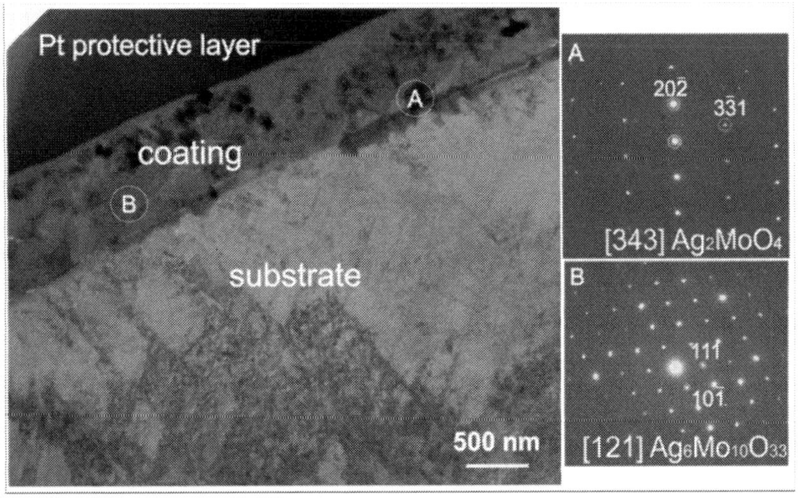

Figure 8. Microstructure of the MoO_3-Ag coating on the model γ-TiAl alloy substrate after ball-on-disc test performed at 450 °C (20000 cycles under a load of 1 N) and SAED patterns taken from areas of the coating marked as A, B and their identification; TEM-FIB lamella.

3.4. Tribological properties

Tribological properties of coated materials were described in detail in our earlier studies [11-15]. The results showed that all investigated nanocomposite coatings improved tribological properties, decreased coefficient of friction (CoF) and increased wear resistance of surface treated materials.

Very low value 0.05 ± 0.01 of the CoF of the nc-TiC/a-C coatings deposited on oxygen hardened Ti-6Al-4V alloy during dry sliding against 5 mm alumina ball at RT was measured in the ball-on-disk tests, in contrast of 0.5 ± 0.1 and 0.85 ± 0.15 values for the annealed and oxygen hardened Ti-6Al-4V alloy, respectively [13]. This low value of CoF is attributed to formation of a self-lubricating film in the sliding contact.

The MoS_2(Ti,W) coating decreased the CoF from 0.85 ± 0.15 for the oxygen hardened Ti-6Al-4V alloy to 0.15 ± 0.02 (at RT) and to 0.09 ± 0.01 (at 300 °C) for the coated one and increased essentially the wear resistance of the alloy [14].

The MoO_3-Ag coating decreased the CoF of the γ-TiAl alloy at elevated temperature of 450 °C and 550 °C from 0.6 ± 0.2 to about 0.2 ± 0.07. The wear resistance of the coated alloy was not satisfactory, both at RT and at elevated temperature [15].

4. Summary

Micro/nanostructure of the low-friction solid-lubricant coatings was examined in detail by various electron microscopy and X-ray diffractometry techniques. It was found that 3.3 ± 0.2 μm thick nc-TiC/a-C coating was composed of TiC nanocrystallities of a size of few nanometres embedded in an amorphous carbon matrix. A 200 ± 10 nm thick graded interlayer consisting of a mixture of Tiα (hcp) and $TiC_{0.62}$ (fcc) nanocrystallities was present between the coating and the oxygen hardened alloy substrate.

The MoS_2(Ti,W) coating (3.1 ± 0.2 μm thick) was dense and of a good quality. A well adhering, intermediate layer composed of Tiα nanocrystallities and some amorphous phase containing Mo and S was present between the coating and the hardened titanium alloy substrate. The MoS_2(Ti,W) coating was composed of MoS_2 nanoclusters embedded in an amorphous matrix. Some Ti α and W nanocrystals were also found in the coatings' microstructure.

The MoO_3-Ag coating (1.9 ± 0.1 μm thick) was dense. The as-deposited coating microstructure consisted of nanocrystalline Ag particles embedded in an amorphous matrix containing O and Mo. A gradient intermediate layer (250 ± 20 nm thick) was present between the coating and the substrate. This layer was composed mainly of Ag nanocrystallities and some amorphous phase containing O and Mo. The content of the amorphous phase increased gradually with to the distance from the substrate. The coating microstructure changed after annealing. Different silver molybdates, Ag_2MoO_4, $Ag_6Mo_{10}O_{33}$, and $Ag_2Mo_2O_7$ were formed in the coating.

It was established that the applied coatings essentially improved tribological properties of the substrate materials used in the experiments. Type 1 coatings are suitable for tribological contact in the temperature range from RT to 200 °C, Type 2 coatings for temperature not exceeding 350 °C, and Type 3 ones for application at elevated temperature 450 - 550 °C.

Acknowledgments

The results presented in this paper have been obtained within the project "KomCerMet" (contract no. POIG.01.03.01-14-013/08-00 with the Polish Ministry of Science and Higher Education) in the framework of the Operational Programme Innovative Economy 2007-2013. The authors appreciate a valuable contribution of Dr. S. Zimowski and A. Gruszczyński, MSc (AGH University of Science and Technology) to the tribological properties investigation and FIB specimen preparation.

References

[1] Zhang S, Sun D and Bui X L 2007 in: *Nanocomposite thin films and coatings, properties and performance*. (Zhang S and Ali N; Eds.). (London: Imperial College Press) 1-110
[2] Moskalewicz T, Wendler B and Czyrska-Filemonowicz A 2010 *Mater. Charact.* **61** 959

[3] Pei Y T, Galvan D, De Hosson J Th M and Cavaleiro A 2005 *Surf. Coat. Technol.* **198** 44
[4] Martínez-Martínez D, López-Cartes C, Fernández A and Sánchez-López J C 2009 *Thin Solid Films* **517** 1662
[5] Moskalewicz T, Wendler B, Smeacetto F, Salvo M, Manescu A and Czyrska-Filemonowicz A 2008 *Surf. Coat. Technol.* **202** 5876
[6] Moskalewicz T, Grogger W and Czyrska-Filemonowicz A 2006 *J. Microsc.* **223** 195
[7] Renevier N M, Fox V C, Teer D G and Hampshire J 2000 *Mater. Design* **21** 337
[8] Bülbül F, Efeoglu I and Arslan E 2007 *Appl. Surf. Sci.* **253** 4415
[9] Renevier N M, Lobiondo N, Fox V C, Teer D G and Hampshire J 2000 *Surf. Coat. Technol.* **123** 84
[10] Gulbiński W and Suszko T 2006 *Wear* **261** 867
[11] Moskalewicz T, Wendler B, Zimowski S, Dubiel B and Czyrska-Filemonowicz A 2010 *Surf. Coat. Technol.* **205** 2668
[12] Włodarczyk K, Makówka M, Nolbrzak P and Wendler B 2009 *J. Achiev. Mater. Manufac. Engng.* **37** 354
[13] Zimowski S, Moskalewicz T, Wendler B, Kot M and Czyrska-Filemonowicz A 2013 *Mater. Design*, submitted
[14] Wendler B, Moskalewicz T, Rudnik D, Zimowski S, Kot M, Grobelny M, Pawlak W, Rylski A, Czyrska-Filemonowicz A, Makówka M, Nolbrzak P and Włodarczyk K 2013 *Composites Theory Prac.* **1** 52
[15] Makówka M, Pawlak W, Wendler B, Sielski J and Kozanecki M 2011 *Mater. Engng.* **4** 553
[16] Stadelmann P 2004 *JEMS Java electron microscopy software.* http://cimewww.epfl.ch/

High-resolution EPMA X-ray images of mother liquid inclusions in a Pd_2Ga single crystal

D Müller[1], J Schwerin[2], P Gille[2] and K T Fehr[1]

[1] Ludwig-Maximilians-Universität, Department of Earth and Environmental Sciences, Section for Mineralogy, Theresienstrasse 41, DE-80333 Munich, Germany
[2] Ludwig-Maximilians-Universität, Department of Earth and Environmental Sciences, Crystallography Section, Theresienstrasse 41, DE-80333 Munich, Germany

E-mail: dirk.mueller@min.uni-muenchen.de

Abstract. During crystal growth from solution inclusions of different compositions were trapped at the rim of a Pd_2Ga single crystal. Their fine-grained (< 5 μm) internal structure demands special requirements for electron microprobe analysis, realized by low-voltage (5 keV) element mapping applying a step size of 0.138 μm for each pixel. It can be shown, that these inclusions represent an isolated chemical system, and that crystallisation upon cooling follows the expected thermodynamic phase relations. Thus the final composition in the centre of the inclusion consists of a small-scale mixture of PdGa and Pd_5Ga_3 evolved out of a solid-solid decomposition of Pd_5Ga_4.

1. Introduction

Pd_2Ga has a high potential to be utilized as a next generation catalytic material in the production process of polyethylene, where pure Pd or alloyed catalysts are presently in use [1]. In view of a polyethylene production of 50 megatons per year, it is quite a huge business in the chemical industry [2]. This makes it necessary to improve catalysts to increase the output, reduce production cost and save the environment [3].

The semi-hydrogenation of acetylene is a necessary step in removing minor amounts of acetylene from the ethylene feed during the production of polyethylene. It is important to decrease the acetylene content from 1 % to a very low ppm range; otherwise it would deactivate the catalyst [1, 2]. Today, commercial catalysts show a high hydrogenation-activity but a limited selectivity to ethylene. In recent years the intermetallic compound Pd_2Ga (space group Pnma, Co_2Si structure type [4]) has been explored as a highly selective heterogeneous hydrogenation catalyst. Compared to currently used catalysts, the intermetallic compound shows an improved long-term stability as well as a high selectivity and activity. These outstanding properties can be explained by covalent bonding interactions and the well-defined crystal structure with fixed sites of the catalytically active element Pd [1-3, 5], also known as active-site isolation concept [6].

Due to higher specific area these intermetallic compounds will finally be used as fine-grained powders or even nanoparticles. To determine their intrinsic properties and basic processes of catalysis, however, it is necessary to investigate well-defined single crystals of cm^3-size. In order to grow such large single crystals the Czochralski method was used.

According to the phase diagram (figure 1), the orthorhombic solid-solution (ss) phase Pd_2Ga can be grown from a congruent melt or from Ga- or Pd-rich high-temperature solutions. Due to the Ga

Content from this work may be used under the terms of the Creative Commons Attribution 3.0 licence. Any further distribution of this work must maintain attribution to the author(s) and the title of the work, journal citation and DOI.
Published under licence by IOP Publishing Ltd

vapour pressure, crystallisation from Ga-rich solution at lower temperatures is preferable. During post-growth cooling this could lead to precipitations of the orthorhombic solid-solution phase Pd_5Ga_3 (space group Pbam, Ge_3Rh_5 structure type [7]), which makes it is necessary to check the crystal composition by electron probe microanalysis (EPMA).

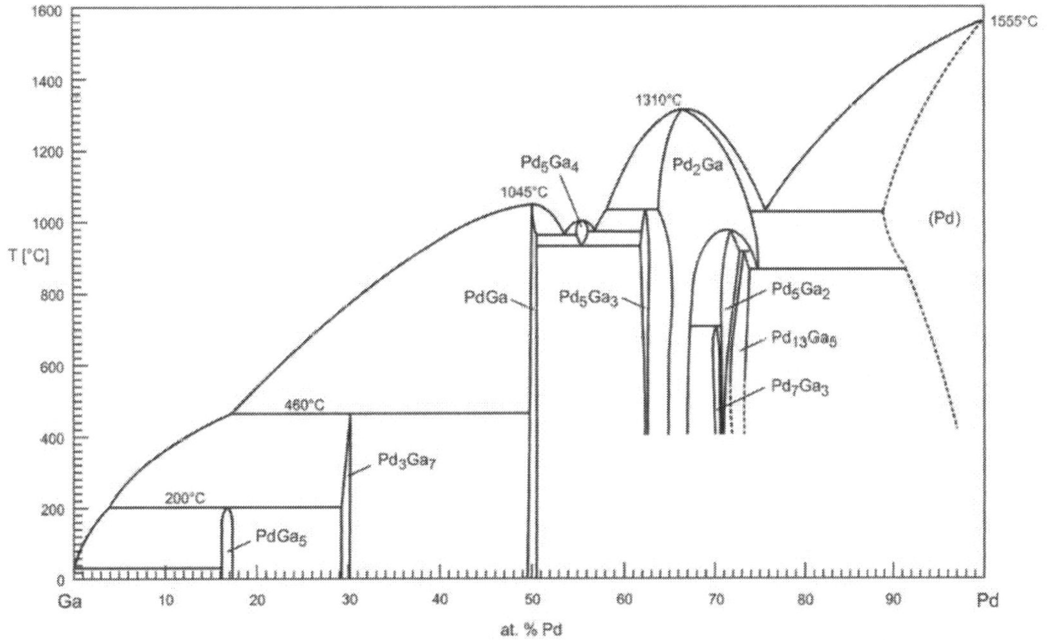

Figure 1. Phase diagram of Ga-Pd [9]

The challenge for EPMA investigations of these inclusions is their fine-grained mixture of different phases. To increase X-ray resolution, beam voltage has to be reduced as low as possible and practicable. Applying a low overvoltage ratio would further improve the spatial resolution of the analysis [8]. Using a LaB_6 electron source the measurement presented here was restricted to the low beam voltage method at commonly used overvoltage ratios of 1.6 - 4.5. In this paper a fully quantitative, high-resolution element mapping was carried out, working at the limits of a LaB_6-equipped electron microprobe.

2. Experimental

2.1. Crystal growth

The Czochralski method is one of the most famous and best studied methods for the production of large single crystals of different compositions. The working principle is shown in figure 2: a well oriented seed (a) is dipped in the molten source material (c). Due to marginal undercooling crystallisation starts on the thin neck of the seed. Subsequently the seed, together with the attached grown crystal (b), is slowly pulled upward while the crystal is growing. For a homogeneous mixture of the source material it is crucial to rotate the crystal in counter direction to the crucible during the whole experiment. The diameter was controlled by adjusting the temperature, while the pulling rate was fixed to 0.15 mm/h. The initial composition of the solution was $Pd_{61.4}Ga_{38.6}$.

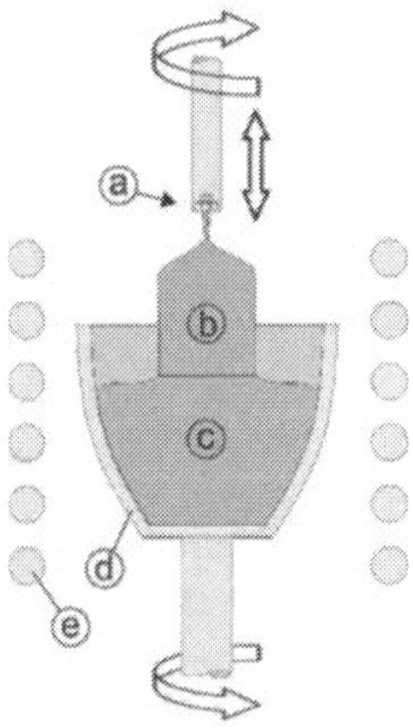

Figure 2. Working principle of the Czochralski method. The arrows on top and bottom symbolize the counter rotation of crystal and crucible. a) seed, fixed in a seed holder; b) grown crystal; c) molten source material; d) inert crucible; e) heating elements.

2.2. Analytical procedure

For EPMA analysis a thin slice was cut from the end of the single crystal (figure 3) and polished using diamond paste of decreasing grain size down to 0.25 μm to obtain a flat surface. Carbon coating was unnecessary as the sample is naturally conductive. The measurements were carried out on a Cameca SX-100 system, equipped with a LaB_6 cathode. An accelerating voltage (E_0) of 5 keV at a current of 20 nA was used. These values were chosen based on the critical excitation voltage (E_c) of the respective elements: Pd Lα - 3.173 keV and Ga Lα - 1.117 keV. To excite X-rays of an adequate amount a sufficient ionisation cross-section has to be reached, depending on the applied overvoltage ratio. This could be calculated from eq. (1), where U is the overvoltage ratio (E_0/E_c) [10]. At 5 keV $U_{PdL\alpha} = 1.58$ and $U_{GaL\alpha} = 4.48$ which gives practicable ionisation cross-sections for both elements of $QE_c{}^2{}_{PdL\alpha} = 2.30$ (10^{-20} cm^{2}keV^2) and $QE_c{}^2{}_{GaL\alpha} = 2.65$ (10^{-20} cm^{2}keV^1), respectively.

$$QEc^2 = 7.92 \cdot 10^{-14} \ln U \cdot U^{-1} \quad (1)$$

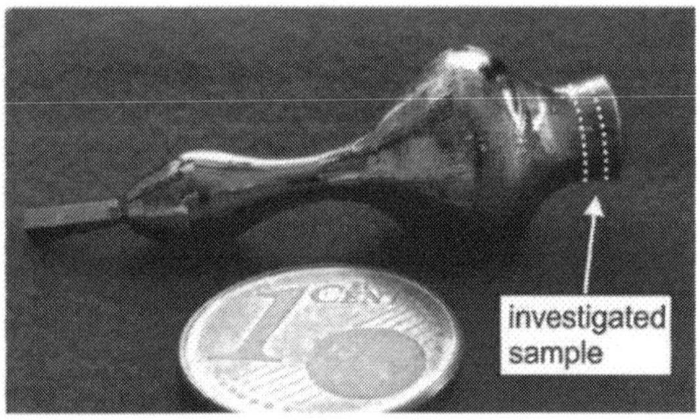

Figure 3. Pd_2Ga solid-solution single crystal grown by the Czochralski method from Ga-rich solution. The marked section represents the investigated slice of the crystal.

Due to the small grain size and heterogeneous distribution of the two phases a quantitative element mapping is preferable in contrast to single point measurements. Therefore, Pd and Ga were analysed simultaneously on two different wavelength-dispersive X-ray spectrometers (WDS). The Pd Lα-line was measured on a LPET and the Ga Lα-line on a TAP crystal. The mapping represents an area of 30 μm x 30 μm, analysed by a grid of 230x230 pixels with a dwell time of 0.07 s for each pixel. This gives a spatial resolution (pixel width) of 0.131 μm, which is in the range of the theoretical resolution for the expected composition of the sample (Table 1), calculated from eqs. (2) - (4). For comparison it should be noted that only eq. (4) accommodates the beam diameter:

Reed 1975 [10] $$d = 0.231\,(E_0^{1.5} - E_c^{1.5}) \cdot \rho^{-1} \quad (2)$$

Gauvin 2007 [11] $$d = k\,(E_0^{n} - E_c^{n}) \cdot \rho^{-1} \quad (3)$$

Merlet & Llovet 2012 [12] $$d = \sqrt{4\,(33\,A\,(E_0^{1.7}\text{-}E_c^{1.7}) \cdot (\rho\,Z)^{-1}\text{-}Z_m)^2 + d_0^2} \quad (4)$$

in which ρ denotes for the density of the investigated phase (ρ_{PdGa} = 9.97 g·cm^{-3}; ρ_{Pd5Ga3} = 10.73 g·cm^{-3}). These values are X-ray densities, calculated on the basis of structure models. *k* and *n* in eq. (3) are variables that were calculated using the equations given in [11]. In eq. (4), *A* is the atomic weight [g·mol^{-1}], *Z* the atomic number, and d_0 the beam diameter, that was assumed to be 100 nm. *Zm* is the depth at which the X-ray distribution curve exhibits its maximum. Mapping was done using the beam scanning method (fixed stage, moving beam). The background was measured on the right side (lower energy) of the peak, applying the same dwell time per pixel of 0.07 s as suggested by [13]. The conversion of raw counts to element concentrations was processed with the PAP matrix correction [14]. The influence of fluorescence would have only a minor impact on the analysis results and was, therefore, not investigated in detail. Calibration was done on pure Pd and stoichiometric PdGa. A peak overlap correction was not necessary.

Table 1. Calculated spatial X-ray resolution d (µm) for Pd and Ga in PdGa and Pd_5Ga_3, derived from different calculation procedures.

	Reed 1975		Gauvin 2007*		Merlet & Llovet 2012	
	d_{PdGa}	d_{Pd5Ga3}	d_{PdGa}	d_{Pd5Ga3}	d_{PdGa}	d_{Pd5Ga3}
Pd	0.128	0.119	0.065	0.061	0.162	0.155
Ga	0.232	0.215	0.105	0.097	0.234	0.221

* to compare these values with results from eqs. (2) and (4) a beam diameter of 100 nm has to be added

2.3. Image processing

Applying low voltage, current and counting times the peak to background ratio and the precision is decreasing, leading to noisy images. A powerful mathematical approach to improve the image quality is the Kriging analysis [15], which was used for the post-measurement processing of the results presented here. First developed for the classification of ore bodies [16] this method became a widespread application for analysing spatially distributed data.

The computed variogram, representing the variance of pairs of variates in relation to their spatial distribution, was modelled applying the nugget effect to remove the noise and a Cauchy function to reduce measurement artefacts. To smooth the image, a neighbourhood filter was used.

3. Results

Inclusions are observed at the rim of the $(Pd_2Ga)_{ss}$ single crystal (figure 4). The maximum diameter of individual inclusions is approximately 300 µm. Most of them show a circular shape and an internal separation, consisting of a homogeneous outer part and a heterogeneous core, comprising a porphyric structure of larger crystals of $(Pd_5Ga_3)_{ss}$ in a fine grained mixture of $(Pd_5Ga_3)_{ss}$ and $(PdGa)_{ss}$, respectively (figures 4 and 5). The homogenous outer region consists of several areas with different backscattered electron (BSE) intensities (different grey levels) suggesting a heterogeneous composition. Quantitative EPMA point analyses, however, indicate that these areas have the same composition, according to $(Pd_5Ga_3)_{ss}$. Thus, this effect could be explained by different crystal orientations (channelling effect) [17].

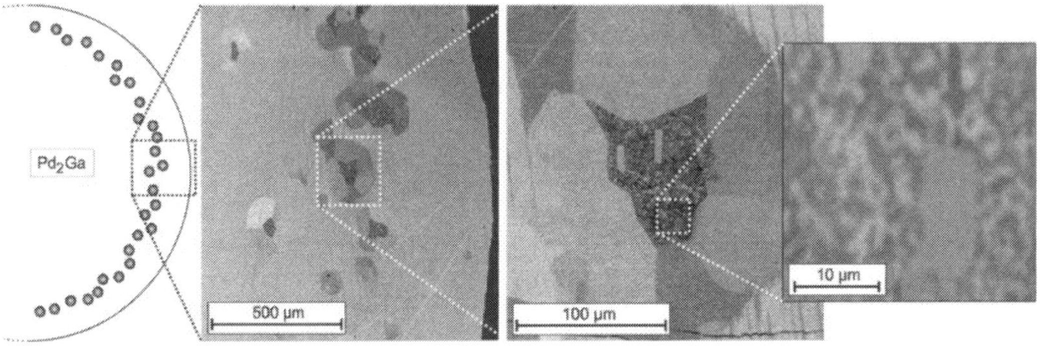

Figure 4. Appearance of Ga-enriched mother liquid inclusions in the $(Pd_2Ga)_{ss}$ single crystal. Their internal structure consists of an outer shell of $(Pd_5Ga_3)_{ss}$ containing a core of $(Pd_5Ga_3)_{ss}$ and $(PdGa)_{ss}$ (visible in the BSE images). The zoomed box on the right shows the mapped area.

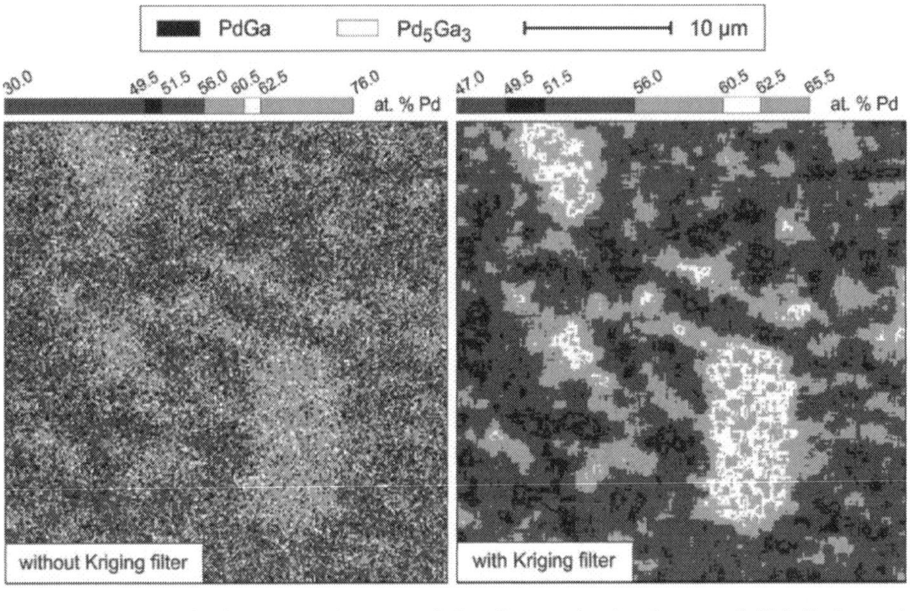

Figure 5. Quantitative X-ray images of the fine grained mixture of $(Pd_5Ga_3)_{ss}$ and $(PdGa)_{ss}$ in the core of the inclusion. Left: without Kriging analysis, showing a noisy, pixelated image. Right: image after Kriging analysis applying a 5x5 neighbourhood filter. Dark grey and light grey colours represent the systematic error of the measurement.

The element mapping was conducted in the core of an inclusion, covering an area which comprises bigger Pd_5Ga_3 crystals as well as the fine grained matrix of $(Pd_5Ga_3)_{ss}$ and $(PdGa)_{ss}$. The brighter regions and crystals in figure 4 represent the Pd-rich phase $(Pd_5Ga_3)_{ss}$ and the darker sections are $(PdGa)_{ss}$. The larger crystals reach a size up to 35 µm in length and 9 µm width, whereas in other inclusions they could be even larger (up to 90 µm). The irregular pattern of the two phases shows different grain sizes below 5 µm.

Figure 5 shows the lateral phase distribution which is in good agreement with the BSE image of figure 4. Regarding the phase diagram in figure 1 the composition of $(PdGa)_{ss}$ is 50.5 at% Pd and for $(Pd_5Ga_3)_{ss}$ 61.5 at% Pd. Taking a random error of ±1 at% into account, these values widen up to 49.5 - 51.5 at% Pd for PdGa and 60.5 - 62.5 at% Pd for Pd_5Ga_3, respectively. Without Kriging analysis a large systematic error is observable resulting in a wide compositional range from 30.0 - 76.0 at% Pd. After applying the Kriging filter the range of composition is limited to 47.0 - 65.5 at% Pd giving a more precise image. The choice of a suitable neighbourhood filter was done empirically by evaluating the area covered from the two phases (figure 6). Since this ratio is no longer changing at values higher than a 5x5 matrix, this filter was chosen as an optimum for the image in figure 5.

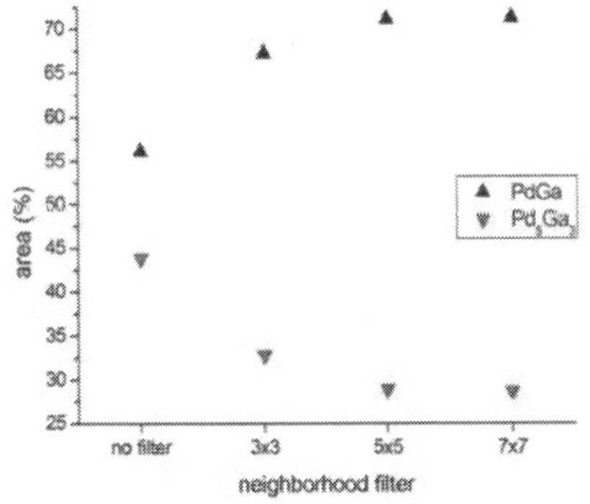

Figure 6. Impact of Kriging analysis on area coverage of $(PdGa)_{ss}$ and $(Pd_5Ga_3)_{ss}$. Neighbourhood filters regarding a matrix of > 5x5 pixels show no significant change anymore.

4. Discussion

The varying sizes of different inclusions and also the presence or absence of bigger $(Pd_5Ga_3)_{ss}$ crystals in the core could be explained by the cutting effect. Presuming a truly spherical shape, the complete content is only visible if the inclusion was cut through the centre (representing the full radius), which depends on their statistical distribution within the single crystal.

The best agreement between BSE and X-ray image could be observed in the central region of the mapped area (figure 5). By applying the Kriging filter the separation of the two phases reveals details more clearly. Based on this EPMA mapping, the internal evolution of these inclusions could be derived. After having been trapped, the liquid inclusion (1) follows the complete crystallisation path, visible in the phase diagram of figure 7: the inclusion represents a closed chemical system. While cooling down (1→2) $(Pd_2Ga)_{ss}$ will form until the temperature reaches the peritectic point (2). Due to the same composition and probably homo-epitaxial growth, this first rim could not be detected with EPMA. At the peritectic temperature $(Pd_5Ga_3)_{ss}$ crystallizes according to the peritectic reaction and simultaneously separates the liquid from the first grown $(Pd_2Ga)_{ss}$ crystal and preserves it from partial dissolution. An on-going cooling along with the crystallisation of $(Pd_5Ga_3)_{ss}$ brings the system to the eutectic point (3), which leads to an abrupt consumption of the remaining melt by the formation of $(Pd_5Ga_3)_{ss}$ and $(Pd_5Ga_4)_{ss}$ crystals. Obeying the lever rule more $(Pd_5Ga_4)_{ss}$ and less $(Pd_5Ga_3)_{ss}$ will be formed. Reaching the eutectoid point at (4) the $(Pd_5Ga_4)_{ss}$ phase decomposes into $(Pd_5Ga_3)_{ss}$ and $(PdGa)_{ss}$ at subsolidus conditions. Due to the relatively low temperatures the diffusivity of the elements during this solid-solid reaction is quite low and leads to the small-scale dissolution structures discussed above.

IOP Conf. Series: Materials Science and Engineering **55** (2014) 012013 doi:10.1088/1757-899X/55/1/012013

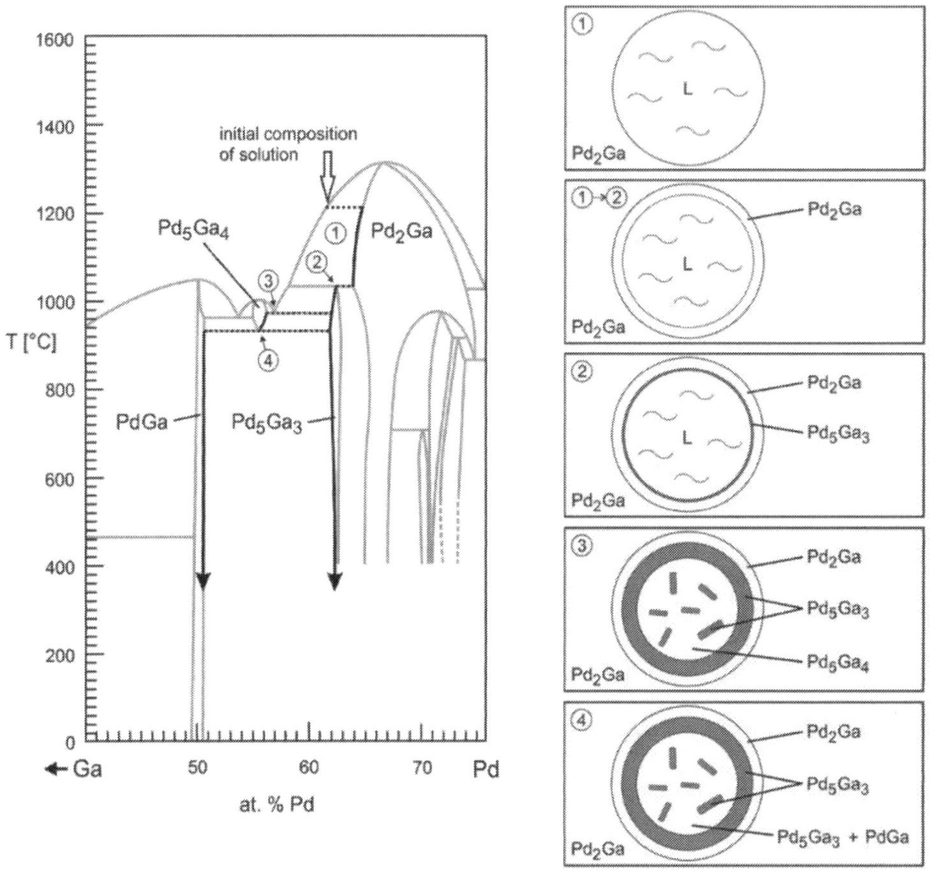

Figure 7. Section of the Ga-Pd phase diagram in the range of 40- 76 at% Pd. Continuous black lines show the crystallisation path of the inclusions during crystal growth while cooling down. Dotted lines represent separation processes. The single steps (1→4) of internal evolution of one single inclusion is shown on the right. L – liquid.

Inclusion formation is a quite common defect in solution growth and can be explained by an insufficient mixing of the solution during the growth process, namely of the rejected solvent adjacent to the growth interface. The rotation of crystal and crucible together with a curved solid-liquid interface leads to the circular arrangement at the rim of the crystal. To avoid this process, the pulling rate has to be lowered, which was successfully done in subsequent experiments.

5. Conclusion

A complete chemical characterisation of the inclusions was possible. On the basis of these results the entire development of the inclusions could be deduced which is in good agreement with the phase diagram. As a consequence, the crystal growth parameters could be optimized to avoid such inclusions in future.

Further improvement of the X-ray resolution might be achieved by reducing the accelerating voltage and the current, which implies longer analysis time and the use of a field-emission (FE) electron-gun. Due to a two-component system it would be possible to measure the Ga-content only and conclude the total composition of the specimen from these values. This would give the possibility to further decrease the accelerating voltage.

Acknowledgement
The authors would like to thank Michael Hahne and Birgitta Meisterernst for helpful discussions and suggestions regarding crystallographic topics. Helpful comments of Xavier Llovet and two anonymous reviewers to improve the paper are gratefully acknowledged. Thanks also to Renate Enders for her assistance in the preparation of the samples.

This work has been partly conducted within the European integrated Centre for the development of new Metallic Alloys and Compounds (European C-MAC).

References

[1] Armbrüster M, Wowsnick G, Friedrich M, Heggen M and Cardoso-Gil R 2011 *J. Am. Chem. Soc.* **133** 9112
[2] Armbrüster M, *et al.* 2012 *Chem. Cat. Chem.* **4** 1048
[3] Kovnir K, Osswald J, Armbrüster M, Teschner D, Weinberg G, Wild U, Knop-Gericke A, Ressler T, Grin Yu and Schlögl R 2009 *J. Catal.* **264** 93
[4] Kovnir K, Schmidt M, Waurisch C, Armbrüster M, Prots Yu and Grin Yu 2008 *Z. Kristallogr.-New Cryst. Struct.* **223** 7
[5] Osswald J, Giedigkeit R, Jentoft R E, Armbrüster M, Girgsdies F, Kovnir K, Ressler T, Grin Yu and Schlögl R 2008 *J. Catal.* **258** 210
[6] Sachtler W M H 1976 *Catal. Rev. Sci. Engng.* **14** 193
[7] Schubert K, Breimer H, Gohle R, Lukas H L, Meissner G and Stolz E 1958 *Naturwissensch.* **45** 360
[8] Newburry D E 2002 *Microsc. Microanal.* **8**, Suppl. 2, 434
[9] Okamoto H 2008 *J. Phase Equilib. Diffus.* **29** 466
[10] Reed S J B 1975 *Electron microprobe analysis.* (Cambridge: Cambridge University Press) 198
[11] Gauvin R 2007 *Microsc. Microanal.* **13** 354
[12] Merlet C and Llovet X 2012 *IOP Conf. Series: Mat. Sci. Engng.* **32** 012016
[13] Reed S J B 2000 *Mikrochim. Acta* **132** 145
[14] Pouchou J L and Pichoir F 1984 *Rech. Aérosp.* **3** 167
[15] Mainy D, Nectoux J P and Renard D 1996 *Mater. Charact.* **36** 327
[16] Matheron G 1965 *Les variables régionalisées et leur estimation.* (Paris: Masson) 306
[17] Meisenkothen F, Wheeler R, Uchic M D, Kerns R D and Scheltens F J 2009 *Microsc. Microanal.* **15** 83

IOP Conf. Series: Materials Science and Engineering **55** (2014) 012014 doi:10.1088/1757-899X/55/1/012014

A new application of SIMS to the analysis of nitrogen in mica minerals: tobelite

L P Ottolini[1], F Scordari[2] and E Mesto[2]

[1] Consiglio Nazionale delle Ricerche (CNR), Istituto di Geoscienze e Georisorse (IGG), Unità di Pavia, IT-27100 Pavia, Italy

[2] Università degli Studi di Bari, Dipartimento di Scienze della Terra e Geoambientali, IT-70125 Bari, Italy

E-mail: ottolini@crystal.unipv.it

Abstract. Muscovite, $KAl_2[AlSi_3O_{10}](OH)_2$, is a common rock-forming mineral in igneous and metamorphic-rocks, sediments, hydrothermal alteration and ore deposits. The site between two adjacent T-O-T (tetrahedral-octahedral-tetrahedral) layers is shared between K and NH_4 in all proportions leading to the building of the "ammonium micas". In this work three tobelite crystals were investigated in terms of major and minor constituents by electron probe microanalysis (EPMA) and in terms of nitrogen, for the first time at authors' knowledge, by secondary ion mass spectrometry (SIMS). The objective was that to gain information on the presence and amount of NH_4 in this matrix. The lack of standards did not allow us to provide quantitative results at the ion microprobe. Nevertheless, the SIMS data agree qualitatively with constraints resulting from EPMA analyses and charge-balance crystal chemical considerations. The results of this study emphasize the capabilities of SIMS in the in-situ analysis of N and point out the possibility to develop reference materials (standards) for quantitative SIMS analysis of nitrogen in mica.

1. Introduction

Nitrogen has been proven to be a powerful tracer in different geochemical system [1]. Urano [2] showed that nitrogen is a reliable indicator of the origin of metamorphic and granitic rocks. In the Earth's crust, N occurs mainly as ammonium ion (NH_4^+), and it is found essentially in clay minerals. The similarity of size and charge of the NH_4^+-ion with that of the K^+-ion, which usually replace each other in several K-bearing sheet silicate minerals (e.g., biotite, phlogopite and muscovite), has led to label NH_4^+ as a "pseudo alkali ion" [3]. Ammonium silicate minerals provide a means of delivering copious quantities of ammonium into the minerals capable of absorbing it [4].

One of the formative processes of the tobelite mineral has been associated with the "tobelitisation of smectite" [5]. Tobelite has been generally found in highly reducing, low-grade environments associated with organic-rich black matter, coal deposits and/or oil deposits. It can be either the product of diagenesis under the highly reducing conditions present in such deposit or a product of hydrothermal alteration of the muscovite from circulating ammonia-enriched ground-water in contact with these organic deposits or both [6]. In the latter case, the natural ammonium mineral is considered to be formed by the substitution of NH_4^+- for K^+-ion in the structure of the primary silicates (muscovite + NH_4^+ = tobelite + K^+).

Content from this work may be used under the terms of the Creative Commons Attribution 3.0 licence. Any further distribution of this work must maintain attribution to the author(s) and the title of the work, journal citation and DOI.
Published under licence by IOP Publishing Ltd

So far, tobelite has never been found as a pure end member, but typically it contains a significant K-component [7-9].

The term tobelite is nowadays used to indicate an illite-like species defined by: (i) $NH_4 > K + \square$ (vacancy) in interlayer site; (ii) Si > 3 atoms per formula unit (apfu) in tetrahedral site(s); (iii) layer charge < 1 vu (valence unit) [10].

Due to difficulties in obtaining a single crystal suitable for X-ray analysis, the structural characterisation of NH_4-rich micas was carried out in the past to more recent times, by means of X-ray powder diffraction [6-9, 11-15]. Occasionally this method was flanked by the Rietveld-based analyses [3, 13] to allow the identification and quantification of the mica polytypes as well as the quantification of mineral phase mixture (illite-tobelite-smectite-vermiculite) (see [3, 5, 13, 15]).

Tobelite structure has been solved recently by single crystal X-ray diffraction [16]. These authors showed that the studied tobelite belongs to $2M_2$-polytype (s.g. C2/c) with the following cell parameters: a = 9.024(1), b = 5.2055(6), c = 20.825(3) Å, and β = 99.995(8)°. The ammonium concentration was semi-quantitatively estimated in two crystals from the absorbance of the OH^--stretching and NH_4^+-bending vibrations in the infrared spectra. An additional estimate was obtained for the NH_4^+-rich muscovite by considering the normalized peak area between $K2p_{3/2}$ and $N1s$ in the X-ray photoelectron spectrum. We remind that the tobelite structure is based on (T-O-T) layers, formed essentially by an Al-octahedral sheet (O) sandwiched by two (Si, Al)-tetrahedral sheets (T). Such T-O-T layers are connected by (K, NH_4, Na, Ca, Ba) interlayer cations (figure 1). Precisely, Mesto *et al.* [16] showed that:

1) the octahedral site is almost filled by Al;
2) the anionic site is substantially occupied by 10 oxygen atoms and 2 hydroxyl groups (no-oxy substitution mechanism is active in the structure).

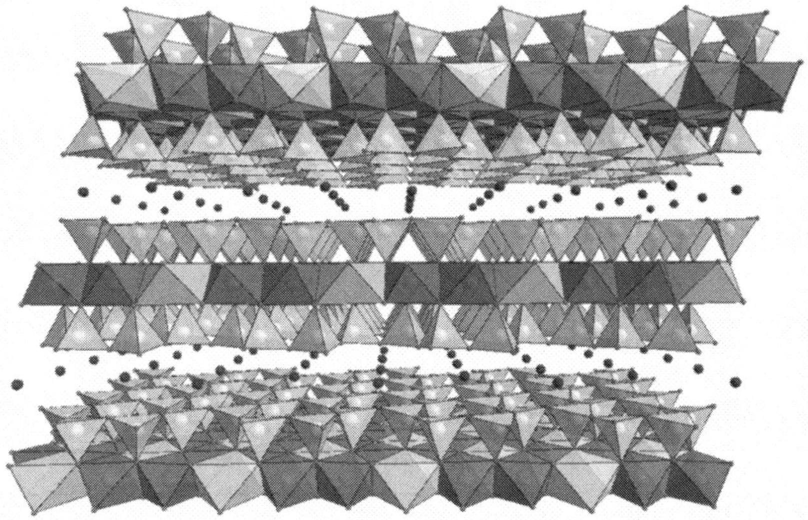

Figure 1. Tobelite structure view along [010]. Tobelite structure is characterized by layers formed by an Al-octahedral sheet (in purple) sandwiched by two Si, Al-tetrahedral sheets (in green). The layers are connected by (K, NH_4, Na, Ba) interlayer cations (in blue colour).

Their values for nitrogen resulted to be in agreement with those obtained from the interlayer spacing in the simulated X-ray powder diffraction spectra, and with charge-balance constraints concerning cations at interlayer and tetrahedral sites after electron probe analysis.

2. Analytical techniques for nitrogen

2.1. Electron probe micro-analysis (EPMA)

Nitrogen is a particularly troublesome element to analyse with the electron microprobe [17], because of extremely low peak-to-background ratios, curved background, and the many higher-order interferences from Zr, Nb and Mo if the lead stearate (STE) diffracting crystals are used. Using the W/Si reflector (layered synthetic microstructure - LSM) with 2d = 60 Å, increases count rates and peak-to-background ratios, and eliminates the higher-order interferences. However, interference of the Ti Ll-peak with the N Kα-peak is still potentially serious in Ti-bearing compounds [18].

Wilson *et al.* [6] demonstrated the effectiveness of the W/Si LSM in a study of tobelite-bearing veins from black shale. They measured the N- and O-contents of NH_4-bearing silicates directly by EPMA, using conditions similar to those of Nash [19]. AlN was used as the standard for N. Because the Ti-contents in these samples are negligible, interference with the N Kα-peak was ignored.

2.2. Fourier transform infrared spectroscopy (FTIR)

Nitrogen element abundances may also be determined by combustion techniques, but more usually both N-content and aggregation state are determined by FTIR [20, 21]. This is a transmission technique that collects data from a cylindrical sample volume with a depth equivalent to the plate thickness. In the paper by [22], the FTIR sample volume had a diameter of 80 μm and a depth in the range 0.5 - 1.6 mm. In the study of NH_4-bearing micas, the presence of ammonium ions in the interlayers was supported by FTIR [23]. Such an approach, however, is unsuitable for N analysis in minerals that have a growth zonation on a scale of less than 50 μm.

2.3. Scanning nuclear microprobe (SNM)

This technique has been employed for the first time by Ager *et al.* [24] to determine low nitrogen content (~ 0.3 wt%) in mica from Amorgos, an island situated at the South-east margin of the Cyclades Islands (Greece). The method used was the deuteron-induced γ-ray emission (DIGE). In that work the $^{14}N(d,p/\gamma)^{15}N$ nuclear reaction at the deuteron energy of 1.4 MeV was employed by means of DIGE to measure the nitrogen content, while the overall composition of the mica was determined by deuteron-induced X-ray emission (DIXE).

2.4. Secondary ion mass spectrometry (SIMS)

Secondary ionisation of nitrogen is generally pretty low both in positive and negative mode, i.e., analysing N as both $^{14}N^+$ and $^{14}N^-$; moreover, N does not readily forms stable ions. Nitrogen abundance was extensively investigated in diamonds using a Cs^+-beam (20 - 80 nA), collecting negatively charged NC^- molecular ions at high mass resolving power, with no energy filtering and 50 eV bandpass [25]. A similar set up was used by [22] and [26], which included: +14.5 kV accelerating voltage, 5.5 - 8 nA primary beam current, ~ 8 μm diameter of the area sampled, 10 – 25 eV energy window, $^{14}N^{12}C^-$ secondary ions at a mass resolution (M/ΔM) ~ 7500 to distinguish $^{14}N^{12}C^-$ from the principal interference $^{13}C_2^-$. The level of adsorbed N on the sample surface was limited by operating at high vacuum in the sample chamber.

In the present work, we describe the results of an investigation of nitrogen in tobelite crystals, carried out by SIMS.

IOP Conf. Series: Materials Science and Engineering **55** (2014) 012014 doi:10.1088/1757-899X/55/1/012014

3. Geological setting and material

The three tobelite samples investigated in this study were picked up in the Saint-Aubin-Châteaux aggregate quarry, in the Central Armorican Domain (CAD), at about 6 km north from the Northern Branch of the South Armorican Shear Zone (for more details see figures 1a and b and 2a in [27]).

This zone and the North Armorician Shear Zone bounded to the CAD, are the two main geologic structures of the Hercynian Armorican Massif. Gloaguen *et al.* [27] give the paragenetic sequence of the Saint-Aubin-des-Châteaux mineralisation and, among the silicate minerals, they include a white mica, which a year before Moëlo *et al.* [28] hypothesized to be tobelite. This mineral is a pure low hydrothermal product, hosted in quartz veins, related to Variscan (or Hercynian) hydrothermalism of the Ordovician iron ore. The temperature of mica formation is about 275 °C [29] and the growth occurred under very reducing conditions in abundant organic matter. Tobelite was found associated to chlorite, dolomite, quartz, tourmaline, chalcopyrite, galena, some sulphosalts (like boulangerite, bourmanite, tetraheidrite), and electrum.

4. Experimental methods and results

4.1. EPMA

Chemical composition was determined by a Cameca SX-50 electron microprobe after embedding in epoxy resin and polishing. Operating conditions were 15 kV accelerating voltage, 15 nA specimen beam current, and 10 μm beam diameter. The analyses were performed in full WDS (wavelength-dispersive X-ray spectrometry) made at CNR-Istituto di Geologia Ambientale e Geoingegneria (IGAG), Rome. The following standard were used: wollastonite (Si, Ca), periclase (Mg), magnetite (Fe), corundum (Al), orthoclase (K), rutile (Ti), jadeite (Na), baritine (Ba), metallic Mn (Mn), metallic Ni (Ni), F-phlogopite (F), silvite (Cl). Conversion from X-ray counts to oxide wt% was accomplished using the Pouchou and Pichoir (PAP) matrix correction [30]. The chemical composition and the atoms per formula unit (apfu) obtained for three samples of tobelite, labelled Tob_m2, Tob_m3 and Tob_3, are shown in Table 1, where the average values over five, five and four analyses, respectively, are shown. A significant difference among the samples is related to the concentration of K_2O (5.3(3), 2.93(4)), whereas their tetrahedral Al-content is nearly unchanged. Therefore, in order to preserve the charge neutrality of the structure, the decrease of K is balanced by an equal increase of NH_4^+ at the interlayer site. The heterogeneous distribution of NH_4/K ratio has been pointed out also by Nieto [12]. This author found that in tobelite formed during the diagenesis at very low-grade metamorphism, as in our samples, the intergrowths of NH_4 and K in micas are on the nanometre-scale, and NH_4-contents range from 30 to 59 % of the interlayer site occupancy, which is in good agreement with this study (42 - 53 %). The quantitative data for NH_4 in Table 1 were obtained by applying the charge-balance principle according to which the decrease of the charge resulted by the number of Al cations per formula unit, replacing Si at the tetrahedral site, must be balanced by the increase of the charge in the interlayer. This allows us to calculate the amount of NH_4 present in the interlayer. For example, let's consider the atomic proportion without NH_4^+ of the sample Tob_m2 (Table 1). The chemical analysis, on the basis of 11 oxygen atoms, gave a deficit of charge, i.e., 0.63 vu (valence unit). The structural formula quoted in Table 2 shows that the charge in interlayer is K + Na = 0.25 and that there are 0.88 tetrahedral Al cations. To achieve the electrostatic balance we need 0.63 v.u., i.e., 0.88Al - 0.25(K+Na). This missing charge has been attributed to ammonium (NH_4). The estimated content of nitrogen, derived from Table 1, for Tob_m2, Tob_m3 and Tob_3 is thus 2.32, 1.90 and 1.53 (wt%), respectively.

4.2. SIMS

SIMS measurements were performed with a Cameca IMS 4f ion microprobe installed at CNR-IGG, Pavia (Italy). A -12.5 kV accelerated $^{16}O^-$ primary-ion beam was used with a current intensity of 4 nA and < 10 μm beam diameter. Nitrogen was detected at the electron multiplier through its most

Table 1. Average electron microprobe data (wt%) and atomic proportion (apfu) for the investigated tobelite crystals. Standard deviations (1σ) are given in parentheses.

Sample name	Tob_m2 (av. of 5 spots)	Tob_m3 (av. of 5 spots)	Tob_3 (av. of 4 spots)
SiO_2	48.28(2)	47.9(2)	46.9(4)
Al_2O_3	37.86(2)	38.0(4)	37.2(2)
MgO	0.037(4)	0.06(3)	0.082(9)
FeO	0.30(5)	0.35(5)	0.33(5)
TiO_2	b.d.l.	b.d.l.	b.d.l.
MnO	b.d.l.	b.d.l.	b.d.l.
NiO	b.d.l.	b.d.l.	b.d.l.
K_2O	2.93(4)	4.2(4)	5.3(3)
Na_2O	0.08(4)	0.22(4)	0.19(5)
BaO	0.04(6)	0.06(7)	0.2(1)
CaO	0.02(2)	0.0(1)	0.007(5)
$(NH_4)_2O$*	4.24	3.50	2.77
F	b.d.l.	b.d.l.	b.d.l.
Cl	b.d.l.	b.d.l.	b.d.l.
H_2O**	4.65	4.64	4.54
Total	98.2	98.7	97.2
Atomic proportion (apfu)**			
Si	3.12	3.09	3.09
Al	2.88	2.90	2.90
Mg	0.00	0.01	0.01
Fe^{2+}	0.00	0.00	0.00
Ti	-	-	-
Mn	-	-	-
Ni	-	-	-
K	0.24	0.35	0.44
Na	0.01	0.02	0.02
Ba	0.00	0.00	0.01
Ca	0.00	0.00	0.00
NH_4^+	0.63	0.52	0.42
F	-	-	-
Cl	-	-	-

* derived by charge balance.
** derived on the basis of 22 negative charges.

Table 2. Comparison between the structural formulas of the studied micas and of literature samples.

	Interlayer site	Octahedral site	Tetrahedral site	Anionic site
Tob_m2	$[K_{0.24}(NH_4^+)_{0.63}Na_{0.01}]_{\Sigma=0.88}$	$Al_{2.00}$	$(Si_{3.12}Al_{0.88})_{\Sigma=4.00}$	$O_{10.00}OH_{2.00}$
Tob_m3	$[K_{0.35}(NH_4^+)_{0.52}Na_{0.02}]_{\Sigma=0.89}$	$(Al_{1.99}Mg_{0.01})_{\Sigma=2.00}$	$(Si_{3.09}Al_{0.91})_{\Sigma=4.00}$	$O_{10.00}OH_{2.00}$
Tob_3	$[K_{0.44}(NH_4^+)_{0.42}Na_{0.02}Ba_{0.01}]_{\Sigma=0.89}$	$(Al_{1.99}Mg_{0.01})_{\Sigma=2.00}$	$(Si_{3.09}Al_{0.91})_{\Sigma=4.00}$	$O_{10.00}OH_{2.00}$
Tob_7 [a]	$(K_{0.18}Na_{0.01}NH_4^+{}_{0.62})_{\Sigma=0.81}$	$(Al_{1.98}Fe^{3+}{}_{0.02})_{\Sigma=2.00}$	$(Si_{3.19}Al_{0.81})_{\Sigma=4.00}$	$O_{10.00}OH_{2.00}$
Tob_Tobe *	$(K_{0.19}Na_{0.01}NH_4^+{}_{0.53})_{\Sigma=0.73}$	$(Al_{1.97}Mg_{0.05}Fe^{3+}{}_{0.03})_{\Sigma=2.05}$	$(Si_{3.17}Al_{0.83})_{\Sigma=4.00}$	$O_{10.00}OH_{2.00}$
Tob_Horo *	$(K_{0.27}NH_4+_{0.57})_{\Sigma=0.84}$	$(Al_{1.95}Mg_{0.01}Fe^{3+}{}_{0.05}Ti_{0.01})_{\Sigma=2.02}$	$(Si_{3.09}Al_{0.91})_{\Sigma=4.00}$	$O_{10.00}OH_{2.00}$
Tob †	$(NH_4^+{}_{1.02})_{\Sigma=1.02}$	$(Al_{2.00})_{\Sigma=2.00}$	$(Si_{2.98}Al_{1.02})_{\Sigma=4.00}$	$O_{10.49}$
Tob §	$(K_{0.15}Na_{0.01}Ca_{0.09}NH_4^+{}_{0.75})_{\Sigma=1.00}$	$(Al_{1.70}Mg_{0.13}Fe^{2+}{}_{0.26}Ti_{0.01})_{\Sigma=2.10}$	$(Si_{2.99}Al_{1.01})_{\Sigma=4.00}$	$O_{10.00}OH_{2.00}$
Tobelitic material ±	$(K_{0.36}Na_{0.03}NH_4^+{}_{0.36})_{\Sigma=0.75}$	$(Al_{1.91}Mg_{0.13}Fe_{0.03})_{\Sigma=2.07}$	$(Si_{3.21}Al_{0.79})_{\Sigma=4.00}$	$O_{10.00}OH_{1.88}F_{0.12}$

Notes: Tob_m2, Tob_m3 and Tob_3 = crystals from this work
[a] [16]; * [7]; † [3]; § [9]; and ± [6].

abundant isotope (^{14}N = 99.63 %) as secondary positive ions, $^{14}N^+$, under +4500 V accelerating voltage, and 250 μm image-field. The analysis was done at a mass resolution (M/ΔM) ~ 1250, adequate to discriminate the contributions of the interfering ions $^{28}Si^{2+}$ and $^{12}CH_2^+$ at the nominal mass number 14. The mass difference ΔM between interfering ions and $^{14}N^+$ resulted to be comparable, within analytical error, to the expected theoretical values in atomic mass units (amu) [31]:

$$\Delta M\ (^{14}N^+ - {}^{28}Si^{2+}) = 0.0146$$

$$\Delta M\ (^{12}CH_2^+ - {}^{14}N^+) = 0.0126.$$

Representative high-resolution mass spectra for crystals Tob_m2, Tob_m3 and Tob_3 are shown in Figure 2. The ion intensities (c/s) at the peak centre of the SIMS mass spectra relative to all analyses done in the three crystals were the following:

- Tob_m2: 399 c/s, 489 c/s, 564 c/s (average: 484 c/s ± 83 (1σ));
- Tob_m3: 29 c/s, 110 c/s; 116 c/s (average: 85 c/s ± 49 (1σ));
- Tob_3: 62 c/s, 65 c/s, 69 c/s (average: 65 c/s ± 3 (1σ)).

The medians for the three SIMS distributions are: 489 c/s (Tob_m2), 110 c/s (Tob_m3) and 65 c/s (Tob_3). For crystals Tob_m2 and Tob_3, the median and the arithmetic mean are very close or coincide. In the case of Tob_m3, the crystal is characterized by a significant inhomogeneity in term of N, which makes critical to assess the representative count rate for N from the average of the experimental data. In this latter case, in the presence of a skewed data distribution, the median might be seen as a better indication of central tendency of the data distribution.

On the basis of the above SIMS results, the higher signal for $^{14}N^+$ pertains to Tob_m2, whereas the lower signal, to Tob_3. Looking at the median of the data for Tob_m3, the N intensity shows an intermediate value between the two. As for the $^{12}CH_2$-peak evidenced in the SIMS mass spectra, we point out that it is of organic origin and, like N, is heterogeneously distributed in the crystals. This behaviour has been observed by some of us using X-ray photoelectron spectroscopy (XPS). We found that the measurements, performed on two different surfaces after exfoliating the same crystal, resulted in a very different Si/C ratio (0.09 versus 0.42). Therefore, it is reasonable to suppose that the SIMS analysis, with a beam diameter much smaller than that used for XPS (~10 versus 400 μm), may have intercepted areas of higher (Tob_m2 and Tob_m3) or lower (Tob_3) carbon concentrations.

4.3. Structural formulas of tobelite

The structural formulas of the set of crystals here investigated, normalized to 22 negative charges, and other natural ammonium micas and tobelite grains from various localities, are shown in Table 2. A comparison with literature samples is also reported in figure 3, in the form of NH_4^+ versus $^{XII}R\text{-}NH_4^+$ (where R represents the sum of the non-NH_4^+-group per formula unit (gpfu) in the interlayer site). From the inspection of the crystal formula in Table 2, it is apparent that there is a close chemical similarity between the samples investigated here (especially for Tob_m3 and Tob_m2) and the Tobe sample of [7], in terms of NH_4^+ interlayer site. Except for the [9] sample, the Tob_m2 is the natural tobelite with the highest ammonium content (Table 2). Actually, although the Betic Cordillera sample [15] has higher NH_4 (0.75 gpfu) concentration compared with that of the other samples reported in figure 3, it is not, strictly speaking, a typical tobelite but rather an intermediate dioctahedral-trioctahedral NH_4^+-rich white mica: about 2.5 of the 3 available octahedral sites are filled by Al, Fe, Mg in this compound. The NH_4-content in the Tob_m2 and Tob_m3 crystals is also greater than that of NH_4-bearing illites and tobelites reported by [11] and [12], respectively (0.20 - 0.55 and 0.40 - 0.50 gpfu).

By means of SIMS analysis the occurrence of nitrogen was proven. While considering the presence of an inhomogeneous distribution of N in sample Tob_3 and the ensuing experimental data scatter, we can reasonably conclude that the SIMS data agree qualitatively with the constraints derived from EPMA investigations associated with previous information by single-crystal X-ray analysis [16].

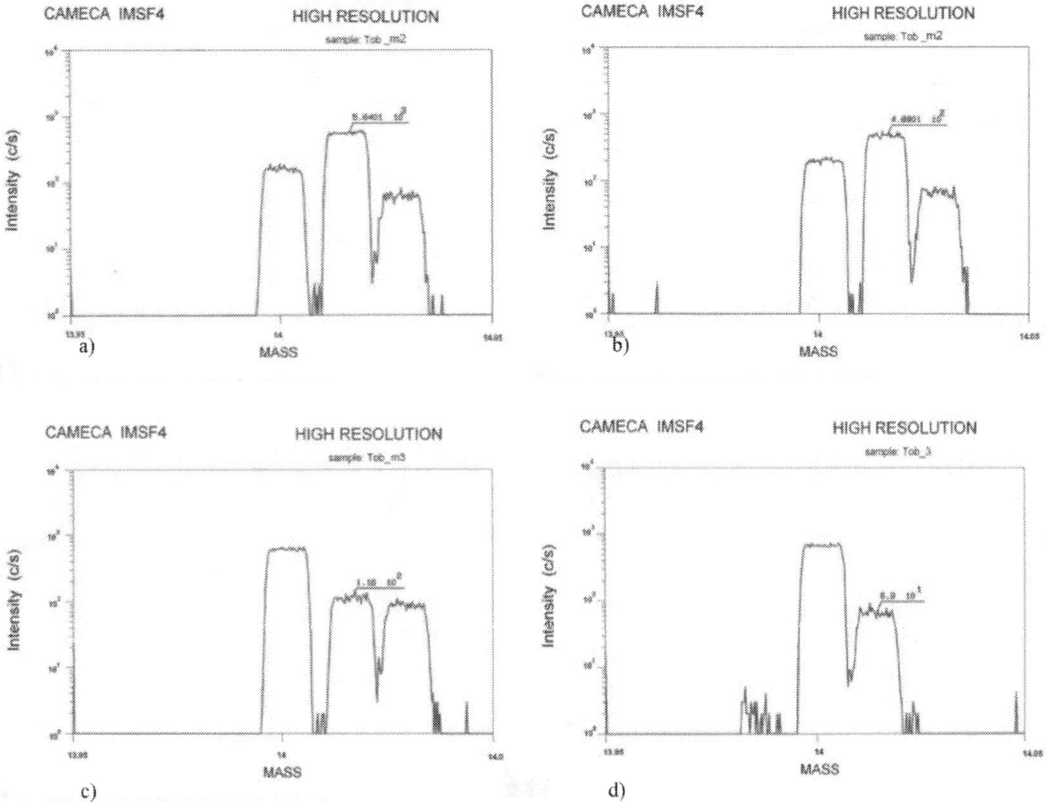

Figure 2. Representative mass scans of peaks of interest at mass 14, obtained at mass resolving power of 1250 (M/ΔM) for the samples analysed: Tob_m2 (figures 2a and b), Tob_m3 (figure 2c), and Tob_3 (figure 2d). We can see the resolved peaks (from next to right): $^{28}Si_2^+$, $^{14}N^+$ and $^{12}CH_2^+$ (this latter in Figures 2a b and). See text for explanation.

5. Conclusion

The detection of NH_4^+ in mica structure is difficult, especially when it occurs in small amount. In this study, by using a multi-analytical approach, it was possible to confirm the presence of ammonium in the interlayer site of the mica structure. In addition, from the analysis of structural formulas, it is feasible to conclude that it is closely and inversely related to K-ion. The present SIMS results match qualitatively with the different NH_4^+-content derived from EPMA analyses and charge-balance constraint, and highlight the prospect of developing reference materials (standards) for quantitative SIMS investigations of nitrogen in mica, and more generally, in phyllosilicates formed in special geological environments.

Acknowledgements
The authors thank G. Toscani (CNR-IGG, Pavia) for improving the quality and sharpness of figure 2. A special thanks to Prof. F. Nieto for providing ref. [24] and an anonymous reviewer. Their constructive criticism and useful comments greatly improved the quality of the paper.

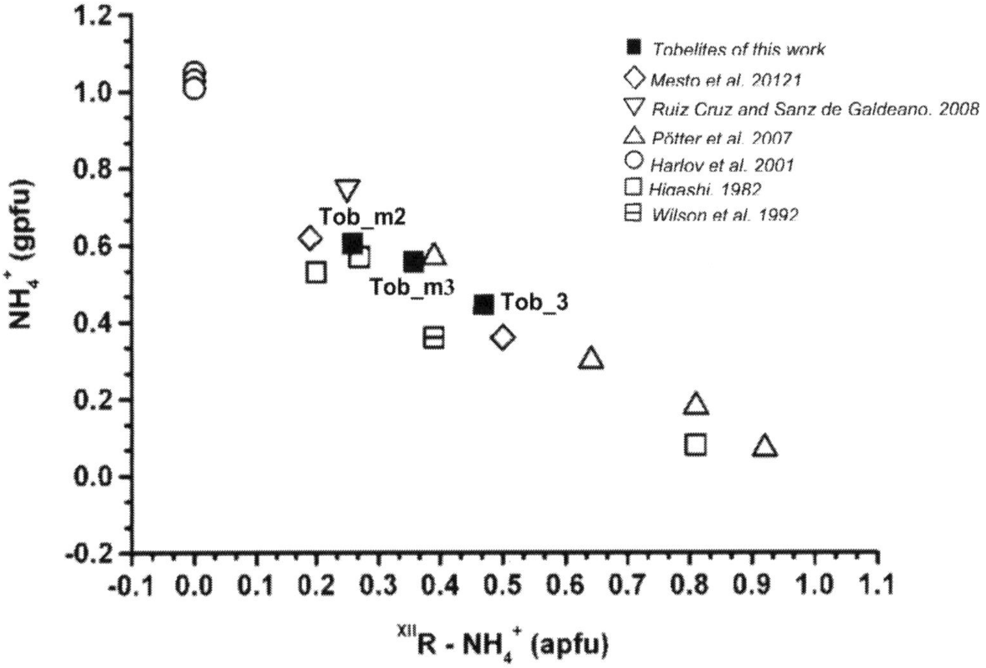

Figure 3. NH_4^+-group per formula unit (gpfu) versus. ^{XII}R-NH_4^+ atoms per formula unit (apfu) diagram modified after [32]. ^{XII}R stands for (K^+, Na^+, Ba^+ ,Ca^{2+}). For comparison, tobelites from literature are included. S ymbols are shown in the legend.

References

[1] Busigny V, Cartigny P, Philippot P and Javoy M 2003 *Chem. Geol.* **198** 21
[2] Urano H 1971 *J. Earth Planet. Sci. Nagoya Univ.* **19** 1
[3] Harlov D E, Andrut M and Pöter B 2001 *Phys. Chem. Minerals* **28** 268
[4] Mookherjee M, Redfern S A T, Zhang M and Harlov D E 2002 *Am. Mineral.* **87** 1686
[5] Drits V A, Lindgreen H, Sakharov B A, Jakobsen H J, Salyn A L and Dainyak L G 2002 *Clays Clay Miner.* **50** 82
[6] Wilson P N, Parry W T and Nash W P 1992 *Clays Clay Miner.* **40** 405
[7] Higashi S 1982 *Mineral. J.* **11** 138
[8] Higashi S 2000 *Appl. Clay Sci.* **16** 171
[9] Ruiz Cruz M D and Sanz de Galdeano C 2008 *Am. Mineral.* **93** 977
[10] Brigatti M F and Guggenheim S 2002 in: *Reviews in mineralogy and geochemistry.* (Mottana A, Sassi F P, Thompson Jr. J B and Guggenheim S; Eds.). (Chantilly, Virginia: Mineralogical Society of America and the Geochemical Society) **46** 1
[11] Juster T C, Brown P E and Bailey S W 1987 *Am. Mineral.* **72** 555
[12] Nieto F 2002 *Am. Mineral.* **87** 205
[13] Pöter B, Gottschalk M and Heinrich W 2007 *Am. Mineral.* **92** 151
[14] Ruiz Cruz M D, Sanz de Galdeano C, Alvarez-Valero A, Rodríguez Ruiz M D and Novák J 2010 *Can. Mineral.* **48** 183
[15] Ruiz Cruz M D 2011 *Miner. Petrol.* **101** 225
[16] Mesto E, Scordari F, Lacalamita M and Schingaro E 2012 *Am. Mineral.* **97** 1460

[17] Bastin G F and Heijligers H J M 1988 *Internal report*; ISBN 90-6819-010-5 (Eindhoven, The Netherlands: University of Technology)
[18] Raudsepp M 1995 *Can. Mineral.* **338** 203
[19] Nash W P 1992 *Am. Mineral.* **77** 453
[20] Evans T 1992 in: *The properties of natural and synthetic diamond.* (Field J E; Ed.). (London: Academic Press) 259-290
[21] Taylor W R, Canil D and Milledge H J 1996 *Geochim. Cosmochim. Acta* **60** 4725
[22] Fitzsimons C W, Harte B, Chinn I L, Gurney J J and Taylor W R 1999 *Mineral. Mag.* **63** 857
[23] Stukalova I E, Petrova V V, Sakharov B A and Pokrovskaya E V 2001 *Lithol. Mineral Res.* **36** 243
[24] Ager F J, Mata M P, Ynsa M D, Respaldiza M A, B Goffé, F. Nieto J A 2006 *Nucl. Instr. Meth. Phys. Res. B* **249** 642
[25] Hauri E H, Wang J, Pearson D G and Bulanova G P 2002 *Chem. Geol.* **185** 149
[26] Harte B, Fitzsimons C W, Harris J W and Otter M L 1999 *Mineral. Mag.* **63** 829
[27] Gloaguen E, Branquet Y, Boulvais Ph, Moëlo Y, Chauvel J J, Chiappero P J and Marcoux E 2007 *Miner. Deposita* **42** 399
[28] Moëlo Y, Gloaguen E and Lulzac T 2006 III. Le minéraux de Saint-Aubin. *Le Cahier des Micromonteurs 1* 13-25
[29] Robert J L 2013 personal communication
[30] Pouchou J L and Pichoir F 1991 in: *Electron probe quantitation.* (Heinrich K F J and Newbury D E; Eds.). (New York: Plenum) 31
[31] Burdo R A and Morrison G H 1971 *Report number 1670.* (Ithaca, NY: Cornell University, Materials Science Center) 1
[32] Tischendorf G, Forster H J, Gottesmann B and Rieder M 2007 *Mineral. Mag.* **71** 285

IOP Conf. Series: Materials Science and Engineering **55** (2014) 012015 doi:10.1088/1757-899X/55/1/012015

Identification of corrosion and damage mechanisms by using scanning electron microscopy and energy-dispersive X-ray microanalysis: contribution to failure analysis case histories

G Pantazopoulos and A Vazdirvanidis

ELKEME Hellenic Research Centre for Metals S.A., 252 Pireaus Street, GR-17778 Athens, Greece

E-mail: gpantaz@halcor.vionet.gr

Abstract. Emphasis is placed on the evaluation of corrosion failures of copper and machineable brass alloys during service. Typical corrosion failures of the presented case histories mainly focussed on stress corrosion cracking and dezincification that acted as the major degradation mechanisms in components used in piping and water supply systems. SEM assessment, coupled with EDS spectroscopy, revealed the main cracking modes together with the root-source(s) that are responsible for the damage initiation and evolution. In addition, fracture surface observations contributed to the identification of the incurred fracture mechanisms and potential environmental issues that stimulated crack initiation and propagation. Very frequently, the detection of chlorides among the corrosion products served as a suggestive evidence of the influence of working environment on passive layer destabilisation and metal dissolution.

1. Introduction

Failure analysis processes aim in determination of the overall damage mechanism (e.g., mechanical overload, corrosion, fatigue, wear) and its specific form and root-cause (e.g., corrosion or damage form and source/initiation) leading ultimately to the failure and to the suggestion of improvement in order to mitigate similar failures in the future. In the present work, the contribution of scanning electron microscopy (field emission SEM) assisted with elemental microanalysis (energy-dispersive X-ray spectrometry) in analysing, determining and identifying mechanisms of corrosion failures in certain metallic components is highlighted. Microfractographic examination using SEM employing secondary and backscattered electron detectors can also characterize the mechanical behaviour and fracture mode of the material under certain environmental conditions which in combination with elemental microanalysis enables outlining the service history towards better understanding of the failure sequence and assisting in the implementation of preventive strategy. Electron trajectories and interaction volume sizes are deduced using statistical modelling (Monte Carlo simulations), see [1]. Quantitative chemical analysis of macroscopic sample volume is usually conducted through specific analytical techniques, such as optical emission and atomic absorption spectroscopy. The utilisation of SEM/EDS microanalysis in fractography and failure analysis provides a unique capability in detection, interpretation and documentation of failure analysis projects; the merits of SEM in fractographic evaluation are highlighted in [2].

Content from this work may be used under the terms of the Creative Commons Attribution 3.0 licence. Any further distribution of this work must maintain attribution to the author(s) and the title of the work, journal citation and DOI.
Published under licence by IOP Publishing Ltd

2. Experimental

Microstructural and morphological evaluation was conducted on mounted cross-sections, after water grinding with up to #1200 grit SiC abrasive papers, followed by fine polishing using diamond and silica suspensions. Immersion etching was performed in appropriate $FeCl_3$ solution followed by rinsing in ethanol and hot air-stream drying. In the present study, immersion etching affects the surface chemistry by deposition of reaction products; therefore it was used only for microstructural evaluation after the performance of microanalysis which was conducted on as polished sections prior to etching. Higher magnification observations on damaged surface and localized chemical analyses were conducted, employing a FEI XL40-SFEG scanning electron microscope using both secondary (SE) and backscattered imaging modes (BSE) coupled with an EDAX energy-dispersive X-ray spectrometer (EDS). Au-sputtered mounts were used for SEM microstructural evaluation under a 20 kV accelerating potential.

3. Case histories

3.1. Stress corrosion cracking of a machineable brass alloy distributor

Water distributor is a component that is used in various plumbing installations, dividing and adjusting the main water supply flow to various stream lines for further use. Fracture surface examination of a brass distributor revealed a characteristic intergranular fracture mode, while the surface was covered by corrosion products (figure 1). SEM observations of a transverse section indicated intergranular cracking with multiple branching, which strongly suggests the occurrence of stress corrosion cracking (SCC) (figure 2). The SCC process is induced by the synergistic action of corrosive environment and stress fields (e.g., residual or applied tensile stresses from component fabrication and/or over torque during installation). The material sensitivity against SCC is enhanced to be result of surface work hardening (due to cold working from machining operations) and local stress concentration imposed by abrupt section/design changes. SCC is initiated on the metal surface and is favoured by three main contributing factors: sensitive material, residual or applied tensile stresses and corrosive environment. SCC is considered as a delayed failure process, where cracks are initiating and propagating at slow rates until stresses applied in the remaining section exceed the fracture strength of the material. SCC is also termed in copper and copper alloys systems as "season cracking" and is mainly promoted by the presence of NH_3 and humidity and it is a more common failure mode for cold worked brasses with more than 15 % Zn. Two main models of SCC general corrosion mechanism are proposed [3]:

(i) Anodic dissolution or chemical model. According to this mechanism, after the passive surface film is fractured, metal anodic dissolution is initiated at the tip of the crack and grows through an active path, which is the least free energy route.

(ii) Mechanical/stress model. According to this specific mechanism the absorbed species alter the stress state at atomic level provoking crack nucleation and facilitating crack propagation.

The entire SCC process comprises several physicochemical processes that involve mostly heterogeneous reaction steps, where the limiting step defines the process control rate. A pit, dent, scratch or other discontinuity acting as stress raiser is a potential SCC contributor, since it may induce a crack to nucleate, especially as the critical notch radius decreases due to stress concentration effects. Intergranular cracking occurs also as a result of hot-shortness which results in fusion of low-melting constituents segregated at grain boundaries causing grain boundary de-cohesion. This type of failure can be found in case of high speed and/or high temperature extrusion [4, 5].

3.2. Dezincification corrosion of a machineable leaded brass component

Dezincification is the result of: Zn selective leaching or both Zn and Cu leaching followed by copper deposition. Characteristic studies of dezincification of brass alloys are presented in Refs. [6-7]. Zinc selective attack leads to the formation of a porous Zn-depleted layer showing poor mechanical strength, toughness and wear resistance.

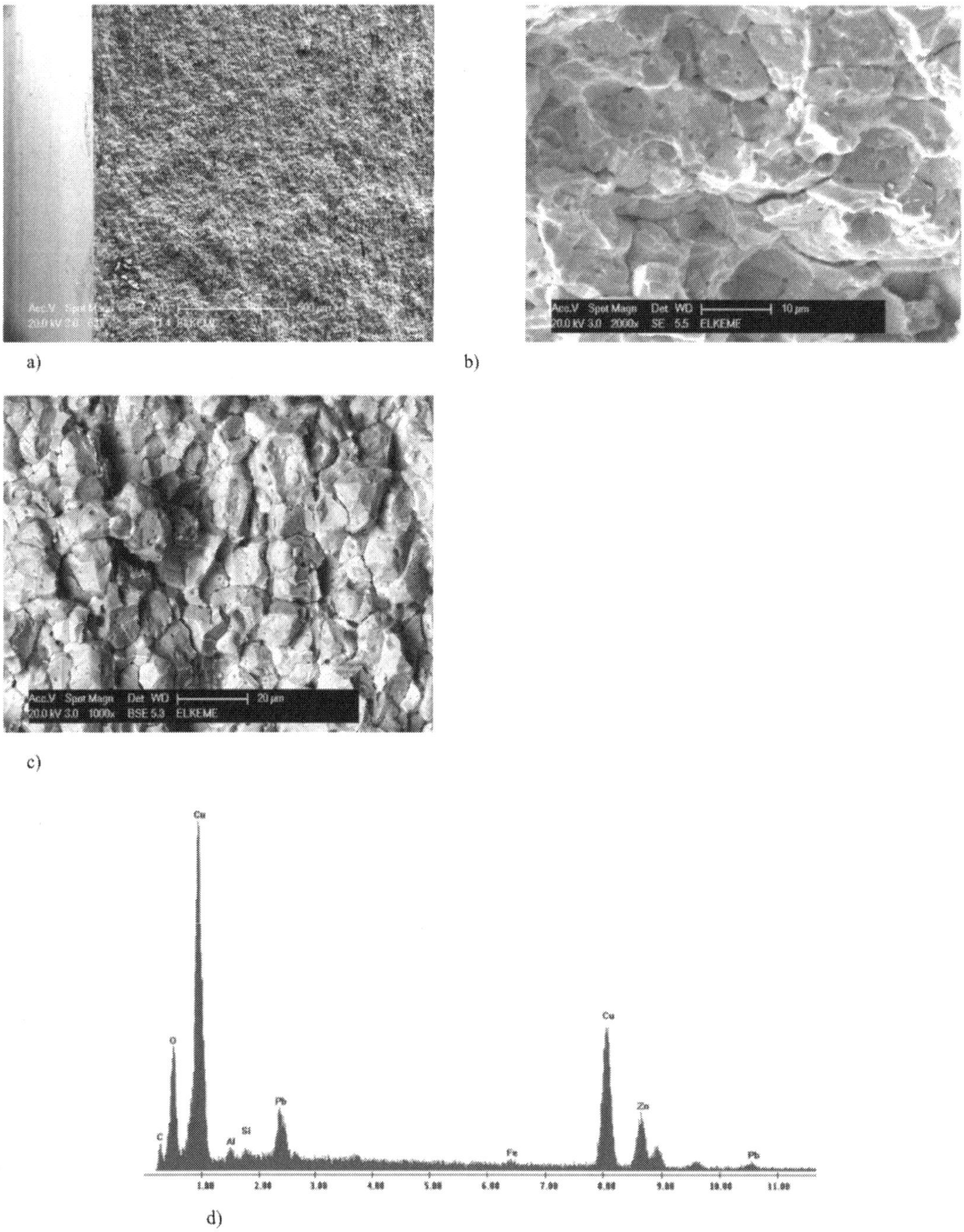

a) b) c) d)

Figure 1. a-c) SEM micrographs (BSE and SE imaging) showing the fracture surface of the ruptured brass component; note the intergranular fracture mode; d) EDS spectrum indicating the elemental chemical composition of fracture surface – the presence of Cu, Zn oxides is evident (ZAF corrected - standardless).

IOP Conf. Series: Materials Science and Engineering **55** (2014) 012015 doi:10.1088/1757-899X/55/1/012015

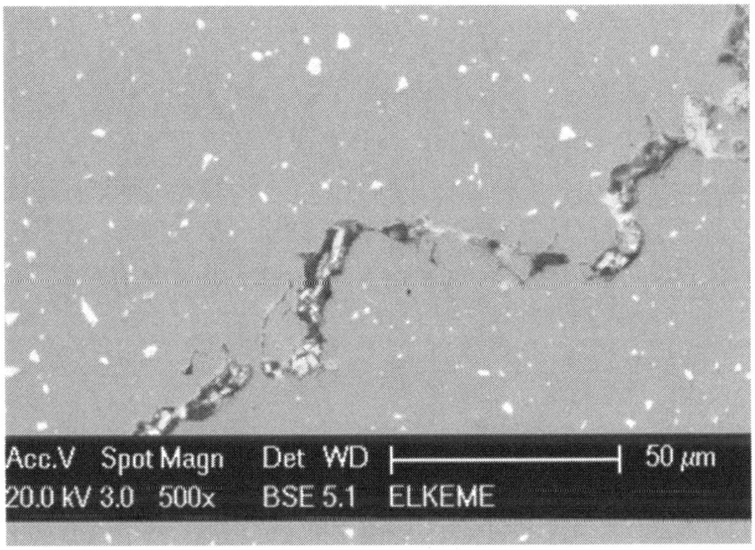

Figure 2. SEM micrograph of as-polished section, transverse to the main crack direction, depicting intergranular cracking and secondary branching remaining also as characteristic indications of SCC in machineable brass. The areas of white contrast in the matrix correspond to lead particles that segregated preferably at grain boundaries.

A threaded brass component demonstrated progressed dezincification, reaching a depth up to ~ 100 μm (figure 3). Dezincification kinetics follows a parabolic rate law controlled by diffusion, see also [7]. In two phase α+β brass alloys the difference in electrochemical potential of the adjacent phases establishes a galvanic coupling, stimulating selective dissolution of Zn ions originating mainly from the β-phase, followed by an equilibrium concentration among neighbouring grains. The degradation of the α-phase occurs at an advanced stage of the dezincification process. In parallel migration through the porous network aggravate the corrosion conditions, accelerating dissolution leading to chemical attack on copper, through the formation of $CuCl_2^-$-complex anions. Sensitivity of brass dezincification is significantly reduced in case of single α-phase brass and especially for Zn contents lower that 15 %, see [8].

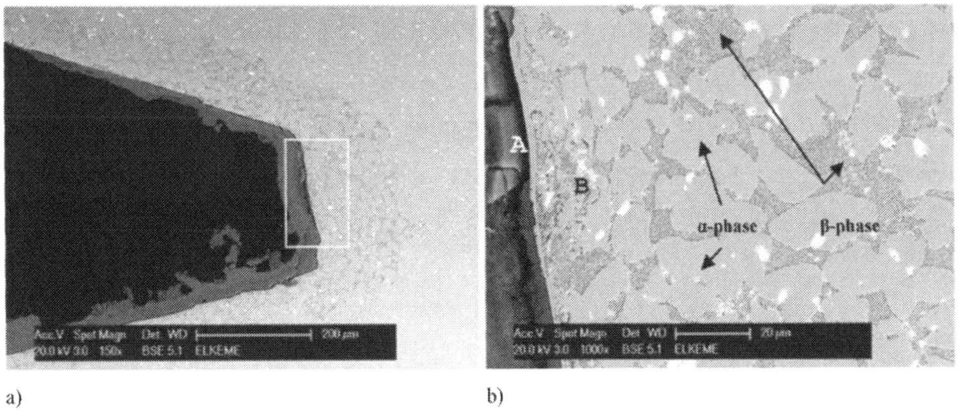

a) b)

Figure 3. a) BSE-image of as-polished section showing a dezincification area in the threaded brass component. b) Detail of the boxed area presented in a). The morphology of corrosion layer suggests that it is the result of layer (uniform) dezincification.

The addition of minor alloying elements, such as Sb and As operating as inhibiting constituents, improves significantly the dezincification corrosion resistance of brasses in aqueous service environments. Depending on service environment conditions (e.g., temperature, corrosive media), two main forms of dezincification are encountered: uniform and plug-type (localized) dezincification, where penetration is steep and leads very frequently to perforation and leakage [9]. SEM micrographs and EDS spectra showing the form of degradation and Zn concentration gradients and selective attack of β-phase, due to differential Zn dissolution are shown in figures 3 and 4.

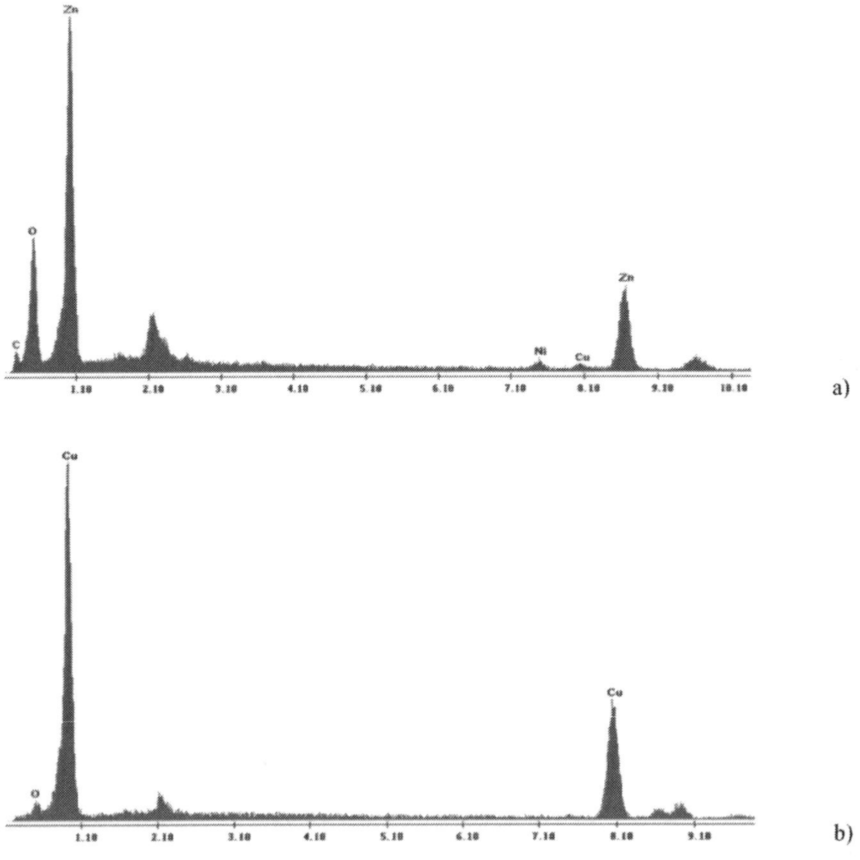

Figure 4. EDS spectra (ZAF corrected - standardless). a) The scale consisting mainly of ZnO (approximate location – A in figure 3b; b) the Zn depleted area showing predominantly the presence of Cu (approximate location – B in figure 3b).

3.3. Pitting corrosion of a deoxidized high phosphorus Cu (CuDHP) water tube

A sample from copper water-tube, which experienced leakage during service after a short-time period, was evaluated using SEM-EDS of 1 mm wall thickness. Deoxidized high phosphorus Cu (P: 0.015 - 0.040 wt%; Cu: 99.90 wt% min) is the main construction material for the fabrication of water and refrigeration pipes. The investigation of failure strongly suggests that the leakage resulted from inside-out driven pitting corrosion. Layers of scale products were massively accumulated on the water side surface. No microstructural abnormality of the CuDHP, that could be connected to the failure was found. The nature and stacking sequence of scale deposits, addressing the operated corrosion phenomena, was evaluated by SEM-EDS analysis. The examination of outer surface scale (green copper patina) indicated the presence of characteristic spherical nodules composed of acicular crystals, corresponding to complex copper carbonates (malachite), see figure 5. Transverse sections adjacent to the leakage area were made to highlight the form of corrosion. The inner tube surface suffered from severe pitting corrosion, while corrosion product inner surface deposits located on that area fell in to two main categories: mixed hydroxides-chlorides on the floor of the pit, copper oxide crystals (Cu_2O) and outer-thick copper hard scale (e.g., carbonates), see figure 6.

a) b)

Figure 5. SEM micrograph (BSE images). a) Pit cap surface showing the presence of characteristic nodular aggregates composed of malachite crystals; b) detail of a) revealing the acicular nature of malachite crystals.

a) b) c) d)

Figure 6. SEM micrograph (BSE images) of a polished transverse section of the corroded inner surface of a CuDHP water tube. a) Layered structure of corrosion products: mixed copper chlorides alternating with copper oxide (Cu2O) and on top copper carbonates; b) the rectangular area outlined in a) at higher magnification; c) nodular malachite crystals on the pit cap at an another location on the transverse section; d) cubic oxide crystals close to the metal/oxide interface.

Perfectly formed cubic copper oxide crystals (Cu_2O) are found underneath the carbonate deposits (see figure 6d) while discontinuous layers of mixed copper chlorides were detected on the floor of the pit. Though the presence of complex hydroxyl-chlorides compounds such as $CuCl_2.3Cu(OH)_2$

initially decreases the corrosion rate, their porous nature and poor adhesion to the base metal frequently leads to detachment and lose of the film sustaining the corrosion process [10]. Selective area elemental analysis conducted by EDS, verified the presence of mixed hydroxy-chlorides and carbonates (figure 7). The presence of carbonates and chloride salts causes cold water type pitting corrosion, referred to as Type I-pitting (see also Refs. [11-12]). Type I pitting is a very severe form of corrosion, leading to accelerated corrosion rates, and it is influenced by the concentration of chlorides and bicarbonates. Corrosion pits can provoke failures through stress corrosion cracking (SCC) as they constitute stress concentration sites. Failure analysis of copper tube affected by the synergy of pitting corrosion and SCC is presented in [13]. Several case histories of copper tube damage in service, including SCC, microbiologically influenced corrosion (MIC), pitting corrosion and fatigue are reviewed in [14, 15]. In the majority of the investigated cases, water supply conditions and quality were the major failure contributors. Water chemistry (physicochemical and microbiological parameters including pH, conductivity, total organic carbon, etc.) should be investigated to pin-point precisely the root cause of the leakage failure. Regular water quality monitoring during installation together with sufficient maintenance and appropriate water treatment (adjusting salinity and hardness as well as dosing corrosion inhibitors and controlling the microbial content) as well as avoidance of stagnation periods are suggested as further actions to prevent corrosion failures.

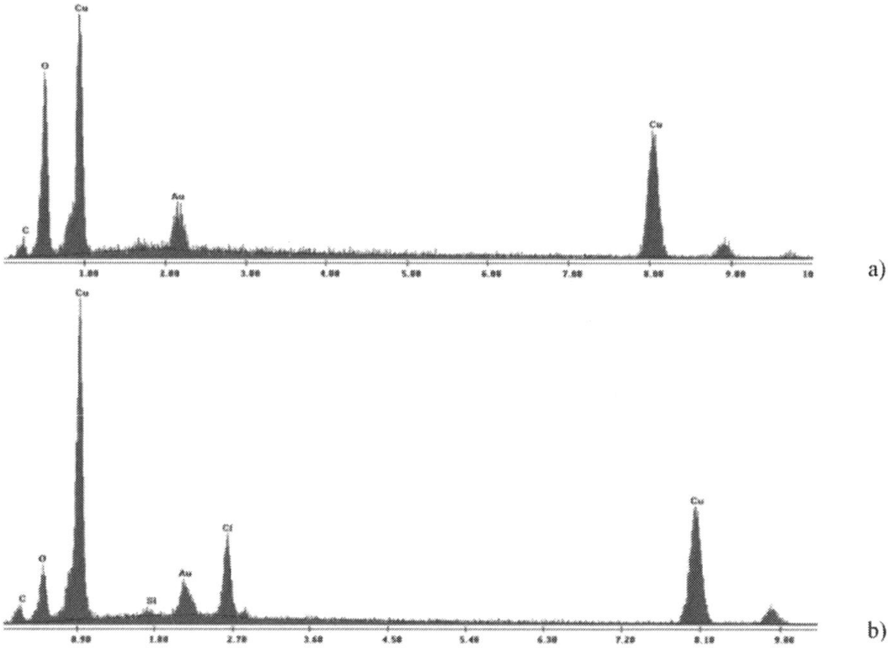

Figure 7. EDS spectra (ZAF corrected – standardless analysis). a) At the surface of the pit confirming the presence of copper carbonate crystals (malachite) – approximate location D in figure 6c; b) Scale attached to the metal substrate consisting of mainly copper (hydro)oxide-chlorides (approximate location – C in figure 6b). Note the presence of significant Cl, considered to be a contributing factor to passive layer destabilisation and pitting initiation. Also note that the energy scale is different on spectra a) and b).

4. Conclusion

Scanning electron microscopy coupled with energy-dispersive X-ray spectrometry constitutes an exceptional analytical technique, offering high spatial resolution and microanalytical capability. The utilisation of SEM/EDS microanalysis provides deep insight and understanding of corrosion and damage mechanisms, outlining the fracture history along with the identification of environmental contributing factors, effectively assisting in failure prevention.

References

[1] Shipley R J and Stevenson M 2011 Interpreting the evidence: elemental analysis in the SEM. *J. Fail. Anal. and Preven.* **11** 274-281
[2] Pantazopoulos G 2011 Damage assessment using fractography as failure evaluation: applications in industrial metalworking machinery. *J. Fail Anal. and Preven.* **11** 588-594
[3] Fontana M G 1986 *Corrosion engineering, 3rd edition.* (New York: McGraw Hill)
[4] Pantazopoulos G 2003 A review of defects and failures in brass rods and related components. *Prac. Fail. Anal.* **3**(4) 14-22
[5] Pantazopoulos G and Vazdirvanidis A 2008 Failure analysis of a fractured leaded-brass (CuZn39Pb3) extruded hexagonal rod. *J. Fail Anal. Preven.* **8** 218-222
[6] Sugawara H and Ebiko H 1967 Dezincification of brass. *Corrosion Sci.* 7 513-523
[7] Trethewey K R and Pinwill I 1987 The dezincification of free-machining brasses in sea water. *Surf. Coat. Technol.* **30** 289-307
[8] Davies D D 1993 *A note on dezincification of brass and the inhibiting effect of elemental additions.* (New York: Copper Development Association)
[9] Jones D A 1996 *Principles and prevention of corrosion, 2nd edition.* (Prentice-Hall NJ)
[10] ASM Handbook 1992 *Corrosion, 4th printing.* (Materials Park, OH: ASM International)
[11] BS EN12502-2:2004 2005: *Protection of metallic materials against corrosion – Guidance on the assessment of corrosion likelihood in water distribution and storage systems Part 2: Influencing factors for copper and copper alloys.* (London: British Standards Institute)
[12] Edwards M, Ferguson J F and Reiber S H 1994 The pitting corrosion of copper. *J. Am. Water Works Assoc.* **87** 74-89
[13] Pantazopoulos G, Vazdirvanidis A and Tsinopoulos G 2011 Failure analysis of a hard-drawn water tube leakage caused by the synergistic actions of pitting corrosion and stress corrosion cracking. *Engng. Failure Anal.* **18** 649-657
[14] Pantazopoulos G and Vazdirvanidis A 2011 A review of characteristic in-service damage of copper tubes in fluid circulation systems: root-cause analysis and failure prevention. in: *Proc. Int. Conf. Copper Metall.* (Krakow, Poland; 26-28 October 2011) 441-449
[15] Pantazopoulos G 2009 Analysis of leakage failure in a domestic hydraulic installation assisted by pitting corrosion of water tube. *J. Fail. Anal. and Preven.* 9 479-485

Improving the quantification at high spatial resolution using a field emission electron microprobe

P T Pinard and S Richter

RWTH Aachen, Gemeinschaftslabor für Elektronenmikroskopie, Ahornstrasse 55, DE-52074 Aachen, Germany

E-mail: pinard@gfe.rwth-aachen.de

Abstract. The capabilities of field emitter electron microprobes to perform quantitative measurements at high spatial resolution are discussed. Using Fe-Cr-C particles in a bearing steel (SAE 52100) as example, a generic procedure was established to find the optimal analytical conditions (beam energy, beam current and acquisition time). The influence of these parameters on the accuracy, precision and spatial resolution was evaluated using experimental measurements and Monte Carlo simulations. A quantification procedure was developed for soft X-ray lines, taking into account the overlap of high order X-ray lines and background anomalies. The accuracy of Kα- and Lα-lines was verified using reference materials. A relationship between experimental and simulated X-ray intensities was determined to evaluate the measurement precision. The spatial resolution of each X-ray line was calculated from the simulated lateral and depth X-ray intensity distribution using simulations integrating experimentally measured beam diameters. The optimal analytical conditions for the studied sample were found to be 5 keV, 10 nA and 10 s acquisition time. Further specialized techniques to improve the spatial resolution are presented: focused ion beam preparation of thin lamella and wedge, and Monte Carlo based reconstruction. The feasibility of the latter to quantify features smaller than the X-ray emission volume was demonstrated.

1. Introduction

By combining the functionality of conventional thermionic gun microprobes with a Schottky field emission gun, field emission microprobes (FE-EPMA) offer new possibilities to characterize smaller features. The higher current density and brightness of this emitter result in a smaller electron probe and extent the analysis range to lower accelerating voltages.

Since their introduction, one decade ago, FE-EPMAs have been used in a wide range of applications from chemical characterisation of micrometeoroids captured on the international space station [1] to partitioning analysis of alloying elements in a cobalt-base superalloy [2]. Several authors combined the improved spatial resolution of FE-EPMA with the spectral resolution of wavelength-dispersive X-ray spectrometers to qualitatively investigate the elemental distribution, including light and trace elements, in various samples by beam and stage mappings [3-8]. Ha *et al.* [9] performed quantitative analyses on different micrometre-size single phase regions of Mo-Si-B alloys under conventional conditions (15 keV and 20 nA). Ohnuma *et al.* [10] measured the chemical composition of iron silicate phases using a lower beam energy (6 keV) and the Fe-Lα. They estimated the spatial resolution to be below 500 nm. Guo *et al.* [11] quantified the carbon concentration inside 0.5 μm to 1 μm Co-rich phases of WC-Co alloys. A beam energy of 8 keV and a beam current of

Content from this work may be used under the terms of the Creative Commons Attribution 3.0 licence. Any further distribution of this work must maintain attribution to the author(s) and the title of the work, journal citation and DOI.
Published under licence by IOP Publishing Ltd

15 nA were chosen as a compromise between spatial resolution and X-ray yield. The spatial resolution, under these conditions, was evaluated to be 400 nm using the electron trajectories of Monte Carlo simulations and the size of the contamination ring produced by a measurement. Two conclusions can be drawn from this literature review: (1) the use of FE-EPMA to quantitatively analyse sub-micrometre feature is limited, and (2) challenges of low beam energy analyses, as described by Merlet and Llovet [12, 13], are often neglected.

The aim of this work is to establish a generic procedure to optimize the analytical conditions in order to improve the spatial resolution, and to evaluate specialized techniques to further improve the spatial resolution. The quantification of Fe-Cr-C spherical particles inside a bearing steel (SAE 52100) is used as a case study throughout the paper. Section 2 details the sample preparation, instrument and software used. The optimisation strategy is described in section 3 where the influence of the beam energy, beam current and acquisition time on the accuracy, precision and spatial resolution is studied using experimental measurements and Monte Carlo simulations. Finally, section 4 discusses how smaller features can be quantified using Monte Carlo based reconstruction approach or advanced preparation techniques such as milling a wedge or cutting a thin lamella.

2. Materials and methods

The bearing steel sample (SAE 52100) and the standards (Cr, Fe, $Cr_{23}C_6$, Fe_3C) were prepared using mechanical polishing up to 0.05 μm colloidal silica, cleaned with ethanol in an ultrasonic bath and dried under hot air. The samples were left inside the specimen chamber for at least one day before starting the measurements. All measurements were performed on a field emission electron microprobe, JEOL JXA-8530F (JEOL Ltd., Tokyo, Japan). The Cr and Fe Kα-lines were measured using LiF crystals on Xe counters, whereas a LDE1 crystal in combination with a P-10 counter was used for their Lα-lines. C-Kα was measured using a LDE2 crystal with a P-10 counter. An oxygen jet was used throughout all measurements to reduce the influence of contamination. The efficiency of this anti-contamination device, especially in the case of carbon quantification, has been presented by Bastin and Heijligers [14]. The quantification of the measurements were performed using in-house scripts and the electron probe quantitation library available as part of the NIST DTSA-II software (version Gemini) [15]. Unless otherwise specified, the XPP simplified model [16] and the NIST-Chantler [17] database were respectively selected as the default matrix correction algorithm and mass absorption coefficient (MAC) database. All Monte Carlo simulations were performed using the NISTMonte programme [18], also included inside the NIST DTSA-II software. New distributions were added to the original code to obtain the results in this work.

3. Optimal analytical conditions

3.1. Accuracy

Depending on the selected beam energy, different X-ray lines must be used for the quantification. Many authors reported large inaccuracies when low energy X-ray lines (< 1 keV) are used (e.g., Lα-lines for the first row of the transition metals). Inaccurate mass absorption coefficients [19, 20], chemical bonding effects [21], oxidation/contamination effects [22], peak shape alterations [14] and interference problems [23, 24] were identified as potential explanations for the large inaccuracies.

In order to assess the reliability of the quantification of the studied sample, two reference samples with homogeneous composition, $Cr_{23}C_6$ and Fe_3C, were quantified using Kα- and Lα-lines of Cr and Fe. To minimize the influence of the aforementioned problems, the quantification was performed using integral intensities (peak areas) obtained from spectrum acquisitions of each X-ray line, detailed background modelling and deconvolution of interfering peaks. Spectra were acquired for each X-ray line with a dwell time of 1.125 s. The step size of each spectrometer was adjusted to keep the total measurement time under 20 min. Pure standards were used for all elements, except for carbon where Fe_3C was used to quantify $Cr_{23}C_6$ and vice versa.

The background of each spectrum was fitted using a single exponential function, except for C-Kα spectra where the absorption edge below the peak significantly changes the shape of the background. Two exponential functions were then fitted for the background values above and below the absorption edge, and juxtaposed with an error function to create a continuous background function. Figure 1 shows the final background modelling under the C Kα-peak acquired on Fe_3C. This method was privileged over the one described by Rémond *et al.* [24] since it does not rely on the knowledge of mass absorption coefficients and can correct for the self-absorption of C-Kα X-rays inside the crystal and detector window.

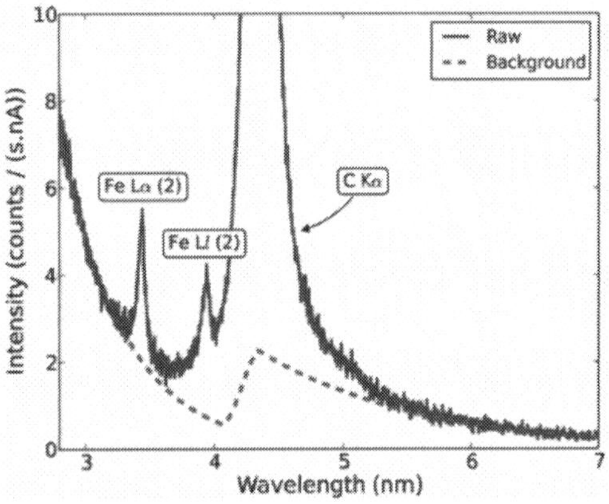

Figure 1. Background modelling below the C-Kα peak acquired on Fe_3C (15 keV, 100 nA, 1.125 s dwell time, 3.5 pm step size, LDE2 crystal, P-10 counter).

Overlapping lines were decomposed using a least-square fitting procedure assuming a pseudo-Voigt function for each peak [25]. The pseudo-Voigt function was assumed to be 57 % Gaussian and 43 % Lorentzian as described by Fournier *et al.* [25]. The optimisation package from the open-source library SciPy [26] was used for the least-square fitting. The initial estimate was manually determined from the peak height, width and position read from each spectrum. For the Cr and $Cr_{23}C_6$ spectra, the Cr-Lα peak was decomposed from the overlapping Lβ- and Ll-lines, whereas for the Fe and Fe_3C spectra, the Fe-Lα was decomposed from the Ll-line. In the case of C-Kα, the overlapping correction of the second order X-ray lines of Cr (Lα and Ll) and Fe (Ll) was more difficult due to the poor resolution of the LDE2 crystal. To simplify the fitting procedure, pseudo-Voigt functions of these peaks were first calculated from the pure standard spectra. These functions were then weighted by the nominal Cr and Fe composition of $Cr_{23}C_6$ and Fe_3C and subtracted from the spectrum acquired from this sample. Figure 2 shows the fitted Cr-Lα (2) and Ll (2) on pure chromium spectrum and figure 3 the raw and corrected $Cr_{23}C_6$ spectrum. No fitting was used for the Kα-lines. The intensity was directly extracted from the sum of the counts under the peak after background removal.

Table 1 gives the quantification results in weight fraction for the $Cr_{23}C_6$ and Fe_3C reference samples measured at 3 and 15 keV. For both reference samples, accurate quantification was obtained using the Kα-lines at 15 keV and the Lα-lines at 3 keV. In the case of Cr, the large errors on the quantification with the Lα-lines at 15 keV are related to the mass absorption coefficients. The accuracy of the MAC is critical due to the large absorption correction at high overvoltage ratio. If the MAC database from Pouchou and Pichoir [16] is used instead of the one from NIST-Chantler, the Cr and C concentration are closer to the nominal values (95.54 % and 5.9 % respectively, total of 101.44 %). Incorrect MAC values could potentially explain the errors for the Fe-Lα, but no improvement was observed by selecting a different MAC database.

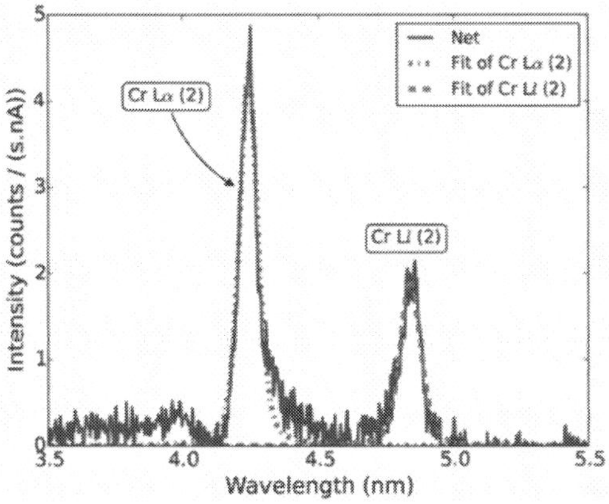

Figure 2. Carbon Kα background-corrected spectrum acquired on Cr showing the pseudo-Voigt fitting of the Cr-Lα (2) and -Ll (2) (15 keV, 100 nA, 1.125 s dwell time, 3.5 pm step size, LDE2 crystal, P-10 counter).

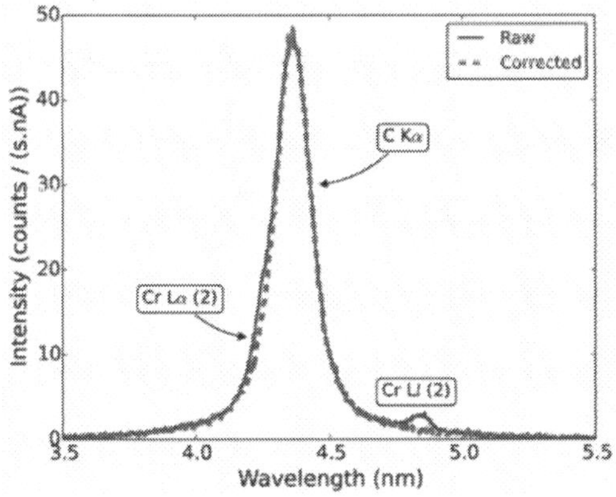

Figure 3. Carbon Kα background-corrected spectrum acquired on $Cr_{23}C_6$ before (raw) and after the correction for the overlap of the Cr-Lα (2) and -Ll (2) (15 keV, 100 nA, 1.125 s dwell time, 3.5 pm step size, LDE2 crystal, P-10 counter).

Table 1. Quantification of $Cr_{23}C_6$ and Fe_3C reference samples (all values in weight percent).

		$Cr_{23}C_6$			Fe_3C		
	Line	Cr	C	Total	Fe	C	Total
Nominal	—	94.32	5.68	100.00	93.31	6.69	100.00
15 keV	Kα	94.30	6.00	100.29	93.13	6.41	99.54
3 keV	Lα	95.12	6.62	101.74	93.40	6.18	99.58
15 keV	Lα	104.89	6.76	111.64	119.12	8.55	127.66

Databases from Henke [27] and Heinrich [28] were tested as well. Other matrix correction models, PAP [16] and Proza96 [29], did not improve the results. Problems with the mass absorption coefficients of the Fe-Lα were also reported by Llovet *et al.* [19] and Gopon *et al.* [20].

Based on these results, both Kα- and Lα-lines of Cr and Fe could be used to quantify the Fe-Cr-C particles, if the overvoltage on the Lα-lines is small. This example, however, illustrates the importance of evaluating the quantification accuracy of low energy lines with appropriate standards before measurements on unknown samples.

3.2. Precision

The precision of a quantitative measurements is determined by the combination of systematic and random uncertainties [30]. Systematic uncertainties can typically be minimized by rigorous calibration of the spectrometers, good sample preparation and use of accurate standards. On the other hand, random uncertainties are unavoidable. Following Poisson statistics, they are equal to the square root of the number of detected X-rays. For an integral acquisition, the on and off peak intensities are obtained from the same spectrum. The uncertainty (σ) is thus equal to the square root of the total intensity (I) under a peak (sum of net and background (I_B) intensities). As the intensity is proportional to the electron dose, increasing the beam current and/or the acquisition time reduces the relative uncertainty (ε) of the measurement. The beam energy has also an influence on the intensities as the ionisation distribution function, $\phi(\rho z)$, is energy dependent. The relative efficiency can be calculated using eq. 1 for an integral acquisition. The total intensity (I) is equal to the area under the peak over a defined interval, whereas the background intensity (I_B) is the area under the fitted background over the same interval.

$$\epsilon = \frac{n\sigma}{I - I_B} = \frac{n\sqrt{I}}{I - I_B} \quad (1)$$

where n is the confidence index.

To evaluate the precision for the studied sample under different analytical conditions, a simple model was developed based on intensities calculated by Monte Carlo simulations and the assumption that the simulated intensities are directly proportional to the experimental ones. The proportionality is only valid for a specific X-ray line as the detector efficiency changes based on the selected crystal and type of counting detector as well as the solid angle, based on the crystal position. The premise also disregards possible inaccuracy in fundamental parameters such as the ionisation cross-section and fluorescence yield [31]. Despite these restrictions, the model allows to quickly estimate the experimental precision from simulated intensities without the need of experimental measurements at each analytical condition.

To calculate the proportionality constant for each X-ray line, the integral intensities measured for the accuracy assessment were compared to the characteristic X-ray and background intensities obtained from the simulations. The background intensities were obtained by summing the X-ray counts over a certain energy range. For the experimental values, the fitted background function of each experimental spectrum was used and for the simulated ones, the simulated energy dispersive spectrum of an ideal detector without characteristic X-ray peaks. Figure 4 shows the relationship between experimental and simulated intensities for the Cr-Kα, Cr-Lα and C-Kα in $Cr_{23}C_6$.

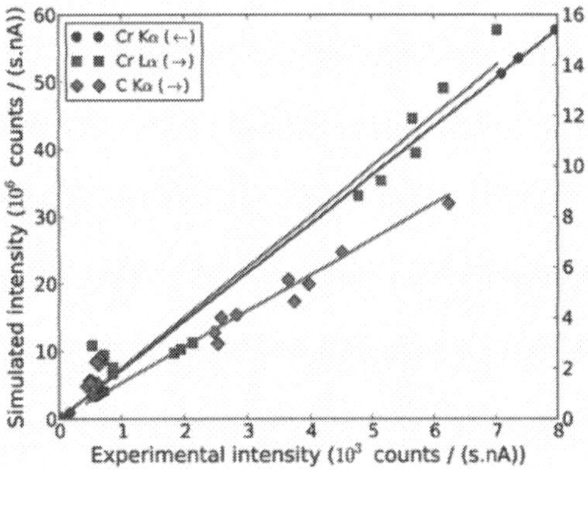

Figure 4. Relationship between experimental and simulated intensities for the Cr-Kα (left axis), Cr-Lα (right axis), and C-Kα (right axis) in $Cr_{23}C_6$. Measurements of net and background intensities at 5, 8 and 15 keV are shown.

Figure 5 displays the variation of the relative uncertainty (2σ) of the Cr-Kα, Cr-Lα and C-Kα in $Cr_{23}C_6$ as a function of the beam energy and dose. The Cr-Kα and Cr-Lα follow a similar trend where the uncertainties decrease as the overvoltage and dose increase. In the case of C-Kα, the uncertainties do not decrease as the beam energy increases. The high absorption of C-Kα X-rays in $Cr_{23}C_6$ limits the increase of the characteristic X-ray intensity at high beam energies, keeps the peak to background ratio constant (≈ 2.3) and prevents the decrease of the uncertainties. From Figure 5, an electron dose of 100 nC would be required to obtain relative uncertainties less than 1 % for all X-ray lines.

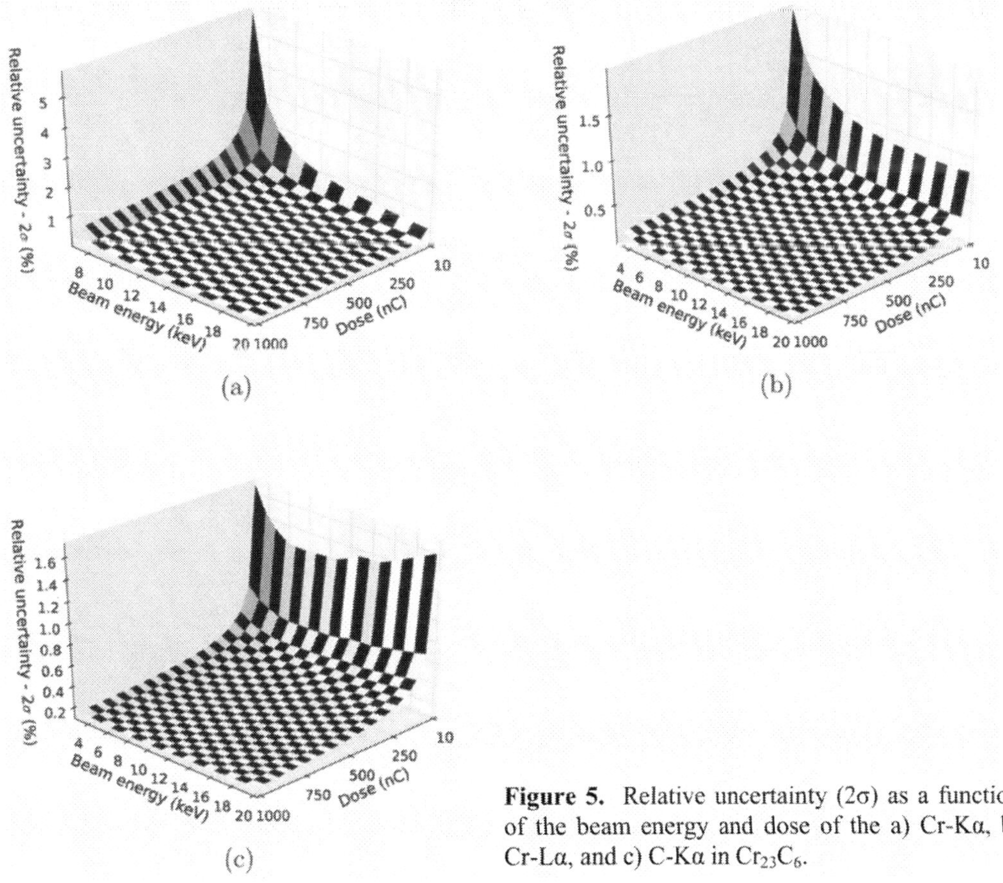

Figure 5. Relative uncertainty (2σ) as a function of the beam energy and dose of the a) Cr-Kα, b) Cr-Lα, and c) C-Kα in $Cr_{23}C_6$.

The composition also affects the precision. A higher dose is required to measure minor and trace concentration of an element. Figure 6 illustrates this effect where the concentration of Cr was varied from 0.1 wt% to 10 wt% in a Fe-Cr-C compound for a fixed beam energy of 10 keV. Doubling the dose from 100 nC to 200 nC has the effect of reducing the relative uncertainty by a factor equal to the square root of 2. Since the composition of a sample is usually unknown prior to the measurement, the dose required to obtain a given precision level could be evaluated from a preliminary semi-quantitative analysis.

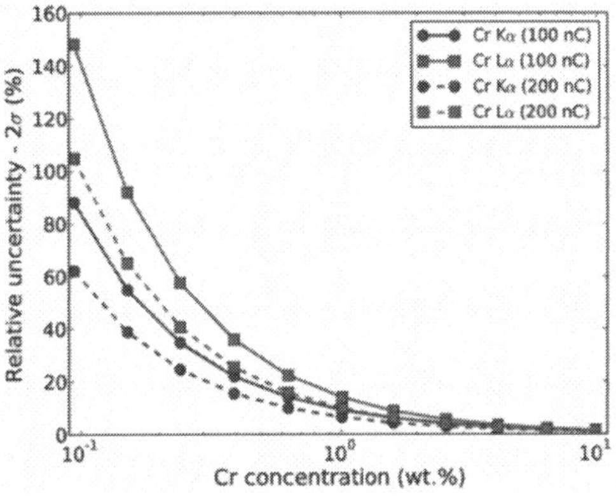

Figure 6. Relative uncertainty (2σ) as a function of the Cr-concentration in Fe-Cr-C where the C-concentration is fixed at 5.68 wt% and the beam energy at 10 keV.

3.3. Spatial resolution

The emission volume of a characteristic X-ray line is defined as the volume from which X-rays are generated and can escape the sample. The dimensions of this volume depend on the beam energy and composition of the sample.

Figure 7 shows the variation of the depth resolution for the Cr-Kα, Cr-Lα and C-Kα in $Cr_{23}C_6$. The depth resolution was calculated by Monte Carlo simulations and is defined as the depth where 90 % of the X-rays are emitted. The density of $Cr_{23}C_6$ was assumed to be 6541.5 kg m^{-3}. The depth resolution of all X-ray lines, except C-Kα, increases approximately linearly with the overvoltage. C-Kα X-rays are strongly absorbed by Cr atoms which limit their escaped depth. Above 10 keV the depth resolution of C-Kα X-rays is smaller than the one of Cr-Kα despite the larger overvoltage (3.5 versus 1.4 at 10 keV).

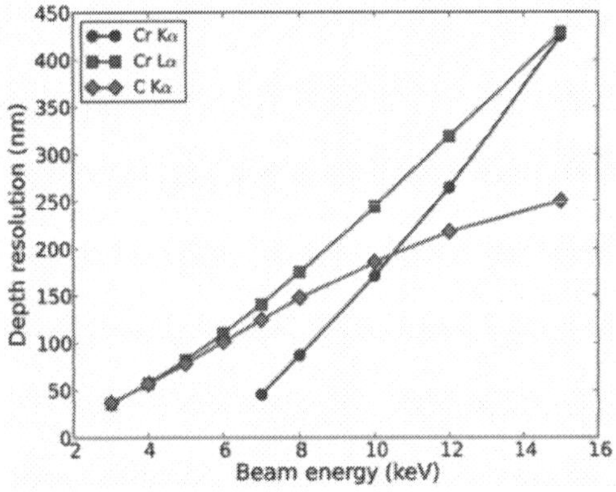

Figure 7. Depth resolution as a function of the beam energy for the Cr-Kα, Cr-Lα and C-Kα in $Cr_{23}C_6$.

Regarding the radial resolution, the beam diameter must be considered as it enlarges the emission volume. One advantage of a Schottky field emission source is the smaller beam diameter [32]. However, the actual width of the beam changes with the analytical conditions. To study this effect and obtain quantitative values for the beam diameter, secondary electron (SE) images of gold flakes on a carbon substrate were taken at different beam energies and beam currents. The resolution of an image was measured using its Fourier transform where it is defined as the extent of the low frequency region [33]. For each image, the Fourier transform was calculated for five regions of interest (512x512 pixels) and the threshold of the low frequency region was manually selected from each power spectrum image following the instructions from Joy [33] and Probst *et al.* [34]. Since the interaction volume of secondary electrons is small in comparison to the measured resolutions (< 5 nm), the beam diameter was assumed to be equal to the resolution.

For all beam energies, the beam diameter increases with the beam current (figure 8). At 100 nA the beam diameter is at least 80 % larger than at 10 nA. Furthermore, for a fixed beam current, the beam diameter decreases with the beam energy. This effect is more clearly shown in figure 9 where the beam diameter was measured as a function of the beam energy for a beam current of 10 nA. The smallest beam diameter was therefore obtained at the highest beam energy (15 keV) and the lowest beam current (10 nA). From figure 9, the broadening of the beam diameter is more pronounced at low beam energies (< 6 keV), which limits the improvement of the resolution. Regardless of the size of the emission volume, the smallest detectable feature can never be smaller than the beam diameter. Reducing the beam current to values lower than 10 nA would result in a smaller beam diameter, but as shown in the previous section, the uncertainties on the measurements would increase unless the acquisition time is increased. Sample drift may however limit the practical increase of the acquisition time.

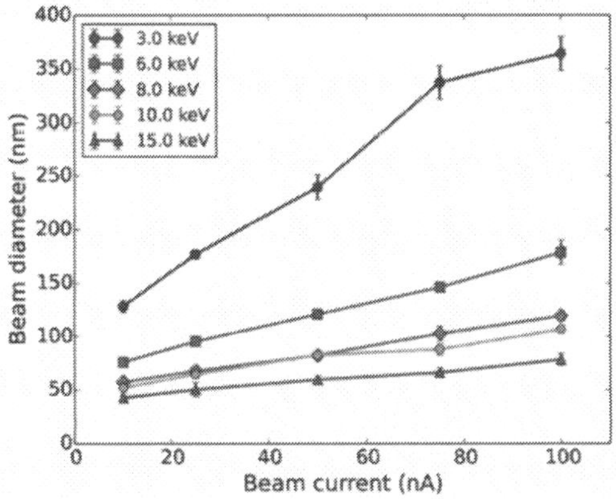

Figure 8. Variation of the beam diameter as a function of the beam current for five beam energies.

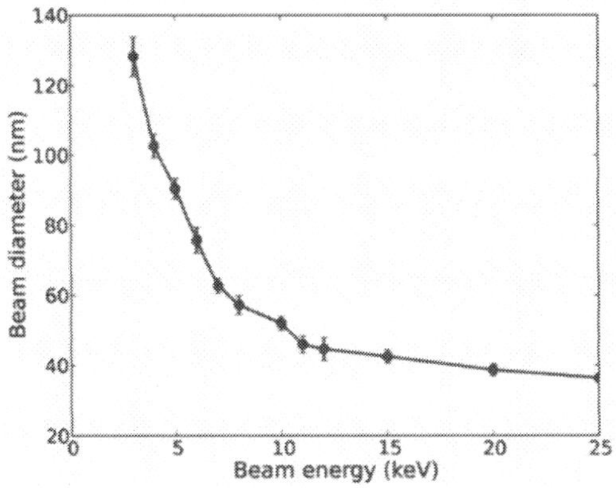

Figure 9. Variation of the beam diameter as a function of the beam energy for a 10 nA beam current.

To evaluate the radial resolution of the different X-ray lines, Monte Carlo simulations were performed at different beam energies using the beam diameters measured at 10 nA. A Gaussian distribution was assumed for the simulated beam with the standard deviation equal to half of the experimental beam diameter. The radial distribution was constructed by collecting the intensities of each emitted X-ray into bins according to the radial distance between the X-ray generated site and the centre of the beam. The width of the bins was calculated to ensure that all bins have an equal area, i.e., an equal probability of collecting an X-ray. As for the depth, the radial resolution was defined as the radius where 90 % of the X-rays have been emitted (figure 10a). For comparison, the same simulations were run with a theoretical beam diameter of 0 nm (figure 10b). The influence of the beam diameter is seen for beam energies less than 10 keV where the decrease of the emission volume is counter-balanced by the larger broadening of the beam. This shows a practical limitation of low beam energy analyses. The comparison of figures 7 and 10a reveals that the spatial resolution is primarily determined by the lateral resolution, especially at beam energies lower than 10 keV. The increase of the beam diameter at low energies does not influence the depth resolution which monotonously decreases with the beam energy. Assuming perfectly spherical Fe-Cr-C particles, the lateral resolution defines the size of the smallest particles that can be measured.

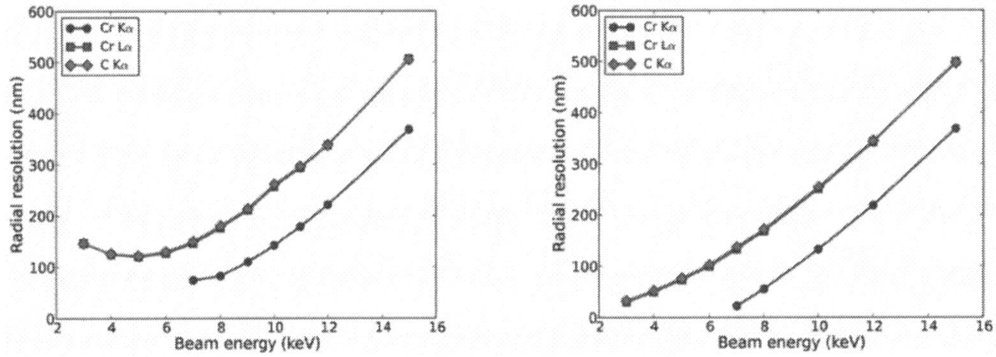

Figure 10. Radial resolution as a function of the beam energy for the Cr-Kα, Cr-Lα and C-Kα in $Cr_{23}C_6$ using a) the experimental beam diameters measured at 10 nA, and b) a theoretical beam diameter of 0 nm.

To evaluate the influence of Fe addition on the radial resolution, simulations were performed using different Fe/Cr ratios from 0.1 to 10 at 10 keV (figure 11). There is a small resolution improvement for Cr-Lα and C-Kα with higher Fe content due to the higher mass absorption coefficients of these X-ray lines in Fe.

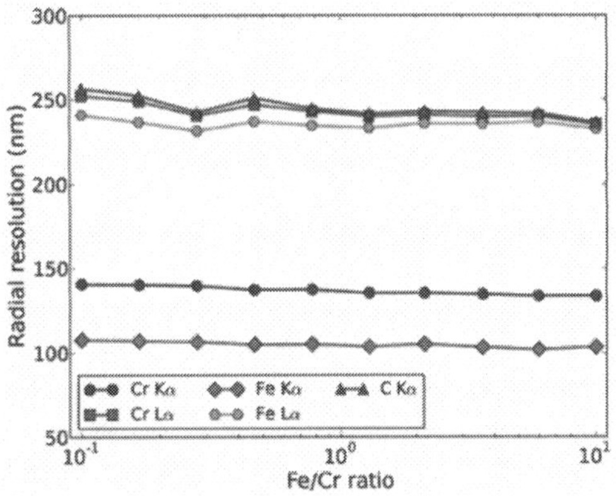

Figure 11. Radial resolution as a function of the Fe/Cr ratio at 10 keV and 10 nA.

Based on these results, Fe-Cr-C spherical particles larger than 250 nm in diameter can be measured using the Cr- and Fe-Lα (assuming a similar emission volume for Fe), the C-Kα and a beam energy of 5 keV. Measurements at 7 keV using the Cr-Kα could also be possible, but as shown in figure 5a, large uncertainties at these low overvoltages would limit the precision of the measurements.

3.4. Quantification of Fe-Cr-C particles

To visualize the distribution and size of Fe-Cr-C particles, qualitative mappings were performed using a beam energy of 10 keV, a beam current of 10 nA and a dwell time of 150 ms (figure 12). The C- and Fe-Kα were acquired using large crystals. The mapping shows no segregation within the particles. Based on the spatial resolution calculations, not all particles can be quantitatively analysed. Sample drift may also limit the measurement of particles with a diameter close to the resolution. The quantification was therefore performed on the largest particles (≈ 1.5 μm).

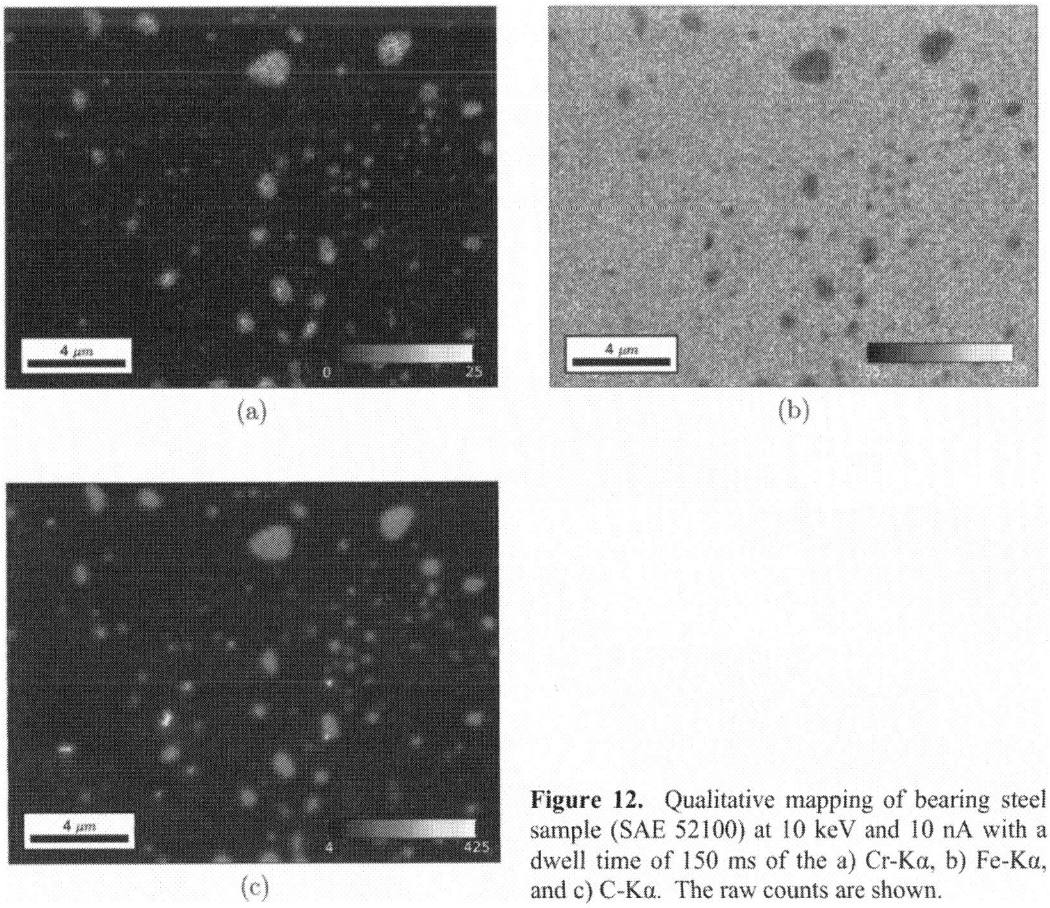

Figure 12. Qualitative mapping of bearing steel sample (SAE 52100) at 10 keV and 10 nA with a dwell time of 150 ms of the a) Cr-Kα, b) Fe-Kα, and c) C-Kα. The raw counts are shown.

From the spatial resolution evaluation, the optimal spatial resolution is achieved with a beam energy of 6 keV and a beam current of 10 nA, which implies the use of the Fe- and Cr-Lα. To further verify the results, the Fe- and Cr-concentrations were also quantified at 10 keV using the Kα-lines. At this beam energy, the radial resolution of the Kα-lines is comparable to the one of the Lα-lines at

6 keV (figure 10a). If the C-Kα intensity is still measured at 6 keV, the overall spatial resolution does not worsen. This example demonstrates an alternative acquisition strategy where different elements are measured at different beam energies to achieve the best spatial resolution and quantification.

A semi-quantitative analysis of the particle revealed a Fe/Cr ratio of approximately 7 and a carbon concentration of a few wt%. For this composition and a beam current of 10 nA, an acquisition time of 10 s is required to obtain relative uncertainties (2σ) less than 1 % on all X-ray lines (figure 5). For integral intensities, an acquisition time of 10 s would approximately be equal to a dwell time of 25 ms per step (assuming 400 steps under each peak). In the present case, the fitting algorithm used to decompose overlapped peaks requires a certain level of precision from the experimental intensities. Otherwise, the accuracy of the fits, and consequently of the quantification, would decrease. A dwell time of 500 ms was selected for the acquisitions to give a total acquisition time of approximately 10 min. The influence of the dwell time and step size on the fitting procedure should, however, be further investigated to increase the speed of integral acquisitions. The MAC database from Pouchou and Pichoir [16] was used.

Table 2 gives the quantification results of two particles: one analysed using the Kα-lines and the other with the Lα-lines. The calculated compositions are comparable. The higher total concentration using the Lα-lines may be explained by inaccuracies in the mass absorption coefficients for Cr and Fe Lα-lines as previously noted. The lower Cr-concentration obtained using the Lα-lines (approx. 15 %) is in line with the findings of Llovet *et al.* [19], which reported an under-estimation of Cr-concentration in stainless steels. The high Fe-Cr ratio correlates with the low intensity measured in the mapping (figure 12). These results confirm the proper selection of the analytical conditions to measure these particles.

Table 2. Composition of Fe-Cr-C particles (all values in weight percent).

Line	Fe	Cr	C	Total
Kα	81.66	11.65	7.58	100.89
Lα	83.72	9.95	7.66	101.32

4. Specialized techniques

4.1. Monte Carlo based reconstruction

Under the analytical conditions used to quantify Fe-Cr-C particles, the emission volume is completely inside the particles and conventional matrix correction algorithms can be used to convert intensities into mass concentrations. Thin film reconstruction algorithms have been successfully developed to calculate the thickness and composition of horizontal layers on a substrate, even when the thickness is much smaller than the emission depth. Can this method be extended to other three-dimensional structures (vertical layers, predefined shaped particles, etc.) in order to quantify smaller features?

To assess the feasibility of the reconstruction, a similar methodology as the work of Peter Statham for multi-layered thin films [35] was applied to the Fe-Cr-C case study. Two unknown parameters, x and y, (e.g., the diameter and the carbon concentration of the particles) were varied and plotted against the residuals between the simulated intensities (I) obtained for each combination of parameters and those of a single reference set of parameters, x_0 and y_0. Residuals were calculated using eq. 2, where m is the number of elements present in the sample (particle and matrix):

$$\text{residual} = \sqrt{\sum_{i=1}^{m} (I_i(x, y) - I_i(x_0, y_0))^2} \quad (2)$$

The two-dimensional parameter space was discretized into a 20 by 20 grid. Monte Carlo simulations were used to obtain the simulated intensities since no analytical model exists for this geometry. The simulations were run with a beam energy of 10 keV, a beam diameter of 50 nm and 5000 electron trajectories to achieve reasonable precision. Only the intensities from Kα-lines were considered. Table 3 summarizes the three test cases studied.

Table 3. Parameters of feasibility study.

	Parameter # 1	Parameter # 2	Composition Matrix	Composition Particle
(a)	Particle diameter (d) (d = 200 nm)	Carbon concentration (x) (x = 5.5 wt% C)	Fe	$Cr_{1-x}C_x$
(b)	Fe/Cr ratio (α) (α = 1)	Carbon concentration (x) (x = 5.5 wt% C)	Fe	$(Fe\text{-}Cr)_{1-x}C_x$
(c)	Fe/Cr ratio (α) (α = 0.1)	Particle diameter (d) (d = 200 nm)	Fe	$(Fe\text{-}Cr)_{23}C_6$

The residual plot of the first test case (figure 13a) shows a pronounced minimum at the reference point which confirms the feasibility to determine the composition and diameter of a 200 nm particle at the chosen analytical conditions. This is an improvement over the experimental resolution where only a 550 nm particle could be measured under the same conditions. From figure 13b, the addition of Fe inside the particle results in a sharp minimum along Fe/Cr ratio direction and a less pronounced minimum along the carbon concentration direction. This example, where one element is present in both the particle and matrix, proves that the reconstruction can be successful for a known particle diameter. No minimum is observed for the last case (figure 13c) even if a low Fe/Cr ratio was selected as the reference point. Instead, a saddle is visible around the reference point indicating a non-unique solution. More measurement points (e.g., line scan) could be taken into account to overcome this ambiguity. Nonetheless, these examples prove the feasibility of using Monte Carlo based reconstruction to determine the dimensions and composition of small structures, although certain constraints are necessary, and thus improve the spatial resolution of quantitative microanalysis.

4.2. Thin lamella

In a thin lamella, commonly used in transmission electron microscope (TEM) analysis, the broadening of the electron beam is almost negligible due to the lower amount of scattering processes. The emission volume is thus reduced. To demonstrate this effect, a multi-layered structure consisting of alternating Si and Al layers with thicknesses of respectively 390 nm and 300 nm was prepared in two different ways: (1) a bulk sample where the cross-section was polished using a focussed ion beam (FIB) after cleavage, and (2) a thin lamella (≈ 150 nm thick) was cut using a conventional FIB routine for TEM sample preparation. On both samples, a line profile (figure 14) was acquired across the layered structure at 10 keV, 100 nA and a dwell time of 500 ms. Al-Kα intensities were respectively collected using a TAP crystal in combination with a P-10 counter. Stronger variations in the Al-Kα intensities are visible in the lamella due to the improved spatial resolution. With the larger emission volume, X-rays are also emitted from neighbouring layers in the bulk sample, which worsens the

IOP Conf. Series: Materials Science and Engineering **55** (2014) 012016 doi:10.1088/1757-899X/55/1/012016

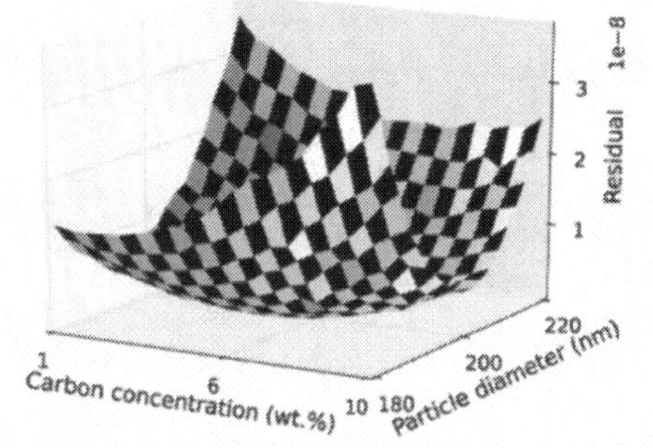

(a)

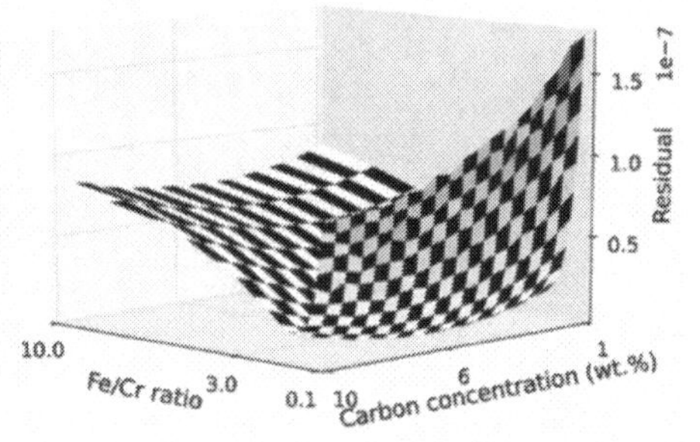

(b)

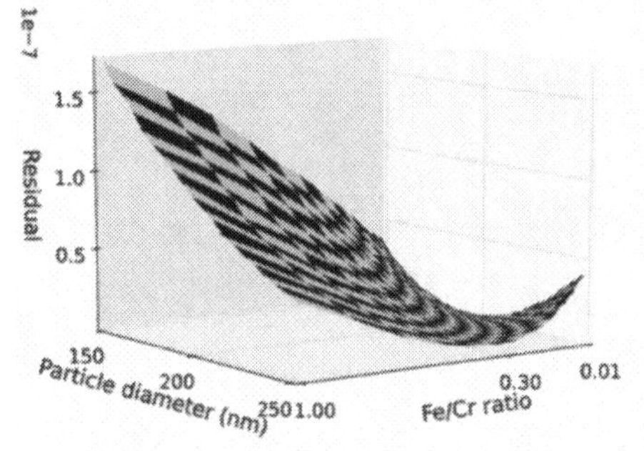

(c)

Figure 13. Residual plots for the three studied cases: a) particle diameter and carbon concentration, b) Fe/Cr ratio and carbon concentration, and c) Fe/Cr ratio and particle diameter.

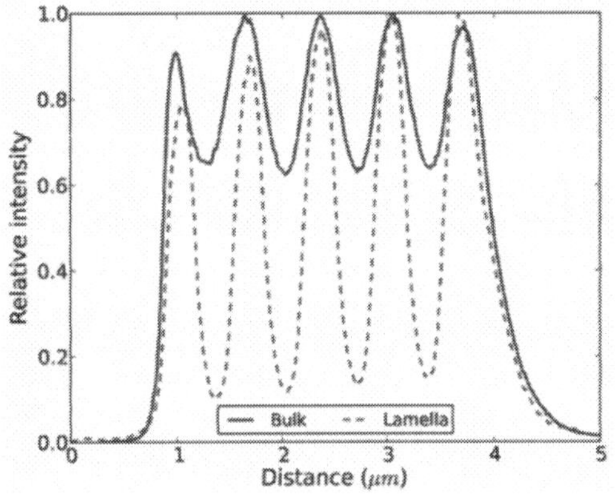

Figure 14. Al-Kα intensity profile across a Al-Si multi-layered structure prepared from a cleaved cross-section (bulk) and lamella (10 keV, 100 nA, 500 ms dwell time and 20 nm step size, TAP crystal, P-10 counter).

Al-Kα signal. Thin samples lead to higher spatially resolved measurements, although the longer specimen preparation, the lower precision and the need for adapted quantification procedures may counterbalance this improvement.

4.3. Bevel surface

Another sample preparation strategy is to create a wedge using the FIB to characterize laterally homogeneous structures inside a sample [36, 37]. With this geometry, the resolution primarily depends on the milled angle (limited by the divergence angle of the ion beam) and to a lesser extent the beam diameter. To study the spatial resolution improvement of the bevel technique over the conventional cross-section analysis, a multi-layered structure consisting of alternating Ni- and Cr-layers with thicknesses of respectively 57 nm and 56 nm was prepared using both sample preparation methods. Intensity profiles of Ni-Kα and Cr-Kα (figure 15) were acquired at a beam energy of 10 keV, 15 nA and a dwell time of 10 s on the bevel, and 10 eV, 10 nA and 2 s on the cross-section. LiF crystals on Xe counter were used for both X-ray lines. The interfaces are more clearly visible in the bevel profile due to the kinks, which are characteristic features of this technique caused by the abrupt change of the concentration at the interface. The oscillations are much smaller in the cross-section profile. This comparison confirms the advantages of the bevel technique to identify interfaces without any knowledge of the beam profile. Using a modified thin film algorithm, quantification of finer structures as a function of depth is therefore possible [36].

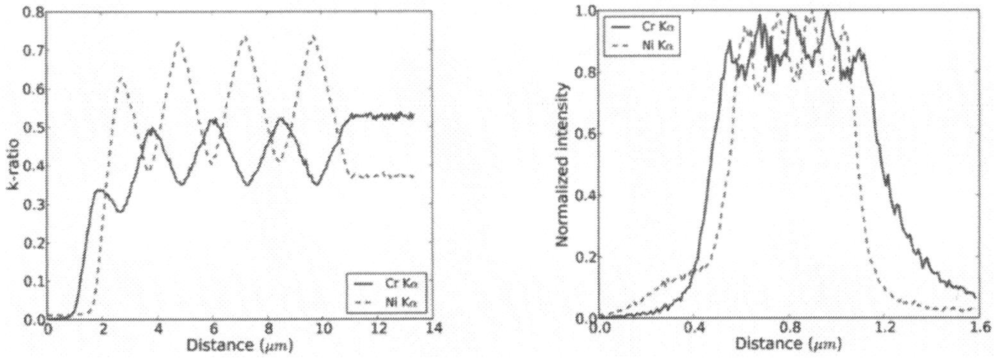

Figure 15. Cr-Kα and Ni-Kα profile across a Ni-Cr multi-layered structure prepared as a) a bevel (10 keV, 15 nA, 10 s dwell time and 52 nm step size, LiF crystals, Xe counters), and b) a cross-section (10 keV, 10 nA, 2 s dwell time and 9 nm step size, LiF crystals, Xe counters).

5. Conclusion

Using Fe-Cr-C particles in a bearing steel as example, a generic procedure was developed to find the optimal analytical conditions to achieve the best spatial resolution. It can be summarized in four points. First, the accuracy of low energy lines should be checked with homogeneous reference materials. If no reference sample is available, verification could be performed on the unknown sample using the high energy lines, as shown for the Fe and Cr Kα-lines. Integral acquisitions, overlapping corrections and advanced background modelling may be required to obtain accurate intensities for all analysed elements. Secondly, the precision can be estimated from simulated intensities and predetermined calibration factors. This evaluation becomes critical in the presence of elements in minor or trace concentration. Thirdly, the beam diameter must be considered in the evaluation of the spatial resolution. The results in this work showed that the improvement of the spatial resolution at low beam energies is limited by the focussing capabilities of the electron column. Once the beam diameter has been measured at different beam energies and currents, the radial and depth resolutions can be obtained from Monte Carlo simulations. Finally, the analytical conditions (beam energy, beam current and acquisition time) cannot be independently determined as they differently influence the accuracy, precision and spatial resolution. One must consider all of them together in combination with practical considerations such as carbon contamination and sample drift to select the optimal conditions for a measurement.

Other strategies can be used to further improve the spatial resolution. A feasibility study demonstrated the possibility to resolve features smaller than the emission volume using a non-linear optimisation algorithm coupled with Monte Carlo simulations. To reduce the risk of ambiguous solutions, either the information content of the measurement (e.g., more measuring points at different conditions) or the number of constraints (prior knowledge of the material structure) have to be sufficiently increased. Both the thin lamella and bevel techniques were shown to improve the spatial resolution due to the shallower emission volume and reduced sensitivity to the lateral X-ray intensity distribution, respectively. The longer and more costly sample preparation is however a disadvantage of these techniques.

Other factors not directly considered in this work, e.g., the reproducibility of the stage movement in serial measurements and automatic beam drift correction, play an important role in defining the analytical spatial resolution. Nevertheless, the present results corroborate the potential of field emission microprobes as accurate and precise instruments for the chemical characterisation of sub-micrometre structures.

Acknowledgments
The authors would like to acknowledge N. Mevemkamp for his support in the reconstruction feasibility study, M. Schiffers and D. Park for the FIB preparations, L. Shang (Metal Chemistry Institute of the RWTH Aachen University) for the Al-Si multi-layers preparation, D. Rhede (Helmholtz-Zentrum Potsdam) for the measurement on the Ni-Cr bevel and J. Börder (JEOL Germany GmbH) for providing the gold on carbon sample.

References

[1] Noguchi T, Nakamura T, Ushikubo T, Kita N, Valley J, Yamanaka R, Kimoto Y and Kitazawa Y 2011 *Earth Planet. Sci. Lett.* **309** 198-206
[2] Sato J, Omori T, Oikawa K, Ohnuma I, Kainuma R and Ishida K 2006 *Science* **312** 90-91
[3] Engström F, Pontikes Y, Geysen D, Peter Tom Jones B B and Blanpain B 2011 Review: Hot stage engineering to improve slag valorisation options. in: *2nd International Slag Valorisation Symposium*. 231-251
[4] Cho S M, Park J W, Han H S, Seok H K, Moon M W and Kim Y C 2013 *ACS Appl. Mater. Interfaces* **5** 395-403
[5] Lan H F, Du L X and Liu X H 2012 *Steel Res. Int.* **83** 1-10
[6] MacRae C, Wilson N, Torpy A and Davidson C 2012 *Microsc. Microanal.* **18** 1239-1245
[7] Roy S and Suwas S 2013 *J. Alloys Compd.* **548** 110-125
[8] Wu R and Reed R 2008 *Acta Mater.* **56** 313-323
[9] Ha S H, Yoshimi K, Maruyama K, Tu R and Goto T 2012 *Mater. Sci. Engng. A* **552** 179-188
[10] Ohnuma I, Abe S, Shimenouchi S, Omori T, Kainuma R and Ishida K 2012 *ISIJ Int.* **52** 540-548
[11] Guo J, Koopman M, Fang Z Z, Wang X, Fan P and Rowe M C 2013 *Int. J. Refract. Met. Hard Mater.* **36** 265-270
[12] Merlet C and Llovet X 2012 *IOP Conf. Ser.: Mater. Sci. Engng.* **32** 012016
[13] Barkshire I R, Karduck P, Rehbach W and Richter S 2000 *Mikrochim. Acta* **132** 113-128
[14] Bastin G F and Heijligers H J M 1990 *Quantitative electron probe microanalysis of carbon in binary carbides.* (Eindhoven, The Netherlands: Eindhoven University of Technology)
[15] Ritchie N W 2009 *Microsc. Microanal.* **15** 454-468

[16] Pouchou J L and Pichoir F 1991 Quantitative analysis of homogeneous or stratified microvolumes applying the model "pap". in: *Electron probe quantitation.* (Heinrich K and Newbury D; Eds.) (New York: Plenum) 31-75
[17] Chantler C T 2000 *J. Phys. Chem. Ref. Data* **29** 597-1056
[18] Ritchie N W M 2005 *Surf. Interface Anal.* **37** 1006-1011
[19] Llovet X, Heikinheimo E, Galindo A N, Merlet C, Bello J F A, Richter S, Fournelle J and van Hoek C J G 2012 *IOP Conf. Ser.: Mater. Sci. Engng.* **32** 012014
[20] Gopon P, Fournelle J and Llovet X 2012 *Microsc. Microanal.* **18** (Suppl 2) 1728-1729
[21] Willich P and Bethke R 1996 *Microchim. Acta* **13** 631-638
[22] Wendt M 1980 *Krist. Tech.* **15** 1367-1375
[23] Newbury D E 2002 *J. Res. Natl. Inst. Stand. Technol.* **107** 605-619
[24] Rémond G, Myklebust R, Fialin M, Nockolds C, Phillips M and Roques-Carmes C 2002 *J. Res. Natl. Inst. Stand. Technol.* **107** 509-529
[25] Fournier C, Merlet C, Dugne O and Fialin M 1999 *J. Anal. At. Spectrom.* **14** 381-386
[26] Jones E, Oliphant T, Peterson P et al. 2001 *SciPy: Open source scientific tools for Python.* http://www.scipy.org/
[27] Henke B, Gullikson E and Davis J 1993 *At. Data Nucl. Data Tables* **54** 181-342
[28] Heinrich K 1986 in: *11th Int. Congr. X-ray Optics and Microanalysis.* (Brown J and Packwood R; Eds.) 67
[29] Bastin G F, Dijkstra J M and Heijligers H J M 1998 *X-ray Spectrom.* **27** 3-10
[30] Marinenko R B and Leigh S 2010 *IOP Conf. Ser.: Mater. Sci. Engng.* **7** 012017
[31] Gauvin R 2012 *Microsc. Microanal.* **18** 915-940
[32] Kimura T, Nishida K and Tanuma S 2006 *Microchim. Acta* **155** 175-178
[33] Joy D C 2002 *J. Microsc.* **208** 24-34
[34] Probst C, Demers H and Gauvin R 2012 *Microsc. Microanal.* **18** 628-637
[35] Statham P J 2010 *IOP Conf. Ser.: Mater. Sci. Engng.* **7** 012027
[36] Richter S, Bückins M, Kyrsta S, Aretz A, Spähn M and Mayer J 2004 *Microchim. Acta* **145** 187-192
[37] Richter S, Kyrsta S, Schneider J and Mayer J 2006 *Microchim. Acta* **155** 257-262

IOP Conf. Series: Materials Science and Engineering **55** (2014) 012017 doi:10.1088/1757-899X/55/1/012017

Prospects for higher spatial resolution quantitative X-ray analysis using transition element L-lines

P Statham and J Holland

Oxford Instruments Nanoanalysis, High Wycombe, Bucks HP12 3SE, Great Britain

E-mail: peter.statham@oxinst.com

Abstract. Lowering electron beam kV reduces electron scattering and improves spatial resolution of X-ray analysis. However, a previous round robin analysis of steels at 5 - 6 kV using Lα-lines for the first row transition elements gave poor accuracies. Our experiments on SS63 steel using Lα-lines show similar biases in Cr and Ni that cannot be corrected with changes to self-absorption coefficients or carbon coating. The inaccuracy may be caused by different probabilities for emission and anomalous self-absorption for the Lα-line between specimen and pure element standard. Analysis using Lℓ(L3-M1)-lines gives more accurate results for SS63 plausibly because the M1-shell is not so vulnerable to the atomic environment as the unfilled M4,5-shell. However, Lℓ-intensities are very weak and WDS analysis may be impractical for some applications. EDS with large area SDD offers orders of magnitude faster analysis and achieves similar results to WDS analysis with Lα-lines but poorer energy resolution precludes the use of Lℓ-lines in most situations. EDS analysis of K-lines at low overvoltage is an alternative strategy for improving spatial resolution that could give higher accuracy. The trade-off between low kV versus low overvoltage is explored in terms of sensitivity for element detection for different elements.

1. Introduction

The excitation volume for X-rays is essentially determined by the incident electron beam energy and specimen density. In a steel specimen at 20 kV beam voltage, X-rays emerge from a region about 2 μm in diameter whereas at 5 kV, the region is about 300 nm across so there is a big incentive to reduce beam kV to improve spatial resolution. However, at 5 kV and below, the K-lines for many of the first row transition elements are not excited and the lower energy L-line emissions have to be used for x-ray analysis.

Two years ago, a report on an inter-laboratory comparison of electron microprobe analysis (EPMA) analysis of alloy steel using 5 - 6 kV beam voltages demonstrated that quantitative analysis using the transition element L-lines could give rise to large relative errors [1]. Using wavelength-dispersive X-ray spectrometry (WDS), one laboratory discovered a systematic trend with composition in a range of 15 steels where Cr analysis was systematically low by about 14 % relative whereas Ni was 22 % too high. For analysis with energy-dispersive X-ray spectrometry (EDS) there was a wide range of results from the different laboratories with much less consistency and even bigger relative errors. The problems of using these L-lines have been reported by other authors and we have done our own research to corroborate these findings and consider new approaches to obtaining more reliable quantitative results at high spatial resolution.

Content from this work may be used under the terms of the Creative Commons Attribution 3.0 licence. Any further distribution of this work must maintain attribution to the author(s) and the title of the work, journal citation and DOI.
Published under licence by IOP Publishing Ltd

2. Analysis using WDS

2.1. Conventional analysis using Lα_1-lines

We first investigated WDS analysis of steel using pure element standards. We chose a steel, SS63 with high Cr and Ni content (Cr: 18.7 %; Fe: 70.5 %; Ni: 9.49 %) and used 5 kV accelerating voltage, 20 nA beam current and acquisition times of 100 seconds for each major element peak. Carbon coating on the standards was measured to be 12 nm by INCA ThinFilmID (Oxford Instruments NanoAnalysis) and k-ratios were processed by Stratagem (SAMx) to give Cr: 15.4 %; Fe: 70.3 % and Ni: 13.0 %. This bad result was in line with the trend observed in [1] with Cr too low and Ni too high and we looked for possible reasons. Figure 1 demonstrates the number of overlaps that can occur with L-line analysis and we took care to choose background measurement points free from peak interference.

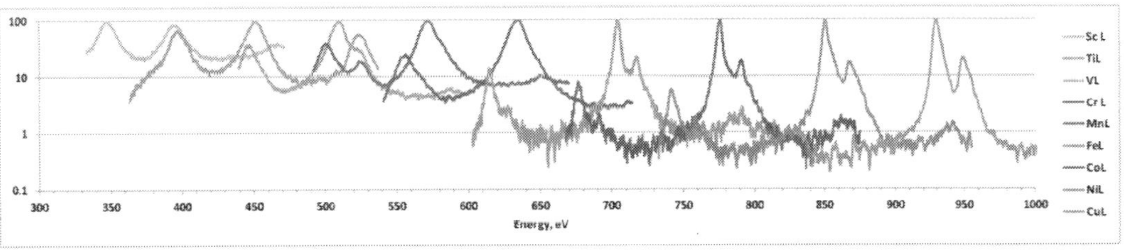

Figure 1. WDS scans (log scale) of transition element L-line emissions. Note change of resolution between scans for Mn-L and Fe-L where crystal changes from a synthetic multilayer, LSM60, to TAP.

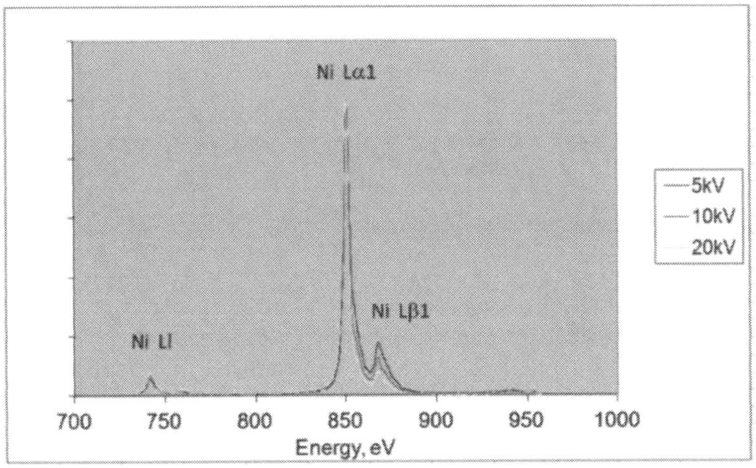

Figure 2. WDS scans of pure Ni standard. Note increasing absorption of Ni-Lα_1 at higher kV.

For the elements Ni and Cr, WDS scans recorded at 5, 10 and 20 kV (figures 2 and 3) show some of the anomalous absorption effects that occur with these L-line series. Both Lα_1(M3-M5) and Lβ_1(L2-M4) are very close in energy to the L3-edge and experience strong self-absorption so that the intensity profile varies as a function of accelerating voltage. Both excitation and absorption may be influenced by the atomic environment which may be different in specimen and standard. Table 1 shows typical values for the mass absorption coefficient, MAC, that are for the most part close to those tabulated by Heinrich [8]. The first three columns are used for conventional analysis using

$L\alpha_1$-lines. The self-absorption coefficients Cr-Lα / Cr, Fe-Lα / Fe and Ni-Lα / Ni are those subject to most discussion in the literature and we have replaced the Heinrich values with our own empirical values. For pure elements, empirical MACs can been derived from $L\alpha_1$ measurements at different kV [2] but for elements Sc-Cr, WDS cannot resolve the $L\alpha/L\beta_1$ overlap and a single MAC is insufficient to represent the composite effects of two lines where one line is much more strongly absorbed.

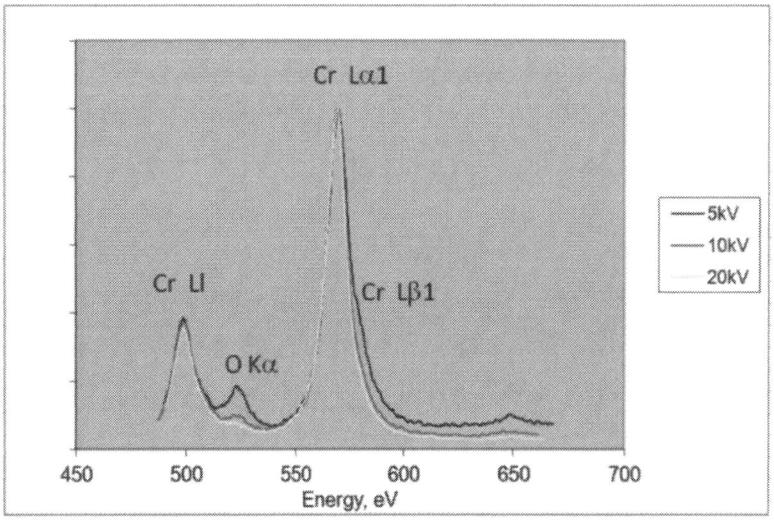

Figure 3. WDS scans of pure Cr metal standard. Note the appearance of oxygen peak from superficial oxide at low kV.

Table 1. MAC (cm^2/gm) for emitters and absorbers involved in the steel analysis.

Emitter energy keV	**Cr $L\alpha_1$** 0.573	**Fe $L\alpha_1$** 0.705	**Ni $L\alpha_1$** 0.851	**Cr $L\ell$** 0.500	**Fe $L\ell$** 0.615	**Ni $L\ell$** 0.743
Cr MAC	3100	17122	11139	3479	19904	15208
Fe MAC	3317	3200	13726	4394	2865	15979
Ni MAC	4221	2713	3500	5623	3632	2424

It is possible that this uncertainty in MAC values and the potential for carbon contamination build up during analysis could be responsible for the systematic bias in results. Therefore, the effects on results of changing the $L\alpha_1$ MACs and also the estimated thickness of carbon on the specimen are shown in Table 2. Column C uses empirical values from ref. [2] and Column H from ref. [1]. Column G is an attempt to use the values from various sources to give the best analysis. Although there is no combination here that overcomes the huge under-estimate in wt% of Cr and over-estimate of Ni, it should be noted that while the self-absorption MACs have been studied in some detail, there are still very high and questionable values for absorption of Fe- and Ni-lines in Cr and Fe.

The physics of L X-ray emission from nickel is complicated and an anomalous Lα absorption resonance has been demonstrated and shown to be less prevalent in some Ni compounds [3]. Such anomalous absorption occurs when the incident energy is close to that required to move an electron from one shell to another unfilled shell and this produces line type absorption in addition to the normal absorption step across the L3-edge. Thus, the composite absorption effect on Ni-Lα depends on what fraction of emission is subject to the much higher resonance absorption. The overall intensity measured for Ni-Lα is therefore going to be affected by spectrometer resolution and whether the

Table 2. Weight% results for SS63 analysis obtained in different tests where the same set of $L\alpha_1$ k-ratios was corrected using Stratagem with different MACs and carbon thickness estimates.

Test	A	B	C	D	E	F	G	H
Carbon nm	12	12	12	15	15	15	15	15
Cr Lα / Cr	3100	3850	3850	3850	3850	3850	3100	3612
Fe Lα / Fe	3200	3200	3350	3350	3350	3350	3200	3559
Ni Lα / Ni	3500	3500	3500	3500	3500	3500	4227	4227
Fe Lα / Cr	17121	17121	17121	17121	15409	17121	18833	17121
Ni Lα / Cr	11123	11123	11123	11123	11123	11123	11123	11123
Ni Lα / Fe	13707	13707	13707	13707	13707	12336	11103	13707
Cr (18.7) %	15.8	15.4	15.4	15.5	15.6	15.5	15.5	15.7
Fe (70.5) %	70.0	70.3	70.3	70.3	70.0	70.8	71.6	70.4
Ni (9.5) %	12.8	13.0	13.0	12.9	13.0	12.4	11.6	12.5

Lα-peak intensity or the total peak area is measured. Furthermore, since the overall absorption factor is a non-linear function of the MACs, the determination of an empirical MAC based on measurements at different kV will also be affected by the range of kVs used. These effects may account to some extent for the different empirical MAC values for pure Ni reported in references [1], [2] and [3]. In the calculations used for Table 2, the same MAC is assumed for both the pure element standard and the SS63 specimen. In the case of Ni it has been shown that the effective self-absorption MAC for Ni-Lα in NiAl is much less than for pure Ni [3]. However, using such a different Ni MAC for the specimen has virtually no effect in the current case because Ni absorption in SS63 is dominated by Fe. For Ni at least, it is difficult to see how the analysis can be improved by any further adjustment of MACs. It has been suggested that Ni-Lα intensity in compounds is enhanced due to increased occupancy of the 3d band [3] and the effects of bonding on the partially filled M4,5-shells may be apparent in any of the first row transition elements [4]. Such effects add to the potential problems that may occur with Lα-line analysis of transition elements.

2.2. Analysis using Lℓ-lines

A vacancy in the L3-shell can be produced by direct ionisation. At 5 kV there are no Auger transitions that could give rise to additional L3-vacancies following a K-shell ionisation but internal Coster-Kronig transitions from L1- and L2-shells are possible. When there is a vacancy in the L3-shell there are a number of possible transitions that could produce an X-ray. As suggested above, the probability of a transition that gives rise to $L\alpha_1$(L3-M5) or $L\alpha_2$(L3-M4) may be affected by the different occupancy of the M4,5-shells. However, the more tightly bound M1-shell is fully occupied so the probability of emission of an Lℓ(L3-M1) transition should be much less affected by the local atomic environment. Furthermore, the Lℓ-line is well away from the L3-absorption edge and does not experience any anomalous resonance absorption effects. Therefore, we tried using the Lℓ(L3-M1)-lines to see if that would improve the accuracy in our SS63 example. As is clear from figures 2 and 3, the Lℓ-line is at energy well below the range of Lα and Lβ spectrometer positions normally made available for analysis with WDS. We therefore had to set new peak and background positions, taking particular note of potential overlaps suggested by figure 1. Furthermore, the Lℓ-intensity can be much less than that for $L\alpha_1$ so we needed to extend analysis times to a few hundred seconds for some lines. Since the depth distribution of Lℓ and $L\alpha_1$ should be the same because they are instigated by ionisation of the same L3-shell, MACs from Table 1 for Lℓ were substituted in Stratagem for the normal $L\alpha_1$-values to provide for absorption of the Lℓ-lines. The new results for analysis with Lℓ-lines are shown in figure 4 and compared with a typical result using $L\alpha_1$-lines. The vertical extent of the symbol has been scaled to correspond to twice the calculated statistical standard deviation and the results of the four repeated analyses with Lℓ-lines appear to be consistent within this expected experimental error. With Lℓ-lines, the Cr results are 11 % higher and the Ni results 27 % lower than the results using $L\alpha_1$-lines. Although the Cr values are still 9 % below the expected value, the Ni results are now very close to the expected concentrations.

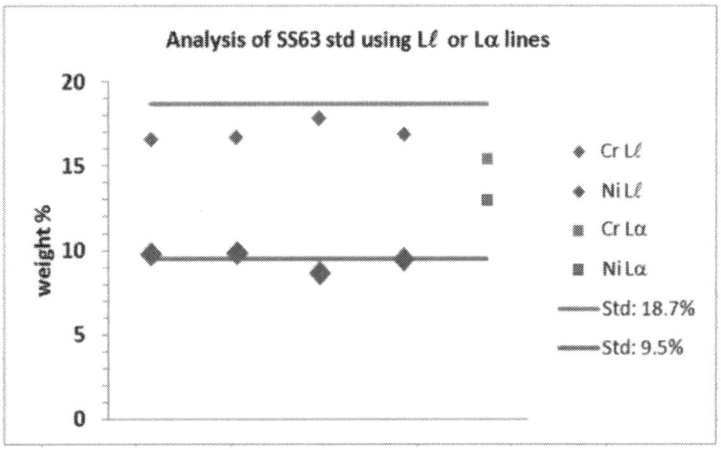

Figure 4. Diamonds show results using Lℓ-lines. Squares use $L\alpha_1$ and conditions as for Test C in Table 2. Horizontal lines show expected concentrations for SS63 standard. Symbol height = 2σ statistics.

Whereas this is a promising result and suggests that analysis with Lℓ-lines may be a route to improved quantitative analysis at low kV, there are some caveats. In our experiments we used 20 nA beam current and still required long acquisition times that made each analysis take over 20 minutes. Achieving high currents at low kV inevitably involves using large probe diameter and in order to exploit the advantages of low kV in reducing the excitation volume, it is essential that the probe diameter does not become the dominant factor in limiting spatial resolution.

3. EDS analysis at low kV compared with low overvoltage

We repeated the SS63 analysis at 5 kV using EDS X-max^N 150 SDD, INCA software) using the same pure element standards. To correct for background and overlaps, filtering and least squares fitting were used to fit the spectrum to L-peak shapes matching those of pure elements at 5 kV. The total L-series intensity was assumed to correspond to $L\alpha_1$. Since the $L\alpha_1$-peak still dominates the L-series, it is perhaps not surprising that the EDS results (Cr: 16.8 %; Ni: 12.0 %) showed similar biases to the WDS analysis using $L\alpha_1$. However, it was immediately apparent that count rates were at least two orders of magnitude higher with the large area SDD detector and analyses could be accomplished in seconds rather than the 10's of minutes required for WDS.

The problems of uncertainty in MACs and difficult peak overlaps can be overcome by using higher beam kV and a low overvoltage to reduce the interaction volume [4, 5]. Figure 5 shows how the range of X-ray generation [6] changes with beam kV for a material with the same density as silicon. The nanometre range will be less for denser material like steel because the range is inversely proportional to the density. It can be seen that similar range and spatial resolution can be achieved with Cu K-radiation using 11 kV or with Cu L-radiation using 5 kV. Spectra under these two conditions for the same beam current are compared in figure 6. It is immediately apparent that the Cu K-peak using 11 kV is considerably smaller than the Cu L-peak obtained using 5 kV with the same current. However, when these spectra are scaled, it is also apparent that the background is higher in the 5 kV case so it is not immediately obvious which condition is best for analysis. Therefore, a few examples have been studied by spectrum simulation to see what EDS spectrum analysis would achieve in terms of peak area and statistics. The ratio of peak counts/statistical sigma ultimately determines the detection limit and is, therefore, one measure of sensitivity. The results are plotted in figure 7 and show two points for each elemental line corresponding to conditions that give 200 nm and 400 nm range in figure 5. As expected, sensitivity is improved at poorer spatial resolution. For Cu, it appears that greater sensitivity is achieved with low kV and L-line analysis, whereas for Cr, K-line and low overvoltage is more sensitive. Whereas this gives an idea of the trade-offs for sensitivity, for quantitative analysis the presence of any superficial layers has a greater effect on the low-overvoltage measurements.

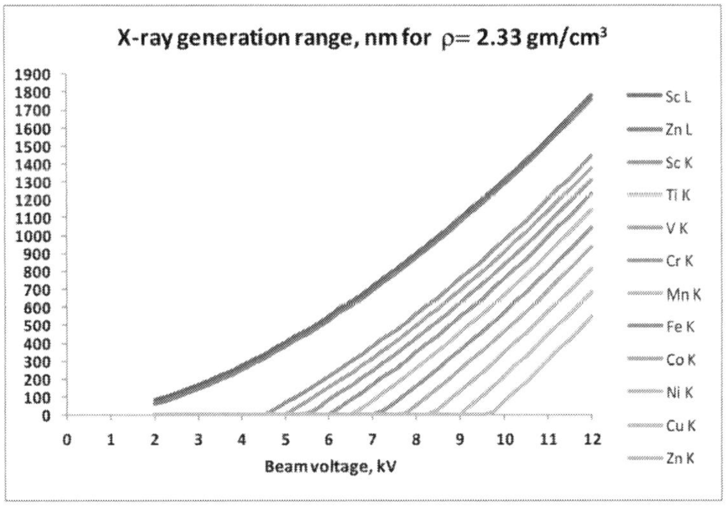

Figure 5. Range of X-ray generation for different element lines in a matrix with density 2.33 gm/cm^3. Note: range scales inversely with density of matrix.

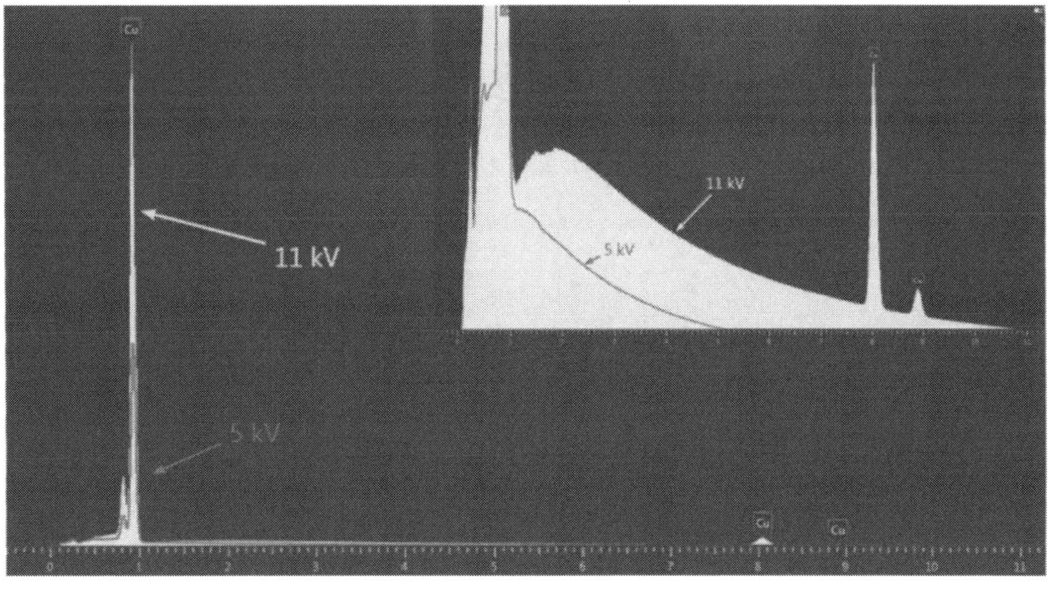

Figure 6. X-ray spectrum from Cu at 5 kV and 11 kV using same beam current and acquisition time.

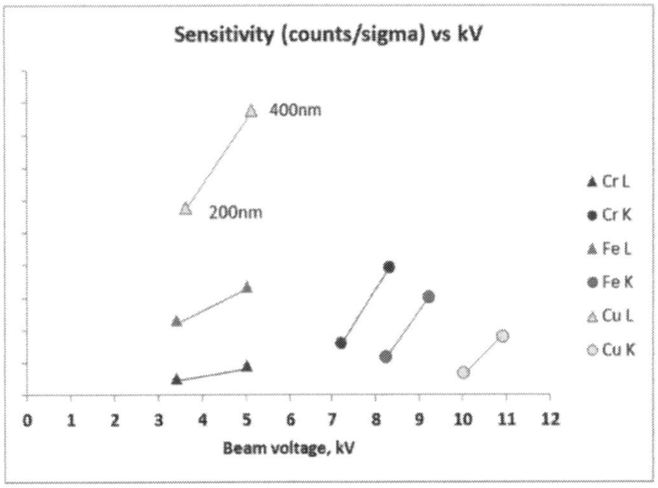

Figure 7. Results from spectrum simulation for SDD (124 eV fwhm at Mn-Kα) with polymer window, take-off-angle 35°, for 1 % of element in a silicon matrix using same beam current and time. Peak counts are obtained by filter-fit method and statistical sigma (standard deviation) is calculated as in ref. [7]. Vertical axis units are arbitrary and scale with time 0.5. In each case, two points are calculated corresponding to kV values that gives rise to 200 and 400 nm range according to figure 5.

4. Conclusions

The inaccuracy in analysis using Lα-lines of transition elements for steel appears likely to be due different probabilities for emission and anomalous self-absorption for the Lα-line between specimen and pure element standard. This is a serious barrier to accurate low kV analysis because it means that each new type of material will require a set of standards showing similar chemical effects if relative errors much less than 20 % are required. Analysis using the Lℓ-line rather than Lα appears to be a promising strategy to improve accuracy because emission is less affected by the atomic environment. However, Lℓ-intensities are very weak and WDS analysis may be impractically slow for some applications. EDS with large area SDD offers orders of magnitude faster analysis and achieves similar accuracy to WDS analysis with Lα-lines but energy resolution precludes the use of individual Lℓ-lines in most situations. For selected elements, EDS analysis of K-lines at low overvoltage is an alternative strategy for improving spatial resolution that could give higher accuracy. However, peak intensity and peak-to-background ratio changes with element so the best approach depends on whether qualitative sensitivity or quantitative accuracy is more important.

References

[1] Llovet X, Heikinheimo E, Núñez Galindo A, Merlet C, Almagro Bello J F, Richter S, Fournelle J and van Hoek C J G 2012 *IOP Conf. Series: Mat. Sci. Engng.* **32** 012014
[2] Pouchou J L and Pichoir F M A 1988 in: *Microbeam analysis 1988* (San Francisco: San Francisco Press) 319-24
[3] Pouchou J L 1996 *Mikrochim. Acta* **Suppl. 13** 39-60
[4] Armstrong J T and Crispin K L 2012 *Microsc. Microanal.* **18** (Suppl 2) 1726-1727
[5] Newbury D E 2002 *Microsc. Microanal.* **8** (Suppl. 2) 434-435
[6] Anderson C A and Hassler M F 1966 in: *Proc. 4th ICXROM* (Castaing *et al.*; Eds.). (Paris) 310
[7] Statham P J 1977 *Anal. Chem.* **49** 2149-2154
[8] Heinrich K F J 1987 in: *Proc. 11th ICXOM.* (London, Canada) 67-119

IOP Conf. Series: Materials Science and Engineering **55** (2014) 012018 doi:10.1088/1757-899X/55/1/012018

Nanoscale characterisation of semiconductors by cathodoluminescence

K Thonke[1], I Tischer[1], M Hocker[1], M Schirra[1], K Fujan[1], M Wiedenmann[1], R Schneider[1], M Frey[1] and M Feneberg[2]

[1] Universität Ulm, Institut für Quantenmaterie / Gruppe Halbleiterphysik, DE-89081 Ulm, Germany
[2] Otto-von-Guericke-Universität, Institut für Experimentelle Physik, Magdeburg, Germany

E-mail: klaus.thonke@uni-ulm.de

Abstract. Cathodoluminescence measurements carried out either in a scanning secondary or a transmission electron microscope allow a direct correlation of structural features with the emission spectra. Applied to semiconductors, numerous material and electronic properties can be determined on a length scale down to 10 nm in favourable cases. In this tutorial, we discuss the nature of the most important light emission processes in semiconductors, and what kind of information can principally be derived. Several examples for the application of this method in studies on bulk and nanocrystalline semiconductor materials are discussed.

1. General

This manuscript is the summary of a tutorial presented at the EMAS 2013 workshop and addresses material scientists working on other classes of materials than semiconductors, or using other electron microscope based methods than cathodoluminescence like EDS, etc. It is not meant as a comprehensive review of all latest achievements, but tries to show some typical cases starting from features and defects in 3D samples, and continues with examples of the analysis of lower dimensional systems. For a broader discussion, the reader is referred e.g., to the book of Yacobi and Holt 0, or for review articles like Refs. [2-8]. A more theoretical analysis of the optical excitation processes can be found in Ref. 0. An overview over data processing procedures and noise reduction of "hyperspectral" CL images can be found in Ref. 0.

Here, we first want to discuss the nature of different electron-hole recombination processes in semiconductors to understand what information these potentially can carry, and then shortly look at some experimental considerations, before we consider several examples and extensions of the method.

2. Luminescence processes in semiconductors

2.1. Fundamentals of radiative recombination processes

To understand what kind of information can be drawn from cathodoluminescence (CL) spectra, we first take a look at the possible radiative recombination processes in semiconductors with so-called direct bandgap, i.e., those with both the valence band maximum and conduction band minimum at the same $\boldsymbol{k}$-value (figure 1). In this type of material, electrons and holes can recombine and – in fortunate

Content from this work may be used under the terms of the Creative Commons Attribution 3.0 licence. Any further distribution of this work must maintain attribution to the author(s) and the title of the work, journal citation and DOI.
Published under licence by IOP Publishing Ltd

cases – emit a photon without any phonon being involved in the process, which in indirect semiconductor is needed to fulfil the ***k*** conservation rule. Drawing just the energies in a simplified band scheme, which shows the conduction and valence band edges (figure 2), several processes can be schematically depicted.

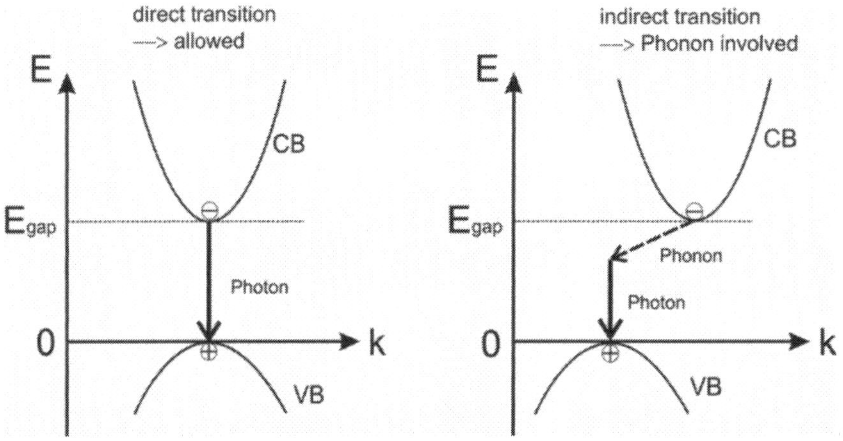

Figure 1. *E(k)* dispersion relation of the single particle electron states in a semiconductor. The band gap separates mostly occupied states in the valence band (VB) from the mostly unoccupied states in the conduction band (CB). In direct semiconductors (like gallium arsenide, GaAs), the maximum of the VB and the minimum of the CB are both at the same ***k***-value, typically $\boldsymbol{k} = 0$, whereas in indirect semiconductors (like silicon, Si) these are shifted. Recombination requires then a phonon to be involved, what makes the process less probable by some 3 or 4 orders of magnitude as compared to the direct case.

Process "a)" is a direct band-band recombination from an electron in the conduction band with a hole in the valence band, which is typically the dominant process at room temperature for defect-free direct semiconductor materials. The emission band starts at the (temperature dependent) band gap energy E_{gap} and extends to higher energies. It is governed in its line shape by the combined density of states *D(E)* in both bands involved, and the occupation probability, which for not too highly doped or highly excited semiconductors can be approximated by a Boltzmann distribution.

At low temperatures, these charged particles (e,h) form electron-hole pairs called excitons ("X"), before they recombine (process "b)" in figure 2). This lowers the emission energy by the exciton binding energy, which is typically in the range of some 5 meV to 50 meV, depending on the effective mass of the particles involved and the dielectric constant of the material.

In the next step, these excitons can diffuse through the crystal and attach by exchange interaction to neutral shallow dopant impurities, forming excitons bound at donors (D^0,X) or acceptors (A^0,X). Depending on the band structure, also binding to ionized impurity cores ((D^+,X), (A^-,X)) might be possible. In direct semiconductors, the ***k*** selection rule leads to relatively sharp excitonic emission lines ($\Delta E \approx 1$ meV) with hydrogen-like excited states, since the ***k***-value of the photon emitted is negligible compared to the k values of the electron and hole states involved.

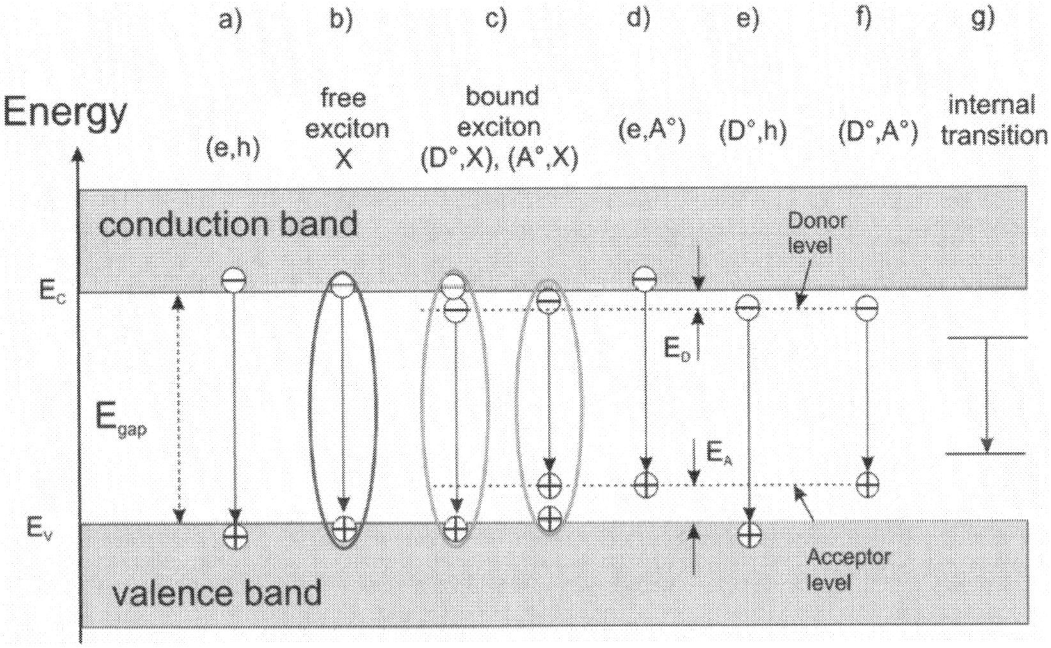

Figure 2. Possible radiative recombination processes in semiconductors in a simplified band model. "e" and "h" stand for free electrons and holes, respectively. "D^0" and "A^0" are neutral donors and acceptors, i.e., shallow dopants, which still have (at low temperature) their excess particle bound. "X" marks electron-hole pairs, so-called excitons.

Processes involving again shallow dopants, but no excitons, are the free electron → acceptor (e,A°) transition or their counterpart, the donor → free hole (D°,h) transition. Since free particles are involved, their density of states and thermal energy distribution is mirrored in the emitted line shape. These bands start at a lower threshold energy (E_{gap} - EA) or (E_{gap} - ED), respectively.

For certain doping ranges, transitions between neutral donors (D^0) and neutral acceptors (A^0) become observable, leading to series of lines merging to broad bands due to Coulomb interaction of the charged donor (D^+) and acceptor (A^-) cores in various distances left over after recombination.

Finally, internal transition between levels localized at point or extended defects might occur. Examples for point defects are transitions in the d-shell of transition metals or the f-shell of rare earth atoms. Extended defects like dislocations often just act as non-radiative sinks for excited particles showing up as dark regions in the sample (like in GaAs), but might also lead to specific (broad) emission bands below the bandgap energy like the D_1-D_4 lines in silicon 0, or the basal plane stacking fault related emission bands in gallium nitride [12, 13].

Generally, all these transitions are affected by several broadening mechanisms. Thermal broadening by shortening the lifetime of the initial or final state of the transitions leads to Lorentzian type line shapes. Broadening by statistical fluctuations of strain in the material, or the influence of charged impurities arranged around statistically, alloy fluctuations in composed materials, etc., lead typically to Gaussian type line broadening. If free particles are involved, their thermal distribution also affects the line shape by introducing exponential high energy band tails.

2.2. Which semiconductor property can be measured via which emission type?

In this section we want to discuss, what kind of information can be drawn from which spectral feature, which might be measurable in advantageous cases. It is clear, that several prerequisites have to be fulfilled to be in a position for getting out interesting information. First of all, the overall luminescence signal level should be high enough to record low noise spectra and low noise "wavelength maps". i.e., charts which show, which signal can be excited on which spot. Unlike optical microscopy, the CL method does not detect where the light is coming from, but where it can efficiently be excited. If for the distinction of close lying narrow lines high spectral resolution is necessary, a monochromator with a long enough focal length and a sensitive detector are required, since in this case the entrance slit of the monochromator has to be set quite narrow. The sample should not spatially drift during the spectra recording, and not pile up charges, which distort the image. Let us first look at ***local properties of bulk semiconductors***.

From the emission energy of either the band-band (e,h) transition (process "a)" in figure 2) or the free exciton X (process "b)" in figure 2), the ***bandgap energy*** can be evaluated (in the latter case the exciton binding energy must be known). For simple materials like elemental or binary III-V or II-VI semiconductor crystals (like GaAs, ZnSe), any deviation from the literature value (at a given sample temperature) contains information about the ***local strain*** in the sample. For most semiconductors of technical interest, the deformation potentials describing the change in bandgap energy as a function of the strain components are known. If the strain situation is clear – what means that the relative contributions of axial and shear strain is known – the shifted bandgap energy gives a direct measure for the strain.

The situation becomes more complicated, when alloys of semiconductor are under investigation. In the case of ternary crystals like (Al,Ga)As, ***a local determination of the composition*** can be undertaken – if the strain situation is known. For the case of more complex materials like quaternary compounds (e.g., (In,Ga,Al)P used for red LEDs), some additional input from other methods (like EDS or EELS) is required to separate the individual atomic ratios.

For indirect semiconductors, the band-band or excitonic recombination is only allowed with an additional phonon involved, as mentioned above. In the case of alloys from elemental indirect semiconductors like the technologically important Si_xGe_{1-x}, additional information can be drawn from the "***phonon replicas***" of the transitions, i.e., those spectral lines, which are shifted from the bandgap or free exciton energy by characteristic phonon energies. For mixed SiGe crystals, different modes of Si-Si, Ge-Ge, and Si-Ge vibrations can couple to the electronic transition and lead to multiple low-energy sidebands. From the relative intensity of these, also information on ***the local composition*** can be drawn. In principle, minor shifts of the phonon energies also give a measure of the local strain – but this is very hard to access for these broader lines.

Another important issue for semiconductor devices is the ***local dopant concentration***. The presence of shallow dopants leads to the occurrence of bound excitons in the spectra (process "c)" in figure 2), which are found at slightly lower energy (some meV) in the spectra than the free excitons. For moderate dopant concentrations, the associated lines are rather sharp, but are becoming broader for high concentrations. Either from their ratio relative to the free exciton or the line shape, the dopant concentration can be estimated. For most semiconductors, different dopants lead to different well separated emission lines –allowing for the simultaneous determination of compensating dopant concentrations.

The information contained in the free-to-bound transitions (processes "d)","e)" in figure 2) is mostly less pronounced, since these lines are broader due to the thermal distribution of the free carriers, and are mostly weaker than the bound exciton lines in the spectra. Nevertheless, there are some examples in literature, where from these bands acceptor energies of (deeper) defects in ZnO or GaN were determined.

The donor-acceptor pair band (D^0,A^0) is often observed for crystals with minor quality, e.g., GaAs substrate material. If the binding energies of the two shallow dopants involved are known, the concentrations of the more abundant dopant can be estimated from the energy position of the emission band maximum.

But also if the defects of interest do not emit light, but just act as efficient ***non-radiative recombination centres***, CL studies reveal the presence of these point-like or extended defects as dark points or lines on an otherwise luminescent background. For not too high defect densities, this allows statistics on the density of such defects 0.

When it comes to ***low-dimensional semiconductor structures*** like quantum wells (2D), quantum wires (1D), and quantum dots (0D), the locally and spectrally resolved CL allows gathering information on quantized states and – in best cases – even their spatial extension. These quantized states lead to maxima in the emitted spectra, which might be compared to model calculations. The local variation of the emission of quantum dots with statistically varying size, embedded in a matrix with higher bandgap, is a typical application of CL.

CL can also provide valuable information on ***optical resonator structures*** built from semiconductors or patterned hetero-layer systems from these. When the detection system is set to a characteristic allowed wavelength of the resonator, spatial mapping yields a landscape, on which spots of the structure excitation is more or less efficient, i.e., where are the maxima and the nodes of the resonant mode electric field distribution. This way, e.g., "whispering gallery modes" in tapered ZnO nanopillars were analysed 0, or of coupled ultraviolet standing waves in ZnO pillars with only ≈ 100 nm diameter 0. Vice versa, fixing the excitation spot to a point of interest, an emission spectrum should result which mirrors the full spectral resonator characteristics.

3. Experimental considerations

3.1. Experimental setup

For spatially resolved CL, two basic types of equipment are of interest: either the implementation in a scanning secondary electron microscope (SEM-CL) – what is the most popular version –, or in a scanning transmission electron microscope (TEM-CL). Whereas in the case of the SEM-CL simultaneous recording of the surface structure can be combined with CL emission maps, in the TEM-CL case one tries to combine a "look through the sample" with emission charts. Both methods have their specific advantages, disadvantages, and experimental requirements. For the SEM version, classically there has to be made a choice between eventually higher image resolution as obtainable with field-emitter-type cathodes at the cost of lower beam current, or higher e-beam brightness as available with LaB_6 cathodes at the cost of reduced image quality. For the newest generation of field emitter type microscopes, this conflict is now overcome with new electron optics providing simultaneously high current (~ 100 nA) and excellent image quality. For most practical cases, a beam current of 0.1 nA … 1 nA is high enough.

For spectroscopy on semiconductors, it is advantageous to have a low temperature cryostat available. Typically, the signal level increases for low temperatures by reducing the efficiency of competing non-radiative recombination channels. Any shallow bound states e.g., like bound excitons, which carry information on doping, are observable only at sufficiently low temperatures, what could mean in cases like donor bound excitons in GaAs temperatures below 10 K. Practically all emission spectra broaden with higher temperature either by thermal filling of states proportional to k_BT (like in the case of quantum wells, locally varying bandgaps in inhomogeneous materials, or free excitons in indirect semiconductors), or by lifetime broadening as a consequence of thermal activation of competing recombination channels. The latter process might even affect sharp intra-3d or 4f transitions in transition metals and rare earth atoms. If one is interested in determining the real radiative lifetime of radiative recombination processes by time-resolved CL, one has good chances to do so at low temperatures.

Although numerous CL investigations can be carried out successfully at room temperature, a liquid nitrogen cooled sample stage, or even better a liquid helium cooled cryostat opens the path for more detailed investigations. But these cooling devices introduce the problem of vibrations – caused by the evaporation of the cooling liquid in the tubes - and special care must be taken in the microscope cryostat design. Also, additional drift problems due to thermal expansion/contraction have to be faced. In any case of cooling, the vacuum system of the electron microscope has to be absolutely oil free, since otherwise the sample – as the coolest spot in the system – will be contaminated rather fast.

3.2. Coupling out the light

In both cases of SEM-CL or TEM-CL, one of the key questions is which way to couple out the light efficiently, without disturbing and shadowing the signal on the electron (or other like X-ray) detectors. Commercial setups for SEM-CL mostly use a scheme, where an elliptic (or parabolic) mirror is placed between the last electron lens and the sample stage (figure 3). A small hole ($\varnothing < 1$ mm) drilled into this mirror allows the electrons to pass. The acceptance solid angle can be made quite large (some 60 % of 2π), but at the cost of losing secondary electron (SE) signal on the typical SE detector placed horizontally apart. Any in-lens detector for backscattered electrons placed in the electron column cannot be used efficiently with this arrangement. Since for high spatial resolution the primary electron energy has to be quite low (e.g., $E \approx 2$ keV for 50 nm in GaN), the working distance WD (i.e., the distance between the pole piece and the sample) has to be reduced to ≈ 3 mm for most electron optics. This cannot be realized with mirror optics squeezed in between. Some minimum height of such mirrors is around 6 mm, and their focal length on the object side ≈ 3 mm. This means that on the detection side a largely magnified light spot is created with typically ≈ 0.5 mm diameter including the blur circle. As long as only low spectral resolution is required, the signal can pass through the entrance slit of the monochromator, but not so if higher spectral resolution is needed.

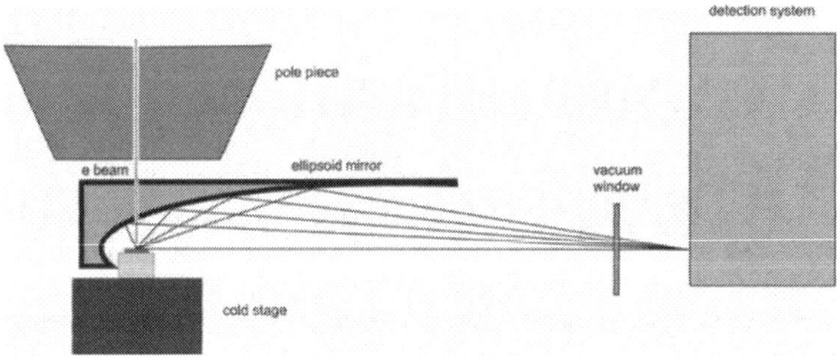

Figure 3. Typical arrangement for SEM-CL: The light is coupled out via an ellipsoid mirror. A small hole allows the electrons to pass. The sample is in one of the focal points of the ellipsoid, and the detection system entrance at the other.

Alternatively, the CL detection system can be attached to an electron probe microanalyser (EPMA) with built-in mirror optics, which collects the light from a certain angle on top and then couples it out sideward into the detection system 0, or to collect the light via a reflecting microscope objective mounted laterally from a sample holder tilted by 45° (see figure 4 and Refs. [10, 18]). In both cases, the collection efficiency is reduced compared to the larger mirror used in the figure 3, but shielding of other detectors is avoided, and perfect electron images, collection of X-rays, etc. are available. The better image quality of this type of optics allows focusing the collected light onto a small (i.e., diameter of ≈ 25 μm 0) spot at the entrance slit of the monochromator.

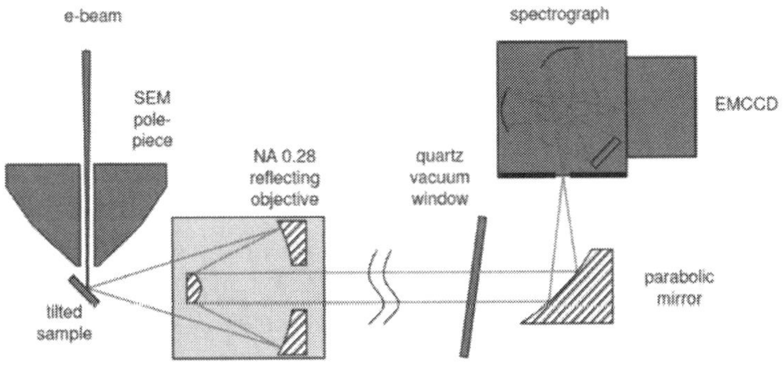

Figure 4. Schematics of a light microscope optics built into a SEM to collect light from a tilted sample without shielding other detection systems, like implemented at the University of Strathclyde. (From [17]; reprinted with permission of the authors).

A further alternative for coupling out the light is to use a glass fibre 0, which is approached from the side under a low angle close to the excited spot (figure 5). For a fibre with 100 μm diameter approached to 50 μm distance, the numerical aperture value is still ≈ 0.7 and thus quite reasonable. The advantage of the latter mentioned setups is that all the electron and X-ray detectors can be used with almost the same signal level as without CL addition – even the in-lens detector. The disadvantage of the fibre approach is that an additional moving part has to be carefully handled, and that not all directions of the emitted light are captured.

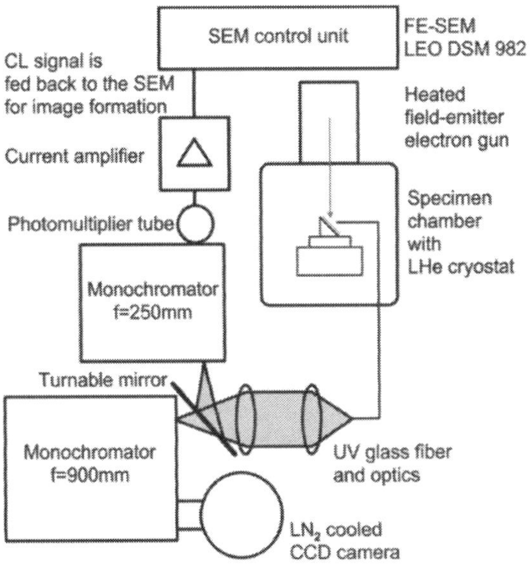

Figure 5. Schematic drawing of the home-made SEM-CL system in the laboratory of the author, based on a Zeiss-Leo DSM 982 field emitter type SEM equipped with a liquid helium cryostat. The light is coupled out through a glass fibre and can be flexibly transferred to arbitrary detection systems sitting on an adjacent optical table.

For the TEM-CL case, the light collection mirror could either be mounted on top of the sample in a similar way as described above, or just underneath the sample to couple out again the light sideward, and also with a hole to let the electrons pass to the rest of the TEM lenses and electron image detector.

3.3. Spatial resolution

The achievable spatial resolution in CL is governed by three main factors: (i) the beam diameter, (ii) the volume, where the electron energy is deposited and (e,h) pairs are generated, and (iii) the diffusion length of minority carriers and excitons. Besides these, vibrations of the sample mounted to a cold finger might be a serious problem.

The beam diameter in modern field-emitter SEMs is ≈ 1 … 2 nm and not limiting the resolution. From the Bohr-Bethe energy-loss relation, phenomenological models for the stopping range s_0 as a function of electron energy E in the form $s_0 = kE^{\alpha}$ were derived, valid in the energy range 5 - 25 keV 0. For SEM-CL, the average depth z_D the electrons reach in the sample (i.e., the centre of the scattering ellipsoid) is $z_D = \frac{2s_0}{1+\sqrt{3}}\left\{1 - \exp\left(-\frac{8}{Z}\right)\right\}$, with the atomic number Z 0. As an example, a rough estimate for ZnO and 2 keV electron energy gives an excited ellipsoid-shaped region with approximate half-axes of $a = 39$ nm and $b = 32$ nm with $z_D = 13$ nm 0. Another approximation for the penetration depth z_D of a sphere-shaped excited region was given by Kanaya and Okayama 0 as $R = \left(\frac{K}{\rho}\right)E^{\alpha}$ with K a material specific constant ~ $Z^{8/9}$, ρ the material density, and $\alpha = 1.67 \ldots 1.75$. There are also several Monte-Carlo type simulation programmes available [25, 26]. The major problem for these is the estimation of the scattering cross-sections and the energy loss at low energies. A simulation for the excited range in GaN for electrons with 2 keV primary energy is shown in figure 6 0. We find most of the scattering events concentrated in a rather small volume, and an overall range of about 50 nm diameter for 90 % of the events for this specific case.

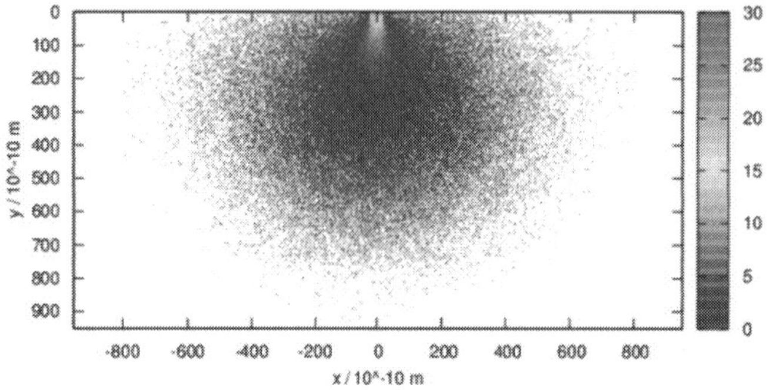

Figure 6. "Scattering bulb" for 2 keV electrons in GaN calculated by a Monte Carlo simulation using the "continuous slowing down approximation". The colour scale encodes the number of inelastic scattering events per volume. (From M. Hocker, Diploma Thesis, University of Ulm, 2012).

The third parameter influencing the resolution, the diffusion length of minority carriers and excitons, is rather material and sample specific. In direct semiconductors like GaN it is at low temperatures in the range of 50 - 100 nm, but can become very large for indirect semiconductors like silicon. For highly doped or defect-rich material this length might be shortened to some few nm. Also high surface recombination rates suppress the diffusion process. Drift movement of excited (e,h) pairs by built-in electrical fields should not be a critical factor, since these fields separate the pairs (or they even can crack excitons), and this way suppress radiative recombination. Our practical experience on ZnO nanostructures or bulk samples with defects is that the scattering bulb diameter gives the right estimate for the characteristic length found for CL contrast changes. We were able to record spectra from GaN structures for energies down to 0.4 keV, what should allow for lateral and depth resolutions on the 20 nm range. Donolato 0 has shown in his calculations on a uniformly excited sphere, that the excess carrier density outside this sphere falls off very steep (faster than exponential) – what means, that the carrier-redistribution by diffusion should not play a big role in most situations.

For STEM-CL, the situation is different, since the very sharply focused high energy (60 - 300 keV) beam mostly penetrates through the very thin (≈ 50 nm) sample prepared for TEM, exciting (e,h) pairs primarily just in a cone shaped volume with ≈ 1 nm diameter on the top. The spread d_b (μm) of the beam at the bottom of the sample is given by 0

$$d_b = 6.25\left(\frac{Z}{E}\right)\left(\frac{\rho}{A}\right)^{1/2} t^{3/2}$$

with the atomic number Z, E the electron energy in keV, ρ (in g/cm³) the material density, A the atomic weight in (g/mol), and the sample thickness t. The excess carriers are mainly concentrated in the central region. Carrier diffusion then also might broaden the excited range, but since the next surface acting as an efficient recombination sink is never far, this broadening effect is expected to be marginal. Actually, there are reports of STEM-CL on nanostructures with ≈ 10 nm lateral resolution [32, 33]. Practically, the preparation necessary for TEM investigation requires sample thicknesses between 50 nm and 100 nm. This creates a defective surface layer on both sides of typically > 30 nm thickness, which has high non-radiative recombination rates. Therefore, additional steps are required for TEM-CL to remove these defects and to get light out of the sample – a rather delicate task. For as-grown nanostructures with suited thicknesses this step is not necessary. Since STEM-CL requires high electron energies, much faster defects in the sample might be created by the electrons.

3.4. Detection system issues

In the simplest case, a photomultiplier tube or any other relatively fast detector suitable for the wavelength range under investigation is enough to record so-called "panchromatic" CL maps. When appropriate colour filters are added, such maps can be recorded wavelength-selective. When complete spectra should be recorded, a monochromator with a grating optimized for the specific task (i.e., the best suited resolving power and grating blaze) has to be inserted. For very efficient collection, a multi-channel detector like a cooled CCD camera is advantageous. This way, for each point of the sample complete spectra may be recorded during the scan forming a so-called "hyperspectral image". These data can then be processed for monochromatic CL images etc. externally. Problems arise with this technique, when large sample drifts or charge pile up occurs during the recording time, or when the sample degrades under the electron bombardment. Alternatively, a single channel detector might be used to record sequentially spectral maps for each wavelength – synchronously with SEM micrographs. To get rid of drifts occurring between the scans, correlation of the SEM micrographs helps to shift the CL maps correctly over each other. For time resolved CL spectroscopy, a fast beam blanker (or a photocathode hit by a pulsed laser) is required on the excitation side, and a fast detection system like a photon counting PMT or a streak camera.

4. Examples of SEM-CL

In this chapter we will take a look at several examples of CL spectra of semiconductors recorded in scanning electron microscopes, either from the top or on cross-sections. For SEM-CL, the samples do normally not need special treatment besides cleaving for access to the desired cross-section.

4.1. Cross-section of a thick GaN layer: hidden defect structures

In figure 7 an example of a cross-section of an epitaxial layer of AlGaN grown on sapphire is shown. The Al content in this layer is slightly below 20 %. Whereas in the secondary electron micrograph no special features are visible, the CL map recorded on the same region clearly shows the traces of several defects, which appear as non-emitting dark horizontal lines in the lower part. In the upper Si-doped range, the emission is shifted to slightly longer wavelength, but relatively homogeneous.

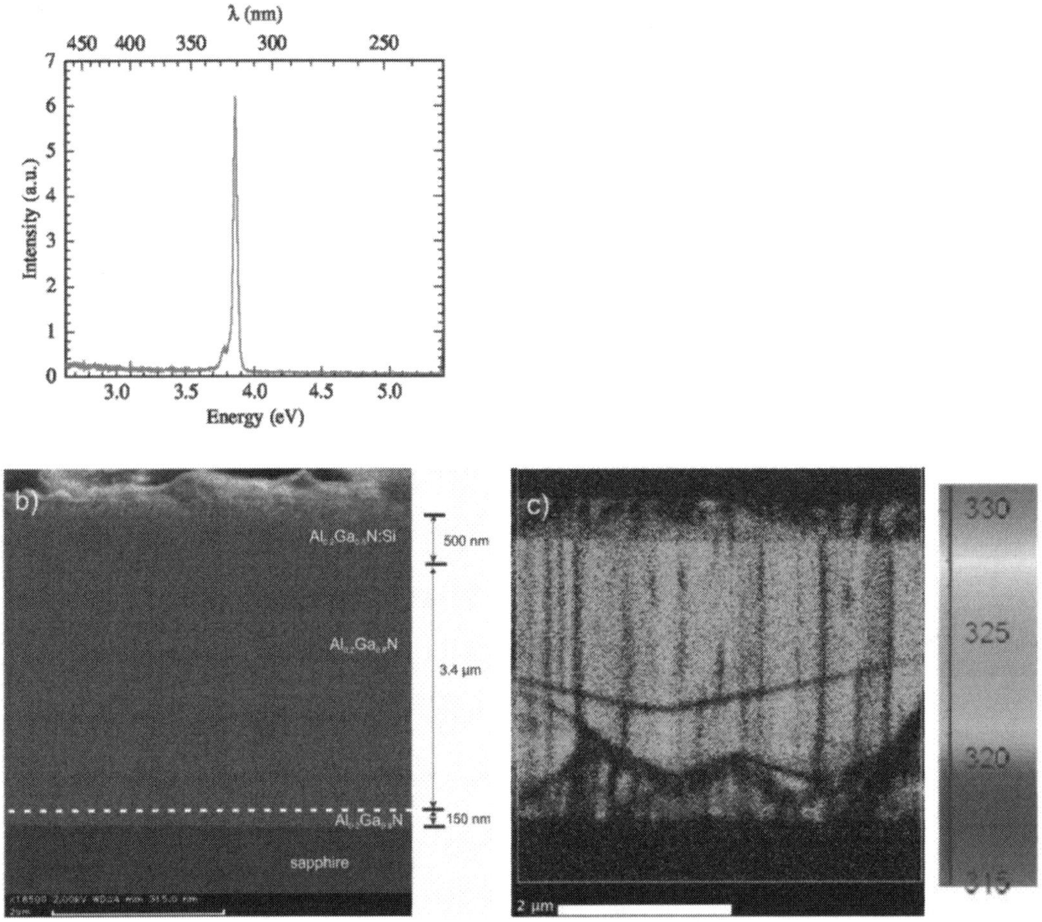

Figure 7. a) Overview CL spectrum recorded on the area of the sample cross section in b), which consists nominally of a layer sequence of sapphire substrate, 25 nm nucleation layer AlN:O, 150 nm $Al_{0.2}Ga_{0.8}N$, SiN_x nanomask, 3.4 µm $Al_{0.2}Ga_{0.8}N$, 500 nm $Al_{0.2}Ga_{0.8}$:Si. b) In the SE picture of the cross-section of the ≈ 3 µm thick Si doped AlGaN sample with low Al content, no special features are observed. c) False colour emission maps composed of emission maps recorded over the narrow range of emissions from 315 nm to 331 nm in 2 nm steps (E = 2 kV, magnification 18500, T ≈ 10 K, spectral resolution 0.4 nm). (Sample courtesy of K. Forghani and F. Scholz, Univ. of Ulm, Inst. of Optoelectronics).

4.2. Non-planar sample with quantum wells: local change of composition

The example shown in figure 8 comprises InGaN quantum wells emitting in the visible range on top of "inverse pyramid" GaN structures. The depressions are approximately 5 µm deep. Whereas in the SE image (left) just a few point-like defects close to the ridges are visible, the colour-coded CL map (right) shows distinctively different emission colours from different facets. Obviously, already minor changes in the orientation of the surface lead to clear changes in the growth rate and composition of the quantum wells, which then change their quantized state energies and emission wavelengths. Remarkable in this example is also the very large depth of focus of the electron and CL image, which allows recording changes in emission wavelength with very high lateral resolution.

Figure 8. Left: SE micrograph of an "inverse pyramid" GaN structure with ≈ 5 µm deep depressions in hexagonal pyramid shape, grown by selective epitaxy on a pre structured sapphire substrate. As top layer, five InGaN quantum wells intended for green emission were grown. Right: CL spectrum recorded at T = 8 K, E = 3 keV reveals multiple facets with different emission wavelengths. (Sample from group of Prof. F. Scholz, Univ. of Ulm, Dept. of Optoelectronics).

4.3. Defect identification in semiconductors: stacking faults in GaN; direct correlation with TEM

Numerous CL studies in literature are dedicated to the local detection of point or extended defects in semiconductors. As an example, we take a look at high-resolution CL spectra recorded on the cross-section of triangular GaN stripe structures (figure 9) 0. Low-temperature CL finds here highly localized horizontal stripes, from which the dominant defect bands at 3.41 eV and 3.32 eV are emitted. Then a thin cross-section lamella was cut out just at the same position, which was then analysed by HR-TEM to localize the exact position and type of stacking faults present there. This way a direct correlation revealed that the band at 3.32 eV is introduced by type I_2 stacking faults, which electronically act as fcc structure GaN quantum wells embedded in hexagonal structure GaN barriers.

4.4. Spectroscopy on V-pit defects in InGaN/GaN

Another example of defect identification by correlation of changes in emission and defect structures visible in SEM micrographs was recently presented by Bruckbauer *et al.* 0. On the surface of c-plane InGaN quantum wells embedded in GaN barriers, they see the typical "V-pits" with diameters in the range of 50 - 100 nm present in high densities, plus chains of such pits and trench-like features. In a detailed room temperature CL investigation over the diameter of typical pits, they found a pronounced change of emission energy and intensity with changes on a scale of 10 - 20 nm, see figure 10. It is not straightforward to distinguish between changes in In-concentration, in strain, quantum well thickness or changes in piezo-field based on CL maps alone, and correlation with data from other methods like TEM, EDS or other is required.

4.5. Local determination of dopant concentration: ZnO nanowires

A CL setup based on a field-emitter SEM (FESEM-CL) allows to record in one synchronous run very detailed images of the nanostructures under investigation together with emission maps. As discussed above, setting the electron excitation spot to a specific region of interest, also spectra with high energy resolution can be recorded, when a larger monochromator and a sensitive detector like a liquid-N_2 cooled CCD camera are attached.

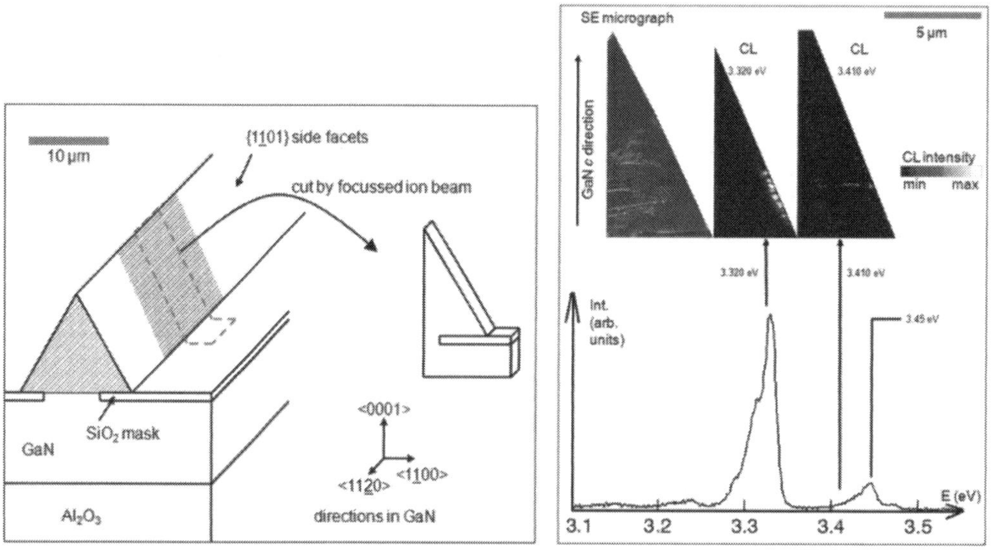

Figure 9. Left: Schematic drawing of GaN stripes with triangular cross-section grown by selective epitaxy on sapphire substrate. By FIB a slice was cut out for high-resolution TEM studies. Right: emission bands at different energies originate from highly localized emission spots and stripes. (Reprinted with permission from 0; copyright American Physical Society).

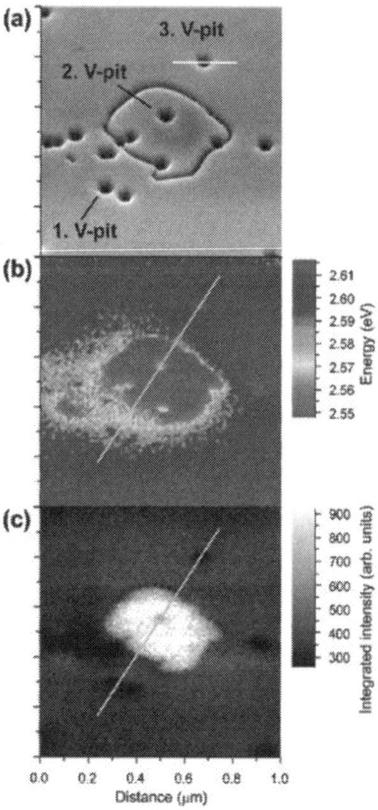

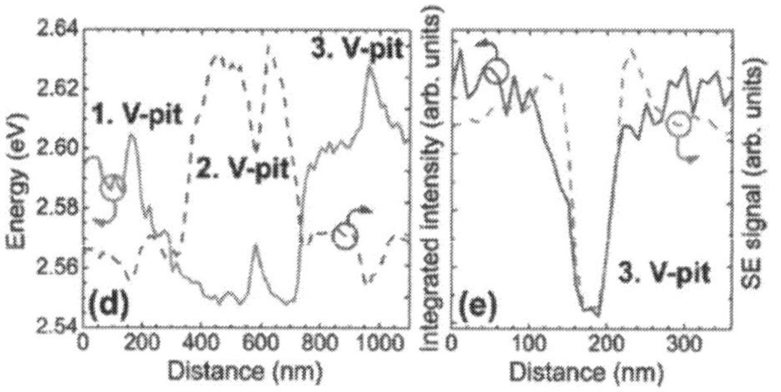

Figure 10. Details of of pits in InGaN/GaN quantum well samples. a) SEM micrograph of a region with different defects like isolated pits, agglomerates of pits, and trenches; b) CL energy map derived from fits to room temperature CL spectra recorded at 5 kV on this area; c) related CL intensity map; d and e) CL line scans through trench-like features with pits and an isolated pit. (From 0; reprinted with permission from Applied Physics Letters; copyright AIP Publishing LLC.).

For the case of high quality ZnO, the half width of the bound exciton emission lines at low temperature can be below 100 µeV. To distinguish different donor dopant species like Ga or In emitting close-lying sharp lines, a spectral resolution of ≈ 200 µeV is required. We carried out such measurements on single ZnO pillars on sapphire substrate in side view under ≈ 45° tilt 0. Indeed, significant changes in the dopant concentration and emission intensity mirroring the crystal quality can be found from bottom to top. Seemingly, a certain amount of Al diffuses from the sapphire substrate into the lower part of the pillar as evidenced by the Al bound exciton (line "I6" in figure 11). The crystal quality becomes best just below the tip (spot #4), whereas it degrades again at the very tip (spot #5), where eventually parts of the catalytic alloy drop, which governed the growth, resides.

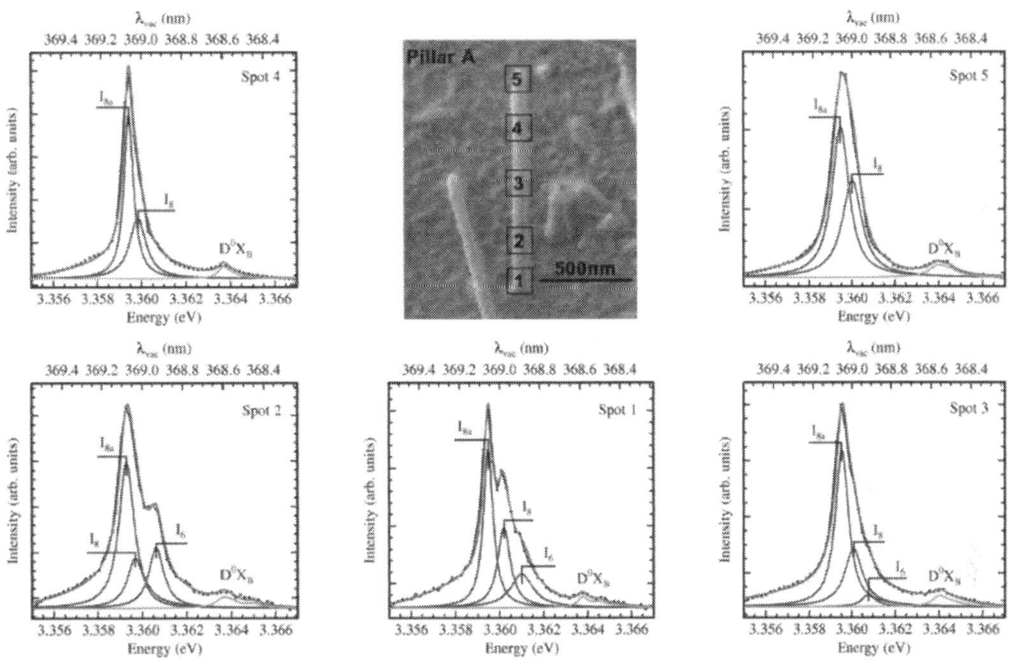

Figure 11. FE-SEM micrograph of single ZnO pillars on sapphire substrate, and low temperature high resolution CL spectra recorded on the marked spots. "I6" corresponds to the aluminium bound exciton, and "I8" to excitons bound to gallium donors. (Reprinted with permission from 0; copyright American Institute of Physics).

4.6. Resolution test for SEM-CL on layered heterostructures

A very nice example to check the resolution of a SEM-CL system was published by Norman 0 (figure 12). He was looking at the cross-section of GaAs quantum wells embedded in AlGaAs barrier layers. There were stacked different sequences on top of each other with well and barrier thicknesses decreasing from 40 nm to 10 nm. When this stack of layers is mapped at 2 keV electron energy, a CL intensity modulation can clearly be seen for the 30 nm structures, and partially still for the 20 nm period ones. Of course, due to the change of quantisation energies in the quantum wells of different thickness, also the emission energy must increase when moving the e-beam to thinner wells. Eventually, the resolution could be even improved applying lower electron beam energy.

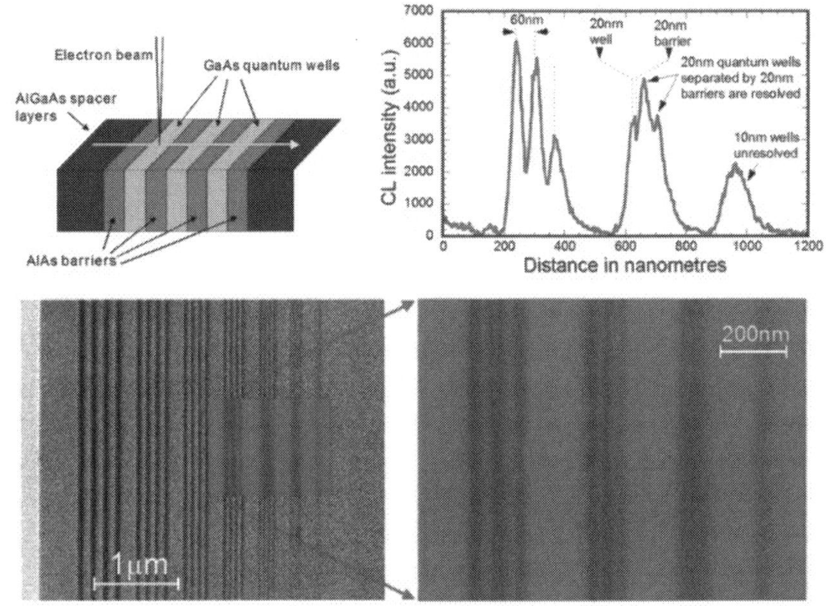

Figure 12. Sample design (top left), SEM micrograph overlayed with pseudo-colour 2 keV CL map of the QWs with 40, 30, 20 and 10 nm thickness (bottom), and CL intensity line scan perpendicular to the QWs and barriers. (Reproduced from [31]; copyright John Wiley and Sons Ltd.).

5. Examples for STEM-CL

Principally, the resolution limit imposed by the volume excited by the electron beam can be made much smaller in STEM-CL, and resolution down to 10 nm has been reported. A 1:1 correlation with high resolution TEM micrographs would be the ideal goal, but is hard to reach due to the vibrations the sample is subject on a cold finger sample holder typically required.

5.1. Triangular III-nitride nanowires

STEM-CL studies on III-nitride core-shell nanowires were reported by Lim *et al.* 0. They obtained a lateral resolution of 20 nm, and recorded either local spectra with a multichannel detector, or monochromatic CL maps at 110 - 300 K sample temperature (see figure 13). Selective area electron diffraction (SAED) patterns recorded on the areas of interest allow to classify the crystal structure and growth direction. Dark-field STEM images were recorded simultaneously with the CL spectral maps.

5.2. Quantum discs in a nanowire

Recently, Zagonel *et al.* 0 have published a study on a nanowire with ≈ 50 nm diameter and a sequence of thin AlN/GaN heterolayers with a period length of ≈ (3 nm (AlGaN) + 1 - 3 nm (GaN)) incorporated, covered by a thin (5 - 10 nm) AlN shell. The structure is shown in figure 14a. The emission wavelength shifts along the wire, what is ascribed to the slowly changing diameter and strain. The authors claim a lateral resolution of 1 nm, what might be correct for the high angle annular dark field (HAADF) signal recorded in parallel, but is not obvious for the CL maps shown in figures 14c-e. A line scan along the NW axis shows clear modulation in the HAADF intensity, and is just resolvable at some spots for the 6 nm periods in the CL signal (see figure 15).

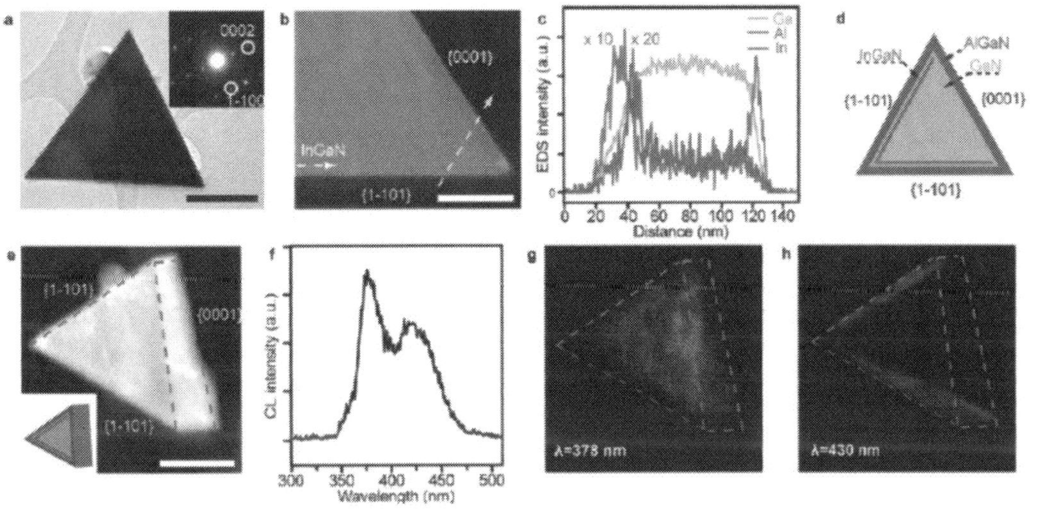

Figure 13. Direct correlation between structural and optical properties on the cross-section of a GaN/InGaN/GaN/ AlGaN SQW nanowire. a) Bright-field TEM image and SAD pattern taken along the [11-20] zone axis (scale bar = 300 nm). b) Dark-field STEM image showing weakly the InGaN SQW (scale bar = 100 nm). c) EDS line scan profiles of Al, Ga, and In along the yellow arrow in (b). d) Cross-section schematic drawing of the SQW nanowire. e) Dark-field STEM image of the same SQW nanowire during the CL measurement (scale bar = 300 nm). f) 110 K CL spectrum of the SQW nanowire. g and h) Monochromatic images. (Reprinted from 0; permission from Nano Letters; copyright American Chemical Society).

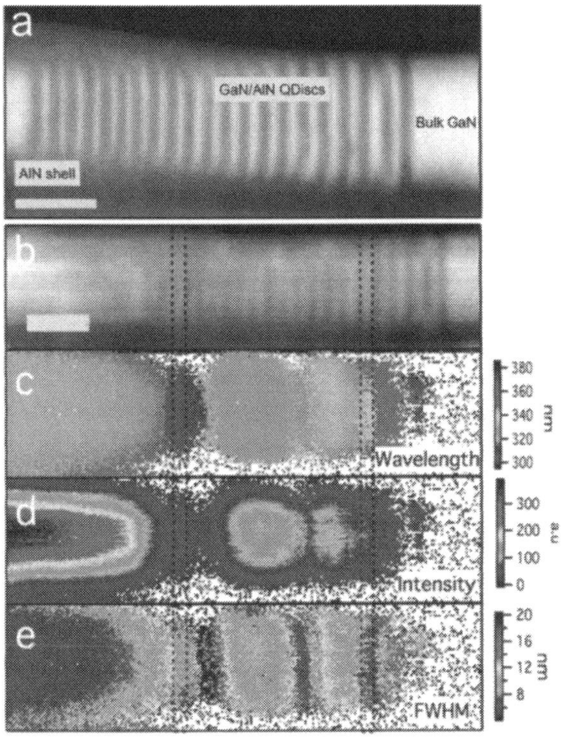

Figure 14. TEM-CL study on a GaN/AlN nanowire with 20 quantum discs embedded. The images have a size of 256x64 pixels with 0.6 nm / pixel, and a colour depth of 256 pixels, $\Delta\lambda$ = 2 nm/pixel, recorded within $\approx$ 6 min. a) HAADF image, where GaN appears whiter, AlN darker, scale bar is 20 nm. b) HAADF acquired simultaneously with the CL, scale bar 20 nm. c) Wavelength position. d) Intensity, and e) FWHM of the most intense peak. Some individual spectral features can be correlated to some individual quantum disks, as emphasized by the black dashed rectangles (E = 60 kV, T = 150 K). (Reprinted with permission from Ref. 0; copyright American Chemical Society).

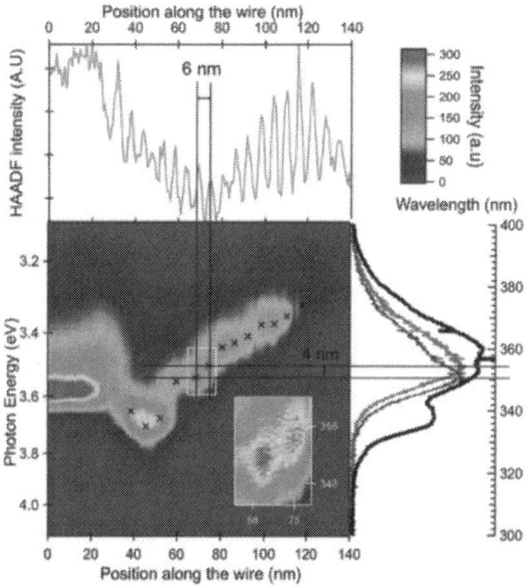

Figure 15. Lateral line scan along the nanowire axis; upper part: HAADF intensity modulation, lower part: spectral CL map (spectral resolution 4 nm). (Reprinted with permission from 0; copyright American Chemical Society).

5.3. Time-resolved CL

Time resolved CL measurements are possible, when the beam can be blanked and a fast enough detection system like a PMT in photon counting mode or streak camera is used. Electronic beam blanking allows for variable excitation pulse width and repetition rates, and allows for time resolution on the scale of 35 psec 0. Even shorter times, but at fixed repetition rate and pulse length are accessible by using a photocathode for the electron emitter excited by a short pulse laser system to produce electron pulses on the 10 psec time scale 0. Figures 16 and 17 show the scheme of the setup, and time resolved spectra recorded with the streak camera on InGaAs/AlGaAs pyramid structures. From the 10 keV electron energy, we would estimate the lateral resolution to be on the order of 100 - 200 nm.

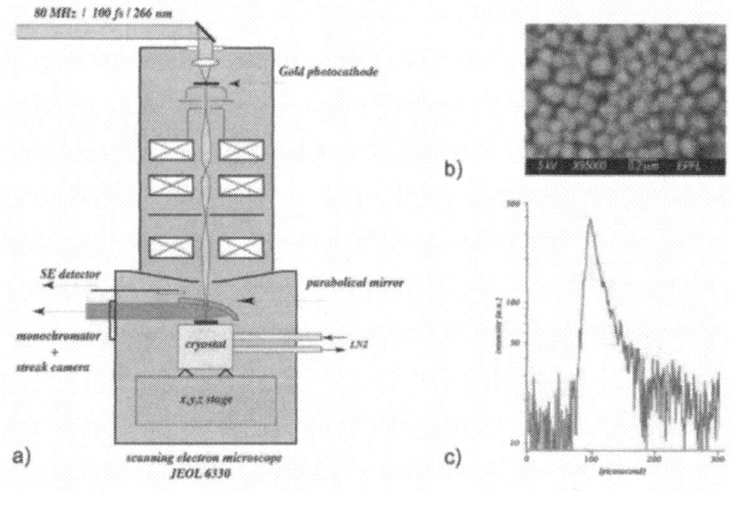

Figure 16. a) Schematic setup for time-resolved PL with a photocathode driven by laser pulses. b) shows an SE electron micrograph of Au particles. c) From the rise time of the CL signal recorded on GaN the authors estimate a width of 10 psec for the electron pulse. (Reprinted by permission from [35]; copyright Macmillan Publishers Ltd.).

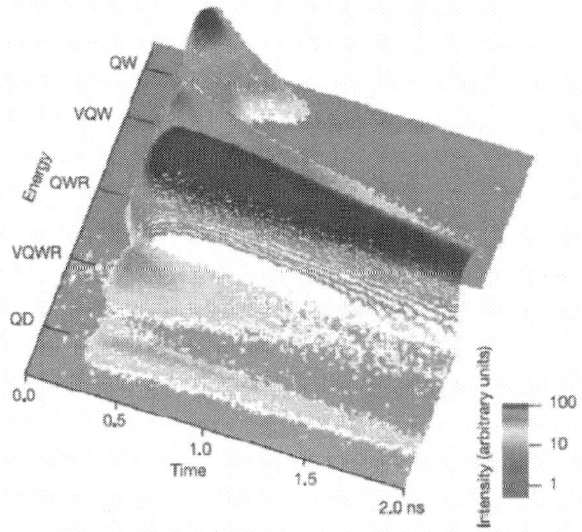

Figure 17. CL signal from InGaAs/AlGaAs pyramids recorded with a streak camera. The sample comprises different quantum structures like quantum dots (QD) on top, vertical quantum wires (VQWR), lateral quantum wires (QWR), vertical quantum wells (VQW), and quantum wells (QW) on the facets of the pyramid (E = 10 keV, T = 90 K). (Reprinted by permission from [35]; copyright Macmillan Publishers Ltd.).

6. Investigation of resonator modes

The CL method can also be utilized to excite specific modes in optical resonator structures made from semiconductors or isolators – or even plasmon modes on metal-nano-antennas. Schirra *et al.* 0 looked at the coupled modes for UV light occurring in very thin ZnO nanowires with ≈ 100 nm diameter and a few μm length. Nobis *et al.* 0 recorded along thicker tapered hexagonal ZnO rods with diameters around 1 μm whispering gallery modes of different order. In a recent publication, Sapienza *et al.* 0 tested the local density of optical states (LDOS) in a photonic crystal cavity (here in Si_3N_4) on the 30 nm scale using the electron beam as a local point source. They were able to resolve the spatial mode structure at a so-called "H1 defect" photonic crystal cavity (see figure 18).

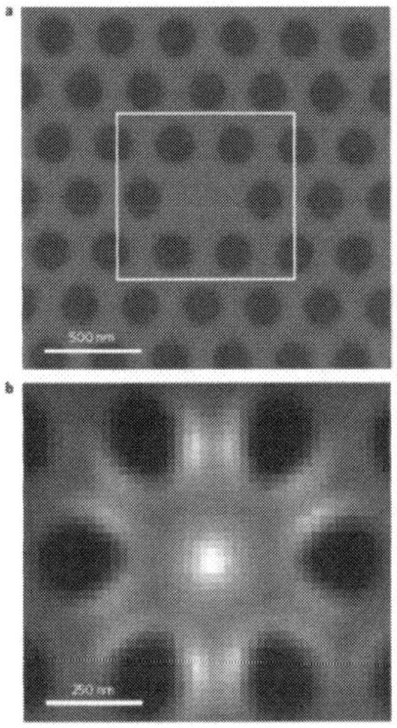

Figure 18. a) SEM image of a photonic crystal membrane with a hole defect (H1 cavity) with a lattice constant a = 330 nm and a hole diameter of 230 nm. b) LDOS image of the cavity in the centre part of a) at a wavelength of 650 nm, with a bandwidth of 9 nm. (Reprinted by permission from [36]; copyright Macmillan Publishers Ltd.).

7. Extension: combined CL characterisation + E-beam lithography (CLL)

By covering the sample with a thin e-beam photoresist, the CL setup can be used in a combined mode for first selecting a region of interest by the emitted CL signal, and then using the e-beam for a writing process (figure 19). Gschrey *et al.* 0 recently suggested and applied this procedure to create well-isolated single quantum dots (QDs) from InGaAs QDs embedded in a GaAs matrix, densely grown in statistical order by the Stranski-Krastanov-Method. For thin PMMA layers, the semiconductor structures underneath can still be excited and their (visible range) emission detected.

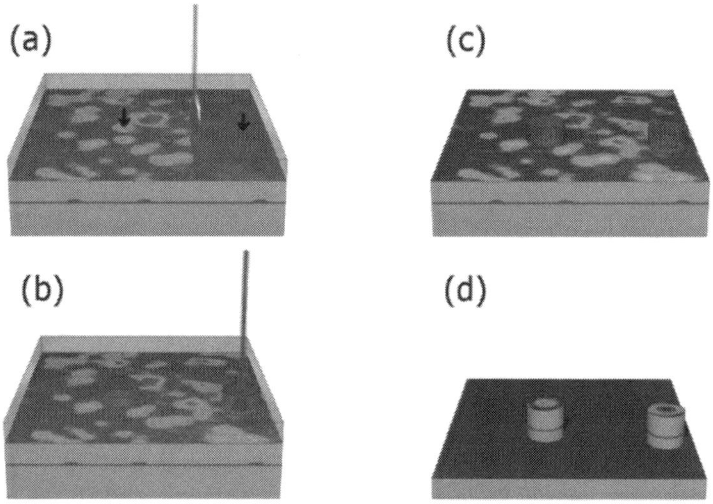

Figure 19. Process steps for combined CL and e-beam lithography: a) Spin coating of the sample by PMMA; b) Selection of an InGaAs QDs by means of the CL signal emitted through the PMMA layer; c) e-beam lithography + dry etching to create sub-μm mesas; d) mesas with single quantum dots result. (Picture courtesy of S. Reitzenstein, TU Berlin).

After having localized special bright spots, the e-beam is used to pattern the PMMA layer. After development of the resist and final dry etching, pre-selected, well isolated QDs embedded in mesa structures are created. Optical experiments on these (by μ-PL) indeed show perfect quantum dot single emitter behaviour.

8. Summary

In this tutorial article, we discussed first the fundamentals of different recombination processes in semiconductors and what information they carry, and looked at some basic experimental considerations. Some typical examples of studies on nanosized defects in bulk samples and on nanorods illustrate the possibilities (and difficulties) of SEM- and TEM-based cathodoluminescence. Extension like time-resolved CL and in-situ e-beam lithography show some further perspectives of this rather powerful method, which can yield very rich information especially when combined on the same sample with methods capable of atomic resolution like TEM.

Acknowledgements

The authors thank the group of Prof. Scholz (Univ. of Ulm, Inst. of Optoelectronics) for the loan of several samples presented here. Anton Reiser, Günter M. Prinz, Tobias Röder and Manfred Madel (Univ. of Ulm, Inst. of Semiconductor Physics, and Inst. of Quantum Matter) grew the ZnO-nanopillars presented. A part of this work was supported by the Deutsche Forschungsgemeinschaft (DFG) in the frame of the SFB 569, and the "PolarCon" research group.

References

[1] Yacobi B G and Holt D B 1990 *Cathodoluminescence microscopy of inorganic solids.* (New York: Plenum Press)
[2] Christen J, Grundmann M and Bimberg D 1991 *J. Vac. Sci. Technol. B: Microelectr. Nanometer Struct.* **9** 2358-2368
[3] Petrov V I 1992 *physica status solidi (a)* **133** 189-230
[4] Gustafsson A, Pistol M-E, Montelius L and Samuelson L 1998 *J. Appl. Phys.* **84** 1715-1775
[5] Gustafsson A 2008 *Scanning* **30** 317-323
[6] Dierre B, Yuan X and Sekiguchi T 2010 *Sci. Technol. Adv. Mater.* **11** 043001
[7] Edwards P R and Martin R W 2011 *Semicond. Sci. Technol.* **26** 064005
[8] Stowe D 2011 *physica status solidi (c)* **8** 1423-1426
[9] Garcia de Abajo F J 2010 *Rev. Mod. Phys.* **82** 209-275
[10] Edwards P R, Jagadamma L K, Bruckbauer J, Liu C, Shields P, Allsopp D, Wang T and Martin R W 2012 *Microsc. Microanal.* **18** 1212-1219
[11] Sauer R, Weber J, Stolz J, Weber E R, Küsters KH and Alexander H 1985 *Appl. Phys. A* **36** 1-13
[12] Liu R, Bell A, Ponce F A, Chen C Q, Yang J W and Khan M A 2005 *Appl. Phys. Lett.* **86** 021908
[13] Tischer I, Feneberg M, Schirra M, Yacoub H, Sauer R, Thonke K, Wunderer T, Scholz F, Dieterle L, Müller E and Gerthsen D 2011 *Phys. Rev. B* **83** 035314
[14] Nobis T and Grundmann M 2005 *Phys. Rev. A* 72 063806
[15] Schirra M, Feneberg M, Prinz G M, Reiser A, Röder T, Thonke K and Sauer R 2009 *Phys. Rev. Lett.* **102** 073903
[16] Nakaji D, Grillo V, Yamamoto N and Mukai T 2005 *J. Electron Microsc.* **54**: 223-230
[17] see a scheme at http://ssd.phys.strath.ac.uk/index.php/File:CL_system.gif
[18] see scheme of P Edwards and Rt Martin at http://nano.strath.ac.uk/local/2009-11/Edwards_Nanometrology_Nov09.pdf, page 8
[19] Hoenk M E and Vahala K J 1989 *Rev. Sci. Instrum.* **60** 226-230
[20] Everhart T E and Hoff P H 1971 *J. Appl. Phys.* **42** 5837-5846
[21] Werner U, Koch F and Oelgart G 1988 *J. Phys. D* **21** 116
[22] Schirra M, Reiser A, Prinz G M, Ladenburger A, Thonke K and Sauer R 2007 *J. Appl. Phys.* **101** 113509
[23] Kanaya K and Okayama S 1972 *J. Phys. D: Appl. Phys.* **5** 43-58
[24] Donolato C 1978 *Optik* **52** 19-36
[25] Drouin D, Couture A, Joly D, Tastet X, Aimez V and Gauvin R 2007 *Scanning* **29** 92-101
[26] Demers H, Poirier-Demers N, Couture A R, Joly D, Guilmain M, de Jonge N and Drouin D 2011 *Scanning* **33** 135-146
[27] Hocker M (University of Ulm) private communication
[28] Tischer I, Feneberg M, Schirra M, Yacoub H, Sauer R, Thonke K, Wunderer T, Scholz F, Dieterle L, Müller E and Gerthsen D 2011 *Phys. Rev. B* **83** 035314
[29] Bruckbauer J, Edwards P R, Wang T and Martin R W 2011 *Appl. Phys. Lett.* **98** 141908
[30] Schirra M, Reiser A, Prinz G M, Ladenburger A, Thonke K and Sauer R 2007 *J. Appl. Phys.* **101** 113509
[31] Norman C E 2002 *Microsc. Analys.* **16**(2): 9-12
[32] Lim S K, Brewster M, Qian F, Li Y, Lieber C M and Gradecak S 2009 *Nano Letters* **9** 3940-3944
[33] Zagonel L F, Mazzucco S, Tenceì M, March K, Bernard R, Laslier B, Jacopin G, Tchernycheva M, Rigutti L, Julien F H, Songmuang R and Kociak M 2011 *Nano Letters* **11** 568-573
[34] Bertram F, Christen J, Dadgar A and Krost A 2007 *Appl. Phys. Lett.* **90** 041917

[35] Merano M, Sonderegger S, Crottini A, Collin S, Renucci P, Pelucchi E, Malko A, Baier M H, Kapon E, Deveaud B and Ganière J-D 2005 *Nature* **438** 479-482
[36] Sapienza R, Coenen T, Renger J, Kuttge M, van Hulst N F and Polman A 2012 *Nature Mater.* **11** 781
[37] Gschrey M, Gericke F, Schüßler A, Schmidt R, Schulze J-H, Heindel T, Rodt S, Strittmatter A and Reitzenstein S 2013 arXiv:1304.3631v1 [cond-mat.mes-hall], April 2013

IOP Conf. Series: Materials Science and Engineering **55** (2014) 012019 doi:10.1088/1757-899X/55/1/012019

Scanning probe - atomic force microscopy: new developments and applications

G Valdrè, D Moro and G Ulian

Università di Bologna, Dipartimento di Scienze Biologiche, Geologiche e Ambientali, Interdisciplinary Research Centre of Biomineralogy, Crystallography and Biomaterials, Piazza di Porta San Donato 1, IT-40126 Bologna, Italy

E-mail: giovanni.valdre@unibo.it

Abstract. Recent developments in scanning probe microscopy (SPM) have widened the spectrum of possible investigations that can be performed at a nanometric level at the surfaces of materials. They range, for instance, from physical properties such as surface potential and electric field topological determination to chemical, nanomechanical, catalytic and spectroscopic analysis in air in liquid or in a gaseous environment. After a brief introduction to new SPM technological developments, we present recent achievements in the characterisation and application of nanomorphology, electrostatic surface potential, dielectric properties, shear force microscopy and radiofrequency measurements, scanning microwave microscopy, nanomechanical properties mapping, frequency modulation SPM in air, liquid, vacuum, and finally high-speed SPM.

1. Introduction

Scanning probe microscopy (SPM) is facing a continuous and rapid development of methodologies thanks to the technological advancements of basic components, innovative detection systems and completely new approaches.

The newly available methodologies allow performing a variety of nano-confined measurements of materials physico-chemical properties. Structure and morphology, electrostatic potential and radiofrequency measurements at sample surfaces, viscoelasticity, adhesion and elastic modulus, and charge density mapping are some of the properties nowadays measurable with nanoscale spatial resolution. Furthermore, true atomic resolution can now be achieved in air, vacuum and liquid environment by implementing frequency modulation atomic force microscopy (FM-AFM).

In this paper we will firstly introduce Kelvin probe force microscopy (KPFM) for the study of surface potential, scanning polarisation force microscopy (SPFM) for liquid films imaging and dielectric mapping, and scanning microwave microscopy (SMM) for the study of carrier behaviour in semiconductor devices. Then, shear force microscopy (ShFM) together with a new radiofrequency (RF) methodology for the simultaneous measurements of viscoelastic/shear properties and impedance characterisation in the RF range of sample surfaces will be presented. Some SPM-based methodologies for nanomechanical property mapping will be reviewed. Finally, a section dedicated to atomic resolution AFM capability in air, liquid and vacuum, and the recent efforts to the development of high speed AFM will be presented and discussed.

Content from this work may be used under the terms of the Creative Commons Attribution 3.0 licence. Any further distribution of this work must maintain attribution to the author(s) and the title of the work, journal citation and DOI.
Published under licence by IOP Publishing Ltd

2. Electrostatic surface potential, dielectric properties measurements and scanning microwave microscopy

2.1. Kelvin probe force microscopy

Kelvin probe force microscopy (KPFM) is a methodology that enables nanometre-scale analysis of surface electrical property of metallic, semiconductor and insulating materials [1]. For instance, KPFM was employed recently to investigate the surface potential of phyllosilicate minerals and the interactions between mineral surfaces and fundamental biomolecules such as DNA, RNA and nucleotides [2-7]. It was found that the nanoscale surface potential distribution of phyllosilicates drive the adsorption, condensation and conformation of the biomolecules.

The "classical" KPFM is a dual pass mode, i.e., during the first pass the surface topography is recorded, then during the second pass the electrostatic interaction is measured by raising the tip above the sample surface following the recorded topographic profile. By this method only the long-range electrostatic interactions are measured and the cross-talk between topography and surface potential measurements is eliminated. However, increasing the tip-sample separation decreases the lateral resolution. The single pass methods allow performing AFM and KPFM imaging simultaneously, decreasing the tip-sample separation and thus increasing the spatial resolution. Furthermore, topographical errors caused by electrostatic forces are minimized because of the active suppression of the electrostatic forces during the simultaneous AFM and KPFM measurement.

In KPFM a conductive tip is biased directly by $V_{tip} = V_{dc} + V_{ac}\sin(\omega t)$. In the case of a metallic sample, the first harmonic of the force, F_ω, is related to the difference between the constant component of the tip bias V_{dc} and the surface potential V_s by

$$F_\omega = \frac{\partial C(z)}{\partial z}(V_{dc} - V_s)V_{ac}\sin(\omega t)$$

where $C(z)$ is the tip-surface capacitance, which is dependent on the tip size and geometry, surface topography and tip-surface separation z. By adjusting V_{dc} to nullify the force, we obtain point by point a surface potential map of V_s.

In the case of a semiconductor sample, F_ω is derived as

$$F_\omega = \frac{Q_S}{\varepsilon_0}\frac{C_I C_D}{C_I + C_D}V_{ac}\sin(\omega t)$$

where Q_s is total charge near the semiconductor surface due to the surface potential of the semiconductor, ε_0 is the dielectric constant, C_I is the capacitance associated with V_{ac} and the air gap between tip and sample, and C_D is the capacitance associated with V_{ac} and the space-charge-layer in the semiconductor [8].

In the configuration of a sample with fixed charges at the surface, placed between a conductive plane substrate and the tip, F_ω can be written as:

$$F_\omega = \left[\frac{\partial C(z)}{\partial z}(V_{dc} - V_s) - \frac{Q_{st}C(z)}{4\pi\varepsilon_r\varepsilon_0 z^2}\right]V_{ac}\sin(\omega t)$$

where $C(z)$ is the tip-substrate capacitance, Q_{St} is the static charge, ε_0 is the permittivity of free space, and ε_r describes the dielectric constant of the materials between the conductive plane substrate and the tip [9, 10].

KPFM have been applied to the characterisation of the electrical properties of metallic nanostructures [1], to the study of surface charges of semiconductor [11], to probe semiconductor devices [1], and study the surface potential of mineral surfaces [12, 13].

A considerable resolution improvement has been achieved in KPFM in the last decade, accompanied by significant advancements in the quantitative understanding of the observed contrast, with the aid of analytical and numerical studies [14, 15]. The electrostatic force F_ω can be detected by employing an amplitude modulation (AM) or a frequency modulation (FM) device [1]. Generally, in single pass KPFM, when electrostatic forces are applied to the tip, the additional oscillating components of the electrical force will be superimposed to the mechanical oscillation. In single pass AM-KPFM, the topography is measured by the oscillation at the first resonance frequency (mechanical excitation), and the surface potential by the amplitude of the oscillation at the second resonance frequency (electrical excitation) to minimize the cross-talk with the topographical signal. In FM-KPFM, F_ω is detected by the frequency shift at ω, and V_{DC} is applied to the AFM tip to nullify the frequency shift. V_{AC} induces a modulation of the electrostatic force detected by the superimposed oscillation at the frequency variation of the mechanical oscillation.

Generally, in FM-KPFM the spatial resolution is mainly related to the dimension of the tip apex, whereas in AM-KPFM the spatial resolution is reduced by an averaging effect between the tip and cantilever. Atomic resolution single pass FM-KPFM of semiconductor surfaces was demonstrated in UHV [16]. Figure 1a shows the topography of a Pb/Si (111)-(7x7) surface with substitutional Pb adatoms as higher protrusions and figure 1b the corresponding FM-KPFM image.

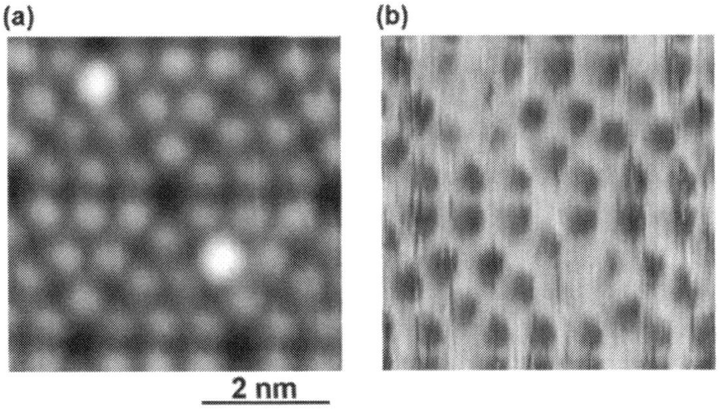

Figure 1. a) Topography and b) simultaneously measured FM-KPFM image of an area of the Si (111)-(7x7) surface with substitutional Pb adatoms (brighter protrusions in a)). (Modified from Sadewasser *et al.* [16]).

Recently, a multiscale approach for simulations of Kelvin probe force microscopy with atomic resolution has been presented [17] in order to obtain a better theoretical understanding of the distance dependence and atomic-scale contrast, recently observed in KPFM measurements. The electrostatics of the tip-cantilever-sample system was treated by a finite-difference method, and the resulting electric field under the tip apex was applied to atomistic wavelet-based density functional theory calculations. The attention was focused on insulating samples and on the interaction of a silicon tip with a NaCl(001) surface. Furthermore, analytical formulae were obtained for the AM- and FM-KPFM signals, taking into account for macroscopic and microscopic bias-dependent forces.

2.2. Scanning polarisation force microscopy

Another interesting approach which involves the polarisation force between an electrically charged AFM tip and a substrate is scanning polarisation force microscopy (SPFM) [18, 19]. SPFM is a non-contact mode particularly useful to study liquid films and droplets, providing information on topography, dielectric properties and contact potential variation. A conductive probe is brought 10 - 20 nm above the sample surface and electrically biased by $V_{tip} = V_{dc} + V_{ac}\sin(\omega t)$. Two lock-in amplifiers are used to detect the forces at the first and second harmonics experimented by the probe. The oscillation amplitude of the second harmonic component is maintained constant by controlling the probe-to-sample distance. T opography and sample polarisability are mixed in images based on detecting the second harmonic term of the electrostatic force. A second feedback loop measures the probe-surface contact potential difference as in KPFM, based on the first harmonic term detection. As a sake of example, SPFM was successfully used to directly image the evolution of the structures of thin water films on $BaF_2(111)$ as a function of relative humidity, RH [20], to study the adsorption of water on alkali halides (KBr, KCl, KF, NaCl) nanocrystals on SiO_2 and their deliquescence as a function of RH [21], and to measure the mobility of muscovite mica surface ions at different humidity levels [22].

2.3. Scanning microwave microscopy

Very recently scanning microwave microscopy (SMM) was developed to investigate carrier behaviour in semiconductor devices. A continuous microwave signal is applied from a vector network analyser to an AFM conductive cantilever with a tall tip, and the reflected portion is analysed [23].

The amplitude of the reflection coefficient was calibrated for quantitative measurements of the capacitance at the tip sample interface, enabling two dimensional mapping of the carrier density across different regions of a semiconductor. SMM is applied for the characterisation and performance modelling of semiconductor devices and failure analysis. SMM was used for calibrated capacitance spectroscopy and dopant profiling measurements in the frequency range from 1 GHz to 20 GHz. As an example, figure 2 shows a topographic and dC/dV image of a bipolar SRAM (static random access memory) integrated circuit device. Several p-n defect structures in the n-doped channels can be identified in the dopant profiling image. A sketch depicts p- and n-doped areas in figure 2a, which corresponds well to the various doping types and concentrations in the dC/dV image.

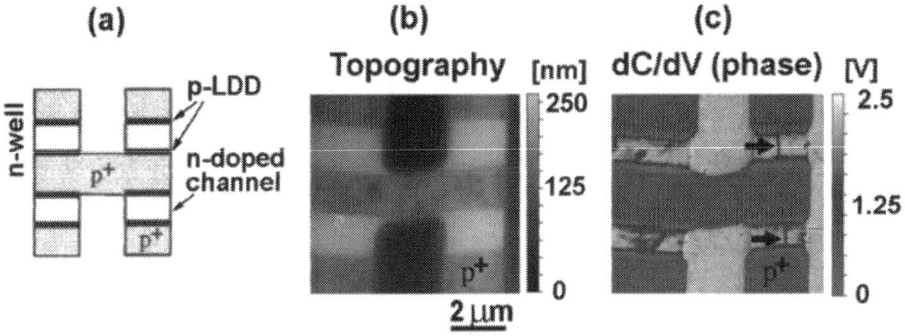

Figure 2. a) Sketch of the p- and n-doped areas of a static random access memory (SRAM). b) AFM topographical image of a SRAM sample. c) dC/dV phase image of the same area as in b) with defect structures (arrowed) in the n-doped channels (modified from Huber *et al.* [23]).

3. Shear force microscopy, radiofrequency measurements and nanomechanical properties mapping

3.1. Shear force microscopy and radiofrequency measurements

Shear force microscopy (ShFM) [24, 25] is a non-contact imaging methodology that uses a cantilever oriented in a vertical configuration with respect to the sample surface, i.e., with the long axis of the cantilever perpendicular to the sample surface. The main advantage of this geometry is the absence of the jump-to-contact. This phenomenon arises in the horizontal configuration when the gradient of the surface attractive force overcomes the cantilever spring constant. As a consequence the lateral stiffness of the cantilevers can be reduced by more than two orders of magnitude with respect to the softest commercially available cantilevers [26]. Furthermore, the probe at the free-end of the cantilever (figure 3), which is oscillated in a direction parallel to the sample surface by a piezoelectric element, is exclusively sensitive to horizontal forces (shear forces) and achieves a very low probe to sample interaction force. This force is considered to be mediated by the water confined between the sample surface and the cantilever apex. When the probe is in the non-contact shear force regime, few nanometres from the surface, the monotonic relation between the cantilever oscillation amplitude and the probe-sample distance is used to image the sample surface by scanning the surface at a constant oscillation amplitude [25, 27]. Thanks to this features ShFM was used, for instance, to investigate molecular layering and layer viscoelastic properties in confined liquid films [25, 28], the effect of surface-adsorbed cations on the viscoelastic response of water to shear in sub-nanometre confinement [29], mechanical properties of biomolecular motors [30], and also as a feedback system in scanning near-field optical microscopy [31, 32].

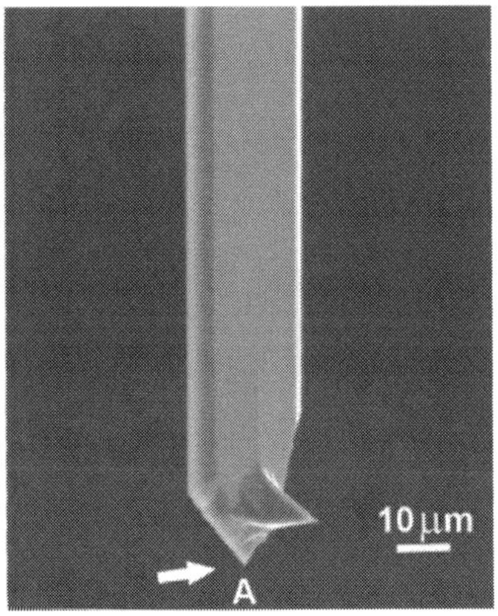

Figure 3. Scanning electron microscopy micrograph of a commercial AFM cantilever oriented in a vertical configuration. The sensing element is the apex at the free-end of the cantilever (arrowed A).

Ulcinas and co-workers [29] measured the viscoelastic response to shear of water in a sub-nanometre confinement by detecting the thermal noise of a vertically oriented microcantilever. This response was related to the effect of surface-adsorbed cations, which act as pinning centres restricting the mobility of the confined water molecules. Measurements were performed in pure water as well as under acidic and high-ionic-strength conditions.

By means of this methodology information about the viscoelastic properties of a fluid can be extracted, being the thermal noise connected to the local mechanical susceptibility. From the fluctuation-dissipation theorem,

$$|u|^2(\omega) = \frac{k_B T}{\pi\omega}\chi''(\omega) = \frac{A\gamma}{(\omega_0^2 - \omega^2) + \gamma^2\omega^2}$$

where u is the displacement, ω is the frequency, χ'' is the imaginary part of the mechanical susceptibility, $\omega_0 = (k/m)_{1/2}$ is the eigenfrequency, k_B is the Boltzmann constant, T is the absolute temperature, k is the effective spring constant, m is the effective mass, $A = k_BT/\pi m$ is the oscillator strength, $\gamma = \zeta/m$ is the drag coefficient, and ζ is the damping. Following the approach of Roters and Johannsmann [33], the power spectral density of the thermal fluctuations around the fundamental vibrational mode is fitted with a Lorentzian with parameters of ω_0, γ, and A, and effective parameters k and m are calculated using fitted values.

As a practical example, the dependencies of the effective stiffness versus the probe-specimen separation (displacement) from the surface in pure water, in a pH 2.9 solution, and in a 55 mM NaCl solution are shown in figure 4.

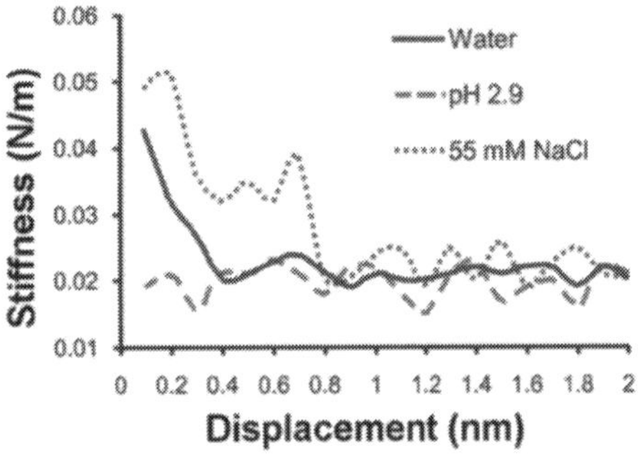

Figure 4. Trend of the effective stiffness obtained from fitting the thermal noise power spectral density spectra as a function of displacement, i.e., at different separations between the probe and the surface. (Modified from Ulcinas *et al.* [29]).

ShFM is usually developed as a prototype at laboratory level. Various microscope setups and detection systems have been proposed based on scattered evanescent electromagnetic waves [34], optical devices [24, 31, 35, 36], piezoelectric tuning forks [37], self-actuated piezo-resistive cantilevers [38], and capacitive sensors [39]. Recently, Valdrè and co-workers [40] developed a simple device that enables to perform ShFM with a commercial AFM head, with no modification of the AFM hardware and software, and using commercial available cantilevers as probes.

Exploiting this favourable probe-to-sample configuration the same research group developed a novel system for simultaneous shear forces/morphology and RF impedance measurements of specimen surfaces [41]. A vector network analyser was connected to the vertically oriented cantilever in order to perform measurements of RF-impedance variation at the nanoscale, minimizing the cantilever parasitic capacitance thanks to the vertical configuration of the cantilever with respect to the sample surface. The complex impedance was obtained by the measurement of the reflected wave amplitude and phase and their comparison to those of the incident wave.

The variation of the phase θ and modulus $|Z|$ of the complex impedance $Z = |Z|e^{i\theta}$ at the probe-to-sample interface were measured in the RF range from 100 kHz to 8.5 GHz, being the impedance variation dependent on the material local properties and on the probe-to-sample distance at the nanoscale.

Preliminary investigations were conducted acquiring RF impedance spectra variation after depositing a micrometre thick synthetic oil film on the atomically flat surface of a pyrolytic graphite substrate (figure 5). The RF impedance modulus, $|Z|$, and phase, θ, signal variation were measured in the frequency range from 100 kHz to 8.5 GHz. A comparison of the spectra acquired on graphite with and without the oil film is shown in figure 5. The measurements were performed with the cantilever apex at forced-contact with the surface (the contact condition which nullifies the free oscillation amplitude). The modulus of the impedance measured for the oil film deposited on graphite reveals higher values at low frequency (< 100 MHz) respect to the same measurement on pure graphite in air (figure 5a). The phase signal shows a pure resistive forced-contact condition at low frequency, < 100 MHz (figure 5b). A strong dependence of the RF impedance phase θ from both the probe-to-sample distance and the material at the interface was observed. In addition, with this device is possible to measure the RF transmitted wave through the specimen to image sub-surface through-specimen defects or inhomogeneities with nanoscale resolution.

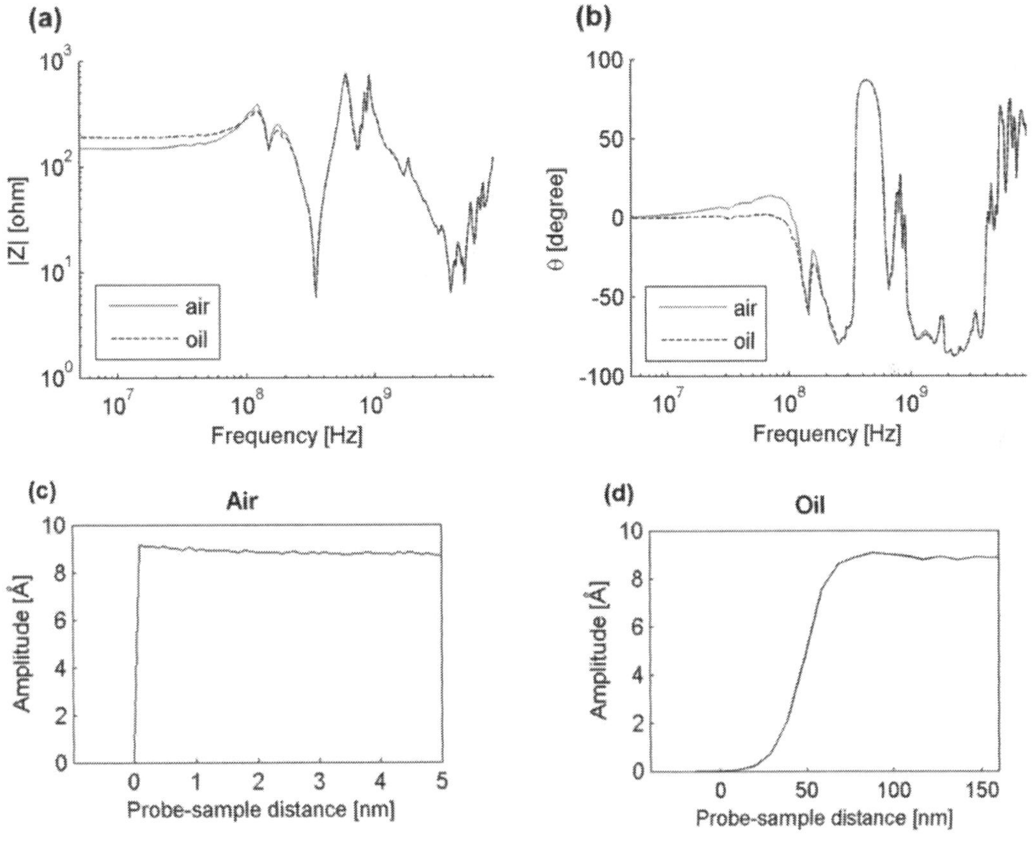

Figure 5. a) RF impedance modulus and b) phase signal variation measured on pyrolytic graphite with synthetic oil film of 1 μm of thickness (dashed line) and without the oil film (solid line). c) Typical ShFM approaching curve in air, showing oscillation amplitude variation as a function of probe-sample distance. d) ShFM approaching curve in oil.

With the same device we have investigated the probe-specimen shear behaviour in different environmental conditions. As an example, we report in figures 5c and d typical ShFM approaching curves in air and oil on graphite, showing the oscillation amplitude variation as a function of probe-sample distance related to different shear forces.

3.2. Nanomechanical properties mapping

A new method developed to measure nanomechanical properties is the peakforce quantitative nanomechanical mapping (PeakForceTM QNMTM, Bruker AXS GmbH, Germany). It allows the acquisition of high resolution images and simultaneous measurement of material mechanical properties [42]. The maximum force applied on the tip is the control parameter of the feedback circuit, and can range down to the picoNewton.

The tip to specimen distance is modulated with a frequency of about 2 kHz and the position is measured as a function of time. The sample properties are calculated by transforming the time function curves in separation function ones. From the curve analysis a wide variety of properties can be obtained, such as tip-sample adhesion, energy dissipation, elastic modulus and the maximum deformation on the sample. To determine the elastic modulus the first linear tract of the withdrawal line in the force-distance curve is fitted using the Derjaguin-Muller-Toporov (DMT) model according to the following equation [42]:

$$F - F_{adh} = \frac{4}{3} E' \sqrt{R(d - d_0)^3}$$

where F - F_{adh} is the force on the cantilever relative to the adhesion force, R is the tip radius of curvature, and d - d_0 is the sample deformation. The DMT contact mechanics model is applied to the withdrawal curve which is not subject to the jump-to-contact and related tip-sample settlements. The reduced modulus E' is calculated by data fitting and, considering the Poisson ratios of the sample, ν_S, and of the tip, ν_{tip} , the elastic modulus E_S is obtained from E' by the following relation:

$$E' = \left[\frac{1 - \nu_s^2}{E_s} + \frac{1 - \nu_{tip}^2}{E_{tip}} \right]^{-1}.$$

By considering the tip elastic modulus $E_{tip} >> E_S$, the previous equation can be approximated to:

$$E' = \left[\frac{1 - \nu_s^2}{E_s} \right]^{-1}.$$

In order to obtain meaningful Young's modulus values, care should be taken that the sample is not plastically deformed. The maximum applied force is controlled by the feedback circuit. Generally, a minimum of 2 nm elastic deformation is necessary for accurate measurements, and the sample surface can be subsequently checked for residual plastic deformation. The Poisson's ratio generally ranges between about 0.2 and 0.5 (perfectly incompressible) giving a difference between the reduced modulus and the sample modulus between 4 % and 25 %. By PeakForce QNM it is generally possible to determine the elastic modulus in a range from 700 kPa to 70 GPa.

For the sake of example, Schön and co-workers [43] applied PeakForce QNM to investigate segmented polyether polyurethanes with varying stoichiometric ratios of the isocyanate and the hydroxyl groups. The morphology was observed to strongly depend on stoichiometric ratio, providing specific ''fingerprint'' phase images that allow the identification of the sample. The elastic moduli of

IOP Conf. Series: Materials Science and Engineering **55** (2014) 012019 doi:10.1088/1757-899X/55/1/012019

stiff and soft segments of the phase separated polyurethane samples were resolved at nanoscale resolution. While tensile testing provided a bulk average value for the elastic modulus of the elastomers, the AFM-based methodology enabled a quantitative investigation of surface stiffness with nanoscale resolution.

The pulsed force mode (PFMTM, WITec GmbH, Germany) is a SPM-based methodology that employs a procedure for nanotopographical measurements without shear lateral forces and simultaneous measurement of materials properties such as superficial adhesion, viscoelastic material behaviour and charge distribution [44].

In PFM a function generator modulates the SPM tip to sample distance with a sinusoidal signal of amplitude in the range 10 nm - 500 nm and frequency from 100 Hz to 5 kHz. The amplitude is re-settled after each contact between the tip and the surface during one cycle.

The maximum tip-sample force is the feedback control parameter and it is set by the user to limit the maximum tip-sample repulsive force. To measure adhesive forces the maximum negative (attractive) force value is recorded when the tip retracts from the surface. Elastic properties are calculated by the force trends before and after the point of maximum tip-sample force in one approach-withdrawal cycle. The difference between a force value in the repulsive part and the maximum force is related to the sample local stiffness. Furthermore, long range electrostatic forces due to surface potential distribution can be mapped recording the fluctuations of the zero force line.

For the sake of example, Roffat and co-workers [45] applied PFM to investigate the pH sensitivity of $Li_{0.30}La_{0.56}TiO_3$ ceramics, which depends on their sintering temperature. These ceramics may be employed as pH sensors and reference electrodes in solid-state combined pH electrode devices. It was observed, by adhesive surface property mapping, that sensitive ceramics have a higher hydrophilic surface properties respect to the insensitive ones. The hydrophilic character of the sensitive ceramics was attributed to surface hydroxyl –OH groups that may display a variable basic character depending on the covalency of their O atoms.

Rezende and co-workers [46] used PFM to evaluate the surface mechanical properties of thin film patterns (less than 100 nm in thickness) obtained by drying an aqueous solution of poly (N-isopropylacrylamide) and sodium dodecyl sulphate on mica. AFM images of dry films showed a morphology formed by droplets distributed on the substrate. PFM adhesion and stiffness mapping revealed heterogeneous features inside the droplets, which were ascribed to be phase separation of the polymer – surfactant complex during the drying/dewetting process. Furthermore, a very thin layer containing polymer and surfactant was observed by a comparison of the adhesion forces measured on dewetted and freshly cleaved mica.

4. Frequency modulation AFM – air, liquid, vacuum

Atomic force microscopy can achieve atomic and also sub-atomic resolution probing short-range chemical forces between a sharp tip and a sample. The tip is positioned with sub-angstrom precision above the surface using piezoelectric actuators.

The most significant recent advancements in AFM high-resolution imaging are based on frequency modulation atomic force microscopy (FM-AFM) [47, 48]. Unlike amplitude-modulation AFM [49], in FM-AFM the cantilever is driven at its resonant frequency. The control signal is the frequency shift of the resonant frequency of the cantilever, which is modified by the tip-surface interaction. The cantilever is kept oscillating at its current resonant frequency, for instance by means of a phase-locked loop (PLL), and the frequency shift induced in the cantilever motion by the tip-sample interaction is used as the source of topographical contrast. In FM-AFM sub-angstrom tip oscillation amplitude enhances the sensitivity to short range forces and it is now possible to directly measure with sub-atomic precision the force of a single chemical bond between two atoms. The image resolution can also be improved by controlling the chemical reactivity of the AFM tip apex picking up adatoms or small inorganic molecules [50].

FM mode, firstly developed for in vacuum experiments, was subsequently applied to ambient conditions to achieve atomic resolution [51].

UHV FM-AFM experiments have involved the chemical identification at the atomic level [52, 53], and also the manipulation of individual atoms on semiconductor surfaces [54].

The research group of Giessibl [55] imaged the silicon (111)-(7×7) surface in ultra-high vacuum resolving fine structures within a single atom. This sub-atomic structure was interpreted as arising from orbitals of the atom at the end of the tip interacting with orbitals of the atom on the surface. It is known that AFM images involve a convolution of tip and sample features, thus comprising sub-atomic information on the electronic structure of the probe.

The extension of FM-AFM to high-resolution operation in liquids has been also achieved. For instance, Fukuma developed a wideband low-noise optical beam deflection sensor with photothermal cantilever excitation system for liquid-environment FM-AFM, applicable to relatively stiff cantilevers (> 20 N/m) [56]. Atomic resolution imaging of mica surface in water was demonstrated (figure 6) by using a stiff cantilever, k = 20 N/m, and an oscillation amplitude of about 0.1 nm. The research group of Garcia [57] combined FM-AFM with bimodal excitation to image and measure protein flexibility with molecular resolution. Mapping of the elastic modulus variation in a single antibody pentamer was achieved in liquid with sub-2 nm spatial resolution.

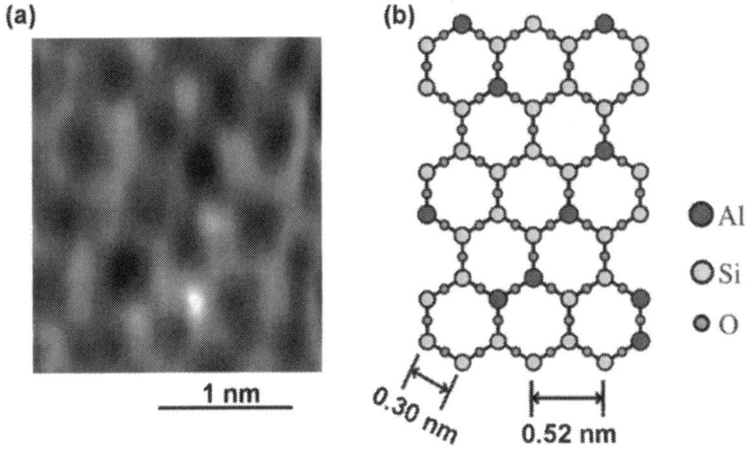

Figure 6. a) FM-AFM image of mica in phosphate buffer solution at RT, pH 7.4. b) Atomic-scale model of the cleaved mica surface. (Modified from Fukuma [56]).

5. High-speed AFM

One of the most challenging advances in AFM involves the combination of high spatial resolution capabilities with improved temporal resolution. Sub-second acquisition timescale is required both in materials science, in biology, and also for quick high resolution investigations of large sample areas. At present AFM is generally a slow process (tens of seconds of acquisition times).

High-speed atomic force microscopy (HS-AFM), developed at laboratory level, allows at present the direct observation of dynamic processes at a sub-second to sub-100 ms temporal resolution, with about 2 nm of lateral and about 0.1 nm vertical resolution [58]. AFM imaging speed is limited by the bandwidth and dynamic behaviour of actuators and mechanical parts. The efforts towards an improved temporal resolution are mainly addressed to the development of small cantilevers and suitable optical beam deflection systems, fast amplitude detectors, fast scanners and new feedback controllers [59, 60]. Picco and co-workers [61] presented a HS-AFM working in contact mode with a flexure stage arrangement allowing imaging video rates of 30 fps (frame per second) and an ultra-high speed version exceeding 1000 fps by employing a tuning fork and a flexure-stage scanning combination.

Recent interest is focused on the development of high-bandwidth flexure-guided nanopositioning systems [62], investigation of the stability of force-feedback applied in HS-AFM imaging [63], on the effect of higher eigenmode flexural oscillations of the cantilever on the signal-to-noise ratio in the topography images [64], and on positioning error mapping to address dimensional metrology issues [65].

Other research groups proposed novel scan trajectories for HS-AFM. For instance, Tuma and co-workers [66] demonstrated a real-time AFM imaging with a frame rate of 1 fps, in a custom-built AFM, employing a two-dimensional Lissajous pattern generated by actuating the scanner with two single-tone harmonic waveforms of constant amplitude and frequency. Yong and co-workers [67] proposed a smooth cycloid-like scan pattern allowing a scan speed up to 156.25 Hz implemented on a commercial AFM with minimal modifications.

References

[1] Melitz W, Shen J, Kummel A C and Lee S 2011 *Surf. Sci. Rep.* **66** 1
[2] Antognozzi M, Wotherspoon A, Hayes J M, Miles M J, Szczelkun M D and Valdrè G 2006 *Nanotechnology* **17** 3897
[3] Valdrè G, Antognozzi M, Wotherspoon A and Miles M J 2004 *Phil. Mag. Lett.* **84** 539
[4] Valdrè G 2007 *Eur. J. Mineral.* **19** 309
[5] Valdrè G, Moro D and Ulian G 2011 *Micro Nano Lett.* **6** 922
[6] Valdrè G, Moro D, Hounsome C M and Antognozzi M 2012 *Nanotechnology* **23** 385301
[7] Valdrè G, Moro D and Ulian G 2012 *IOP Conf. Ser.: Mater. Sci. Engng.* **32** 012027
[8] Hudlet S, Saintjean M, Roulet B, Berger J and Guthmann C 1995 *J. Appl. Phys.* **77** 3308
[9] Nyffenegger R M, Penner R M and Schierle R 1997 *Appl. Phys. Lett.* **71** 1878
[10] Girard P 2001 *Nanotechnology* **12** 485
[11] Valdrè G, Moro D, Lee D, Smith C G, Farrer I, Ritchie D A and Green R T 2008 *Nanotechnology* **19** 045304
[12] Valdrè G, Malferrari D and Brigatti M F 2009 *Clay Clay Miner.* **57** 183
[13] Valdrè G, Tosoni S and Moro D 2011 *Am. Mineral.* **96** 1461
[14] Valdrè G and Moro D 2008 *Nanotechnology* **19** 405501
[15] Valdrè G and Moro D 2008 *Nanotechnology* **19** 405502
[16] Sadewasser S, Jelinek P, Fang C-K, Custance O, Yamada Y, Sugimoto Y, Abe M and Morita S 2009 *Phys. Rev. Lett.* **103** 266103
[17] Sadeghi A, Baratoff A, Ghasemi S A, Goedecker S, Glatzel T, Kawai S and Meyer E 2012 *Phys. Rev. B* **86** 075407
[18] Hu J, Xiao X-D, Ogletree D F and Salmeron M 1995 *Science* **268** 267
[19] Hu J, Xiao X-D and Salmeron M 1995 *Appl. Phys Lett.* **67** 476
[20] Verdaguer A, Cardellach M and Fraxedas J 2008 *J. Chem. Phys.* **129** 174705
[21] Arima K, Jiang P, Lin D-S, Verdaguer A and Salmeron M 2009 *J. Phys. Chem. A* **113** 9715
[22] Xu L and Salmeron M 1998 *Langmuir* **14** 5841
[23] Huber H P, Humer I, Hochleitner H, Fenner M, Moertelmaier M, Rankl C, Imtiaz A, Wallis T M, Tanbakuchi H, Hinterdorfer P, Kabos P, Smoliner J, Kopanski J J and Kienberger F 2012 *J. Appl. Phys.* **111** 014301
[24] Antognozzi M, Haschke H and Miles M J 2000 *Rev. Sci. Instrum.* **71** 1689
[25] Antognozzi M, Humphris A D L and Miles M J 2001 *Appl. Phys. Lett.* **78** 300
[26] Vicari J A, Ulcinas A, Hörber J K H and Antognozzi M 2011 *Ultramicroscopy* **111** 1547
[27] James P J, Antognozzi M, Tamayo J, McMaster T J, Newton J M and Miles M J 2001 *Langmuir* **17** 349
[28] Antognozzi M, Protti A, Miles M J and Valdrè G 2003 *GeoActa* **2** 101
[29] Ulcinas A, Valdrè G, Snitka V, Miles M J, Claesson P M and Antognozzi M 2011 *Langmuir* **27** 10351

[30] Scholz T, Vicary J A, Jeppesen G M, Ulcinas A, Horber J K H and Antognozzi M 2011 *Nanotechnology* **22** 095707
[31] Yang P C, Chen Y and Vaeziravani M 1992 *J. Appl. Phys.* **71** 2499
[32] Betzig E, Finn P L and Weiner J S 1992 *Appl. Phys. Lett.* **60** 2484
[33] Roters A and Johannsmann D J 1996 *J. Phys.: Condens. Matter* **8** 7561
[34] Antognozzi M, Ulcinas A, Picco L, Simpson S H, Heard P J, Szczelkun M D, Brenner B and Miles M J 2008 *Nanotechnology* **19** 384002
[35] Froehlich F F and Milster T D 1994 *Appl. Phys. Lett.* **65** 2254
[36] McMaster T J, Miles M J, Antognozzi M and Humphris A D L 2003 *U.S. Patent US20030160170*
[37] Karrai K and Grober R D 1995 *Appl. Phys. Lett.* **66** 1842
[38] Woszczyna M, Gotszalk T, Zawierucha P , Zielony M, Ivanow T, Ivanowa K, Sarov Y, Nikolov N, Mielczarski J, Mieluarska E and Rangelow I W 2009 *Microelectron. Engng.* **86** 1212
[39] Leong J K and Williams C C 1995 *Appl. Phys. Lett.* **66** 1432
[40] Valdrè G, Moro D and Ulian G 2012 *Meas. Sci. Technol.* **23** 085903
[41] Valdrè G and Moro D 2012 *Rev. Sci. Instrum.* **83** 126103
[42] Pittenger B, Erina N and Su C 2010 *Veeco Application Note* AN128 Rev. A0
[43] Schön P, Bagdi K, Molnár K, Markus P, Pukánszky B and Vancso G J 2011 *Eur. Polym. J.* **47** 692
[44] Gigler A, Gnahm C, Marti O, Schimmel T and Walheim S 2007 *J. Phys.: Conf. Ser.* **61** 346
[45] Roffat M, Noël O, Soppera O and Bohnke O 2009 *Sens. Actuator B-Chem.* **138** 193
[46] Rezende C A, Lee L-T and Galembeck F 2009 *Langmuir* **25** 9938
[47] Morita S, Giessibl F J and Wiesendanger R 2009 *Noncontact atomic force microscopy, vol. 2.* (Berlin: Springer)
[48] García R and Pérez R 2002 *Surf. Sci. Rep.* **47** 197
[49] García R, Magerle R and Pérez R 2007 *Nat. Mater.* **6** 405
[50] Boneschanscher M P, van der Lit J, Sun Z, Swart I, Liljeroth P and Vanmaekelbergh D 2012 *ACS Nano* **6** 10216
[51] Yang G 2009 *Surf. Sci. Rep.* **64** 99
[52] Sugimoto Y, Pou P, Abe M, Jelinek P, Pérez R, Morita S and Custance O 2007 *Nature* **446** 64
[53] Sugimoto Y, Namikawa T, Abe M and Morita S 2009 *Appl. Phys. Lett.* **94** 023108
[54] Custance O, Perez R and Morita S 2009 *Nat. Nanotechnol.* **4** 803
[55] Giessibl F J 2003 *Rev. Mod. Phys.* **75** 949
[56] Fukuma T 2009 *Rev. Sci. Instrum.* **80** 023707
[57] Martinez-Martin D, Herruzo E T, Dietz C, Gomez-Herrero J and García R 2011 *Phys. Rev. Lett.* **106** 198101
[58] Ando T 2013 *Microscopy* **62** 81
[59] Schitter G, Fantner G E, Kindt J H, Thurner P J and Hansma P K 2005 in: *Proc. IEEE/ASME Int. Conf. Advanced Intelligent Mechatronics.* (Monterey, CA, USA; 24-28 July 2005) (New York: IEEE) 261
[60] Ando T, Uchihashi T and Fukuma T 2008 *Prog. Surf. Sci.* **83** 337
[61] Picco L M, Bozec L, Ulcinas A, Engledew D J, Antognozzi M, Horton M A and Miles M J 2007 *Nanotechnology* **18** 044030
[62] Yong Y K, Moheimani SOR, Kenton B J and Leang K K 2012 *Rev. Sci. Instrum.* **83** 121101
[63] Kim B I and Boehm R D 2013 *Ultramicroscopy* **125** 29
[64] Payton O D, Picco L, Miles M J, Homer M E and Champneys A R 2012 *Rev. Sci. Instrum.* **83** 083710
[65] Klapetek P, Picco L, Payton O, Yacoot A and Miles M 2013 *Meas. Sci. Technol.* **24** 025006
[66] Tuma T, Lygeros J, Kartik V, Sebastian A and Pantazi A 2012 *Nanotechnology* **23** 185501
[67] Yong Y K, Moheimani S O R and Petersen I R 2010 *Nanotechnology* **21** 365503

IOP Conf. Series: Materials Science and Engineering **55** (2014) 012020 doi:10.1088/1757-899X/55/1/012020

A review of advances and challenges in EBSD strain mapping

A J Wilkinson[1], **T B Britton**[2], **J Jiang**[1] **and P S Karamched**[1]

[1] University of Oxford, Department of Materials, Parks Road, Oxford OX1 3PH, Great Britain

[2] Imperial College London, Department of Materials, Exhibition Road, London SW7 2AZ, Great Britain

E-mail: angus.wilkinson@materials.ox.ac.uk

Abstract. High sensitivity cross-correlation based analysis of EBSD patterns was introduced by Wilkinson, Meaden and Dingley in 2006 [1, 2]. This paper will describe the basis of the method and the various modifications and improvements that have been made to it over the past few years. Strain sensitivity of ~ 10^{-4} is readily obtained and sensitivity to lower strains is achievable if signal to noise ratio in the patterns is improved by simple integration. The method allows maps of local stress and dislocation density distributions to be generated in parallel with information concerning grain orientation, grain boundary misorientation and the presence of other microstructural features. The method is illustrated by example maps from a Ni-based superalloy and deformed Cu.

1. Introduction

Over the past decade dramatic advances have made in developing models capable of simulating the micromechanical response of complex polycrystalline and even multiphase engineering materials to thermal and mechanical loading. The aim is to develop models which successfully capture the detailed grain to grain interactions, and intra-grain variations in deformation that occur. In tackling failure processes (e.g., fracture, fatigue) the driving stress field and consequent material deformation is clearly localized which forces attention to a process zone near the crack tip or crack nucleation site. Furthermore, even when volume averaged responses such as flow stress are of interest the correct micromechanics is important if the effects of changes in microstructure are to be predicted. Experimental methods allowing measurement of stress, strain, and dislocation content at length scales commensurate with the microstructure are of vital importance in this effort in guiding model development and validating predicted responses.

Synchrotron, and neutron facilities have an important role to play in this respect, however, complementary laboratory-based techniques such as digital image correlation and EBSD are also essential. A step change in the sensitivity of EBSD was realized in 2006 by Wilkinson, Meaden and Dingley [1, 2] who introduced a cross-correlation-based analysis of EBSD patterns allowing elastic strain variations to be probed.

The first attempts to use EBSD to study local deformation made use of the loss of contrast and the blurring of the pattern brought about by the increased dislocation content within the diffraction volume [3-5]. The effect is analogous to peak broadening in X-ray and neutron diffraction data but the analysis remains much cruder for EBSD as the kinematic diffraction theory that can be utilized for the X-ray and neutron cases is not applicable for EBSD where the dynamical effects become prohibitive.

Content from this work may be used under the terms of the Creative Commons Attribution 3.0 licence. Any further distribution of this work must maintain attribution to the author(s) and the title of the work, journal citation and DOI.
Published under licence by IOP Publishing Ltd

The pattern blurring approach was largely dropped after automated analysis of EBSD patterns and orientation mapping became available through the nineties [6-9] as this enabled measurement of the extent of intra-granular lattice rotation. There have now been many studies [10-16] aimed at producing a single simple scalar metric to represent the extent of intra-granular lattice rotations measured over many grains. The metrics are then correlated with macroscopic engineering parameters such as equivalent plastic strain so as to enable assessment of damage development within plant and components extracted from service. It is possible to use the geometrical constructions developed by Nye [17] to estimate the density of geometrically necessary dislocations (GNDs) within the material from the measured lattice curvatures. Coupling this analysis with the higher sensitivity of the cross-correlation based measurements allows GND densities to be found to $\sim \pm 10^{12}$ m^{-2} at 200 nm step size albeit that because the EBSD data is two rather than three dimensional the measured values relate to a lower bound value [18, 19].

Initial studies using the high resolution EBSD technique for strain mapping concentrated on semiconductor structures which offer the advantages of relatively simple well-defined sample geometry, single crystal samples and generally much lower dislocation density than metallics [20-24]. However, applications to deformed metals is now being pursued in earnest and the method has already been used to map stresses near indents [19, 25, 26], at the head of block slip bands [27], near martensite islands in steels [28, 29], and in selected grains of stainless steel during in situ loading [30]. Generally, however where there has been extensive plastic deformation of metals the method has mostly been restricted to assessment of geometrically necessary dislocation (GND) density [1, 18, 19, 27, 31, 32] because large lattice rotations are thought to limit the accuracy with which the smaller elastic strains can be determined [33-35]. Recently, pattern remapping methods have been developed and demonstrated to improve significantly the measurement of small elastic strains in the presence of larger lattice rotations [33, 34] providing the opportunity to study intra-granular stress variations in deformed polycrystalline metals with improved fidelity.

2. Cross-correlation-based analysis of EBSD patterns

The basis of cross-correlation-based analysis of EBSD patterns is as follows. Elastic strain variations and small lattice rotations within the sample crystal create small shifts in the positions of zone axes and other features in the EBSD patterns. The magnitudes and directions of these shifts vary systematically with position across the EBSD pattern and equations can be written to relate the shifts to the size and nature of the strain and rotation state with the diffracting volume [1, 2]. Within each grain one pattern is selected as a reference pattern and all measurements are made relative to this. Absolute strains can be obtained only in situations where the reference pattern comes from a point on the sample for which the strain is known. When the reference pattern comes from a region with unknown strain then the map show only the relative variation in strain from the unknown reference state and knowledge of the absolute strain values is not possible. However, important information such as lattice curvatures and elastic strain gradients can still be determined in either case and the measured lattice curvatures used to recover the GND density distribution. In polycrystals, multiple reference patterns must be used to capture the strain gradients and lattice curvatures within each grain independently. With the larger datasets that are now studied methods for automatically selecting the reference points have been developed [18]. The issues resulting from the 'reference pattern' for EBSD in many ways parallel those resulting from the stress free lattice plane spacing d_0 in X-ray and neutron diffraction based methods.

Cross-correlation functions are calculated at a number of square sub-regions dispersed across the patterns from which the shift at each sub-region in the test pattern with respect to the reference pattern is determined. These calculations are performed via the Fourier domain as this is more efficient than direct calculation for the sub-region sizes generally used. We also take advantage of this and apply a band pass filter in the Fourier transforms to remove high frequency noise and low frequency background intensity variations. More details of the image analysis procedures are given in [1, 2].

The original work [1, 2] claimed that pattern shifts in 256 by 256 pixel sub-regions could be measured to ± 0.05 pixels or better and section 3 below will explore how pattern binning and averaging can improve this.

Measurement of shifts in x and y at 4 non-coplanar sub-regions of the pattern in principle provide all the information that can be obtained and allow eight of the nine possible degrees of freedom in a general strain and rotation to be calculated. Pure hydrostatic strain accounts for the final degree of freedom which cannot be detected by the technique, as strain of this type does not change any inter-zonal angles and thus does not cause shifts in the sub-regions of the pattern. Instead we rely on some simple mechanics to fully separate the three normal strains. The EBSD patterns originate from the topmost few tens of nanometres of material and allow us in most instances to assume that the local stress state must satisfy the equilibrium conditions for a free surface which must remain traction free. The stress normal to the free surface is thus set to zero and can be written as a linear sum of the elastic strains multiplied by the appropriate elastic constants found by transforming the stiffness tensor to the measured grain orientation.

In practice we tend to use many more than four (typical 50) sub-regions in the analysis so as to reduce noise and bias in the measurements. The use of more than four sub-regions leads to an over-determined set of linear equations to solve for the displacement gradient tensor components (strain and rotation). In the original work this was treated using a conventional least squares error solution [1, 2]. More recently [35] we have made use of the large redundancy in the data when 50+ sub-regions are used and have employed robust iterative solutions in which the influence of sub-regions for which the predicted shifts are furthest away from the measured values are 'down-weighted' in the next iterative solution until a convergence criterion is met.

The size and number of datasets obtained using the technique has grown considerably over the last few years and rather rapidly moved to a situation in which it is impossible for the investigator to make a visual check of each EBSD pattern. Thus it has been necessary to implement and use some simple scalar data quality parameters so that the influence of poor quality patterns can be automatically removed from the maps produced. The first of these uses the angular difference between the measured shift and those expected from the best fit solution for the displacement gradient tensor. These differences were simply averaged over all sub-regions in the initial formulation but with the robust statistical analysis the weighted mean angular error is used instead [35]. Where the measured strains and rotations are larger than the mean angular error there is good confidence in the measurement, but where they are smaller they are below the noise limit and must be considered suspect.

The second data quality parameter in use is the geometric mean of the normalized peak heights of the cross-correlation functions calculated for each sub-region. The normalisation is such that cross-correlation of the reference pattern with itself (i.e., auto-correlation) generates a peak height of one. Use of the geometric mean rather than arithmetic mean ensures that poor correlation at only a small number of sub-regions (perhaps due to shadowing) is sufficient to significantly reduce the geometric mean. Typically a threshold value within the range 0.2 to 0.4 is used and points with a geometric mean peak height below this threshold are considered suspect and are ignored in producing subsequent maps and further calculations.

When examining deformed polycrystalline metals Britton *et al.* [35] noted that the apparent elastic strains measured were often larger than expected and echoed the spatial patterning seen in the lattice rotation. They suggested that the apparent strain were likely to be in error as a result of finite deformation effects due to the lattice rotations which were often an order of magnitude larger than the elastic strain variations. When large lattice rotations are present the mapping between sub-regions on test and reference pattern is not so well represented by a simple linear translation, assumed by the cross-correlation process, and this results in significant errors being induced in the lattice strain variation [35]. To overcome this problem 'remapping' algorithms have been developed [33, 34]. Our approach [33] utilizes an initial pass of cross-correlation analysis to estimate the rigid body rotation between the test and reference lattice. This rotation is then used to remap the pattern intensity within the test pattern so as to bring it in closer (crystal) alignment with the reference pattern. A second pass of cross-correlation analysis is then undertaken and the elastic strains and a small correction to the rotation are determined.

3. Measurement sensitivity

The initial 2006 paper reported that using a Si sample the pattern shifts were measured to a precision of ~± 0.05 pixels corresponding to strain and rotation sensitivity of ~ ± 8 x 10^{-5} radians at the pattern centre [2]. The same authors noted an improvement in sensitivity to ~ ± 0.02 pixels and ~ ± 4 x 10^{-5} radians in similar measurement from GaAs which they ascribed to the higher signal to noise ratio obtained from the higher atomic number material [36].

Here we briefly report results of a systematic strain sensitivity study [37] of how the pattern shift sensitivity is affected by data collection, particularly exposure times and camera binning. Patterns were recorded using an 11 point line scan with a 10 μm step size along the tilt axis (x-axis), corresponding to the [110] direction of a Si (001) sample. This was repeated for camera binning levels of 1x1, 2x2, 4x4 and 8x8 for which corresponding exposure times of 1.04 s, 0.24 s, 0.05 s and 0.02 s were used. For each condition the line scan was repeated 5 times so that averaging over different numbers of frames could be used after data collection so as to vary the signal to noise ratio and study its effect on the shift measurements. Each line scan was treated as 10 pattern pairs with identical 10 μm steps between them. The standard deviation in x- and y-pattern shifts were calculated for these repeat measurements and these were then averaged over any repeat line scans available and used to calculate the Euclidian shift which we use as the pattern shift precision. The Euclidian shift is given in terms of fractions of a raw pixel in the unbinned patterns. For a single EBSD exposure using different hardware binning (figure 1a) the highest precision, ± 0.04 pixels, was obtained using no binning (1x1), and as expected the precision worsened significantly as the binning was increased, to ± 0.73 pixels with 8x8 binning.

Software binning was used to generate patterns at different binning levels but similar exposure times. Patterns at the same positions in the five line scans obtained at 1x1 binning were averaged and then software binned and figure 1b shows that at 1x1 and 2x2 binning similar precision of ± 0.018 and ± 0.019 pixels respectively were obtained, while with greater binning the precision deteriorates. Notably even for 8x8 binning the precision is better than ± 0.05 pixels once frame averaging has improved the signal to noise sufficiently. The similarity of the 1x1 and 2x2 binning results suggests that there is in fact little real gain in information content afforded by the additional pixels for this detector. Finally, the detector binning was fixed at the highest resolution 1x1 binning and the effect of software integration for increasing total exposure times was examined. Figure 1c shows that the precision improved with increasing exposure time but the response was non-linear with an ultimate precision of ± 0.019 pixels with five seconds exposure. Sensitivity gains beyond this would require drastic increases in total exposure time and would not realistic for mapping operations. The ± 0.019 pixels reported here for Si corresponds to an angular resolution at the pattern centre of ~ ± 4 x 10^{-5} radians.

4. Two illustrative applications

4.1. Thermal residual strains near a carbide in a superalloy

We have used the cross-correlation-based EBSD method to study the thermally induced deformation around carbide precipitates in a nickel-based superalloy (Mar-M-002). The final step of the heat treatment was ageing for sixteen hours at 870 °C and cooling to room temperature. The difference in thermal expansion coefficients between the matrix and carbides, of order 10^{-5} K^{-1}, should lead to significant thermal residual strains, of order 10^{-3}. A JEOL JSM6500 FEG SEM equipped with TSL digivew II based EBSD system was used to acquire EBSD patterns on a square grid with a 500 nm step size, which is well above the spatial resolution of the method. The strain-free reference point was selected toward a corner of the map in a position remote from the carbide.

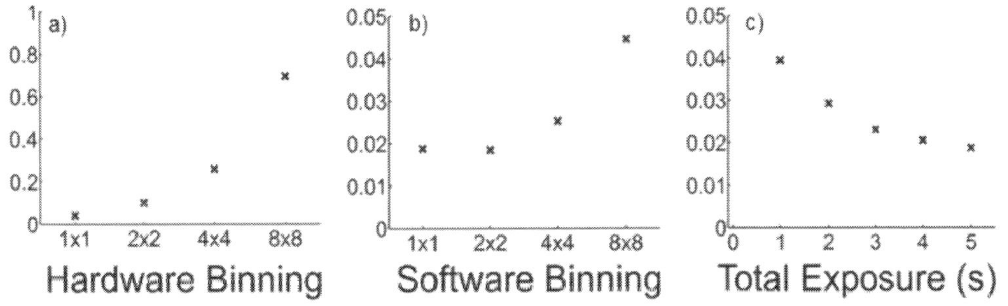

Figure 1. Pattern shift precision in raw unbinned pixels as a function of: a) on camera binning (varying exposure time); b) software binning (fixed exposure time); c) exposure time (fixed 1x1 binning).

Maps of the data quality parameters in the matrix region surrounding the selected carbide are shown in figure 2. The cross-correlation peak height map in figure 2a indicates a good match (peak height above 0.8) between the test patterns and the reference pattern throughout the Ni matrix, while the different pattern generated in the large carbide toward the centre of the frame results in low values of 0.3 and below. There are a number of localized low values toward the bottom right of the map which correspond to a group of smaller carbides. The carbide is also obvious through large values in the mean angular error map (figure 2b) while throughout the matrix regions the angular errors are below ~ 3×10^{-4}. Data within the matrix region was considered good and shown in subsequent analysis if the peak height was above the threshold value of 0.4.

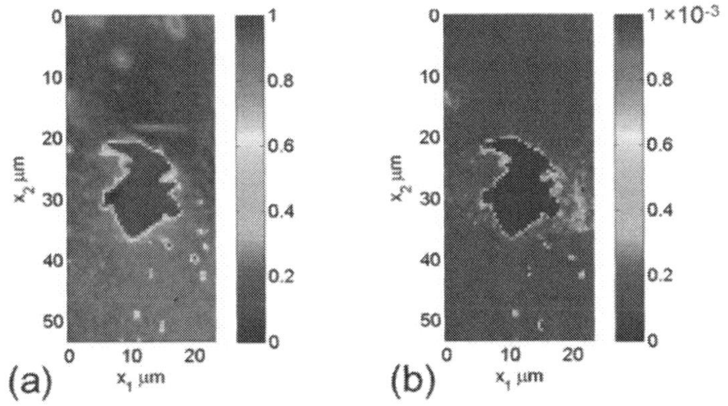

Figure 2. a) Geometric mean of the normalized cross-correlation peak heights. b) Mean angular error (radians).

The thermally induced strain field (figure 3) around the large central carbide particle is far from symmetric presumably due to the complex particle shape and the distribution of fine carbides around it both in and below the section plane. The normal strain ε_{11} along the horizontal x_1 axis is compressive to the left and right of the carbide and tensile above and below it, while the ε_{22} strain along the vertical axis shows strains with the opposite sign. This is broadly consistent with a radial compressive strain and tensile hoop strain that is expected to develop during cooling as the matrix contracts more strongly than the carbide it surrounds. The in-plane shear strain ε_{12} distribution shows maximum positive and negative values at ~ +45° and ~ -45° around the carbide, which is again generally in accord with expectation. The two out of plane shear strains remain close to zero in accord with the expectation that the corresponding out of plane shear stresses should be zero to comply with the traction free surface condition. Fuller analysis and related crystal plasticity modelling is available [38-41].

$$\begin{bmatrix} \varepsilon_{11} & \varepsilon_{12} & \varepsilon_{13} \\ & \varepsilon_{22} & \varepsilon_{23} \\ & & \varepsilon_{33} \end{bmatrix} =$$

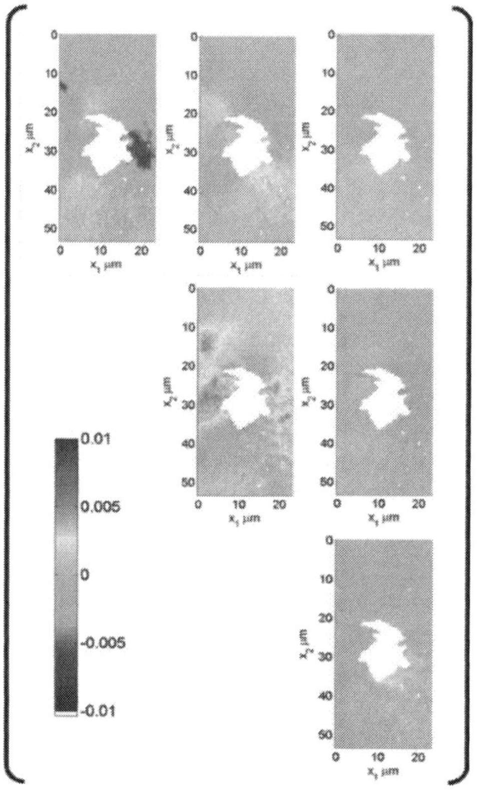

Figure 3. Thermally induced strains near selected carbide particle. One map is shown for each component of the strain tensor.

4.2. Tensile deformation of a polycrystal

The second example concerns residual strains, lattice rotations and dislocation density with a Cu polycrystal deformed in tension to 6 % plastic strain. We focus on the distribution within a selected grain that is close to the cube orientation which is shown with its immediate neighbourhood in figure 4a. The reference pattern was taken from a point toward the middle of the grain (for which the strain is unknown) and where the mean angular error measured for a first pass of the cross-correlation analysis is low (figure 4b). However, the considerable orientation spread within the grain (figure 5a) causes a noticeable increase in the mean angular error toward the left and right hand ends of the grain. The higher mean angular errors were reduced by remapping the intensities in the test patterns so as to affect a virtual rotation of the crystal back to the reference orientation and then applying a second pass of the cross-correlation analysis (figure 4c). This remapping algorithm is described briefly at the end of section 2, and in more detail by Britton and Wilkinson [33].

Problems with the first pass of the cross-correlation analysis are also suggested in figure 5. The lattice rotations (represented by a single component w_{12} in figure 5a) are generally much larger than the elastic strains (again represented here by the in plane shear strain e_{12} in figure 5b). There is also some suggestion that the elastic strains are mimicking the spatial pattern shown by the rotation field, and are taking very large values over appreciable areas. Application of the pattern remapping

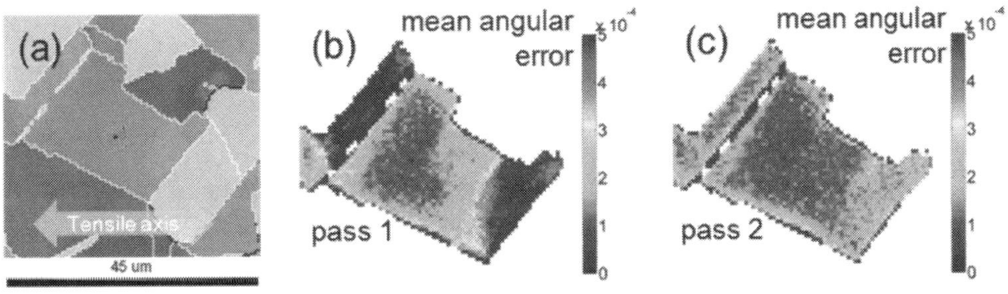

Figure 4. a) Inverse pole figure map showing the selected near to cube orientated grain and its neighbourhood. Maps showing mean angular error (rads) from initial cross-correlation analysis b), and from a second pass after pattern remapping c).

algorithm and a second pass of the cross-correlation analysis reduces many artefacts from the elastic strain map (figure 5c). At the left and right hand ends of the grain it is seen that the large magnitude of the shear strains is significantly reduced and the sense of the shear is inverted. The histogram of shear strain distribution (figure 5d) as calculated at the first and second pass of cross-correlation analysis also show the reduction in shear strain magnitudes brought about by the remapping. On this plot the green line represents the elastic strain calculated after remapping following a simple (infinitesimal deformation theory) linear decomposition of rotation and strain for the second pass of cross-correlation. For the blue line the deformations extracted at both passes of the cross-correlation are combined and a multiplicative polar separation (finite deformation theory) into a finite rotation tensor and a Green's strain tensor. In most instances we have examined including this particular instance there is very little difference between these two results.

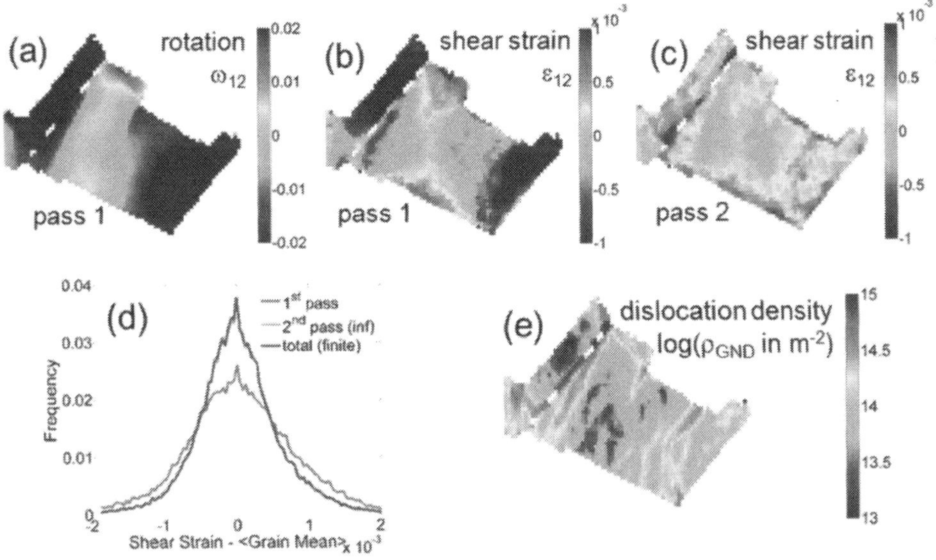

Figure 5. a) In plane lattice rotations in the selected grain. Comparison of in plane elastic shear strain variations from b) single pass of cross-correlation analysis, and c) after pattern remapping and a second cross-correlation analysis. d) Histograms comparing shear strain distributions from first and second pass analyses. e) Map of the GND density.

The gradients of the lattice rotations can be used within Nye's dislocation tensor framework to form a lower bound estimate of the dislocation density at each point in the map. The analysis is given in full in [19] which also includes some discussion of numerical methods employed. Regions of high dislocation density are seen to correspond well to regions in which the lattice rotation varies most rapidly. There is an obvious tendency for dislocations to accumulate in high density regions separated by regions of low dislocation density. For this section through this grain the high GND density regions are arranged in thin wall-like features that span across the grain and tend to emanate from triple junctions or the end of narrow twins. Correlations between the spatial variations of GND density and elastic strains are much less obvious and to understand these will require much more detail analysis, including the dislocation types and the tensor nature of the stress field.0

5. Summary and outlook

High resolution EBSD provides a useful bridge between X-ray and neutron techniques on the one hand and very localized transmission electron microscope methods on the other. The example applications given here illustrate the considerable potential of the method for studying structural metallics. The main strengths of the technique are (i) that all components of the 3D strain and rotation tensors are obtained in parallel, (ii) the use of an electron beam allows very small volumes of material to be probed, and (iii) the equipment required is rather widely available. The detailed information shown in the examples can be acquired with 1 - 2 hours of SEM time and of course much larger, richer and more microstructurally representative datasets can be obtained with overnight scans. A disadvantage is that the depth probing capabilities of X-ray, in particular synchrotron methods, are clearly not available.

The major challenge in further developing the technique is the need for a reference pattern. In situations where the strain is generated at some localized feature (such as the carbide particle in section 4.1) then a reference may be obtained by moving the probe away from the centre of strain. However, in the more general cases (e.g., polycrystal deformation in section 4.2) it is not possible to identify a position on the sample at which the strain state is known. In these cases the strains reported must either be relative to the reference position or show the variation from the grain averaged strain state. The reference pattern issue has many parallels with the need for stress free lattice parameters in X-ray and neutron diffraction studies. Dynamical diffraction based simulations of EBSD patterns such as those undertaken by Winkelmann [42] may eventually resolve this issue, however, current methods of determining the detector geometry are not sufficiently accurate and can induce errors as large as 10^{-2} [43].

Acknowledgements

The authors wish to thank Dr. Graham Meaden (BLG Productions) and Prof. David Dingley (University of Bristol) for continued enlightening discussions on all aspects of EBSD. We are grateful to the Engineering and Physical Sciences Research Council for funding through grants EP/H018921/1, EP/E044778/1, and EP/C509870/1.

References

[1] Wilkinson A J, Meaden G and Dingley D J 2006 *Mater. Sci. Technol.* **22** 1271-1278
[2] Wilkinson A J, Meaden G and Dingley D J 2006 *Ultramicroscopy* **106** 307-313
[3] Quested P N, Henderson P J and McLean M 1988 *Acta Metallur.* **36** 2743-2752
[4] Wilkinson A and Dingley D 1991 *Acta Metallur. Mater.* **39** 3047-3055
[5] Wilkinson A J and Dingley D J 1992 *Acta Metallur. Mater.* **40** 3357-3368
[6] Dingley D J 1989 *Inst. Phys. Conf. Series* **98** 473-476
[7] Juul Jensen D and Schmidt N H 1990 in: *Proc. Recrystallisation '90* 219-224
[8] Krieger Lassen N C, Juul Jensen D and Conradsen K 1992 *Scan. Microsc.* **6** 115-121
[9] Wright S I and Adams B L 1992 *Metallur. Trans. A* **23** 759-767
[10] Brewer L *et al.* 2006 *J. Mater. Res.* **21** 1775-1781

IOP Conf. Series: Materials Science and Engineering **55** (2014) 012020 doi:10.1088/1757-899X/55/1/012020

[11] Kamaya M 2009 *Mater. Character.* **60** 1454-1462
[12] Kamaya M, Wilkinson A J and Titchmarsh J M 2005 *Nucl. Engng. Design* **235** 713-725
[13] Kamaya M, Wilkinson A J and Titchmarsh J M 2006 *Acta Materi.* **54** 539-548
[14] Lehockey E, Lin Y and Lepik O 2000 Mapping residual plastic strain in materials using electron backscatter diffraction. in: *Electron backscatter diffraction in materials science.* (Schwartz A J, Kumar D P F M and Adams B L; Eds.). (New York: Kluwer-Plenum) 247-264.
[15] Wu X, Pan X and Stubbins J 2007 *J. Nucl. Mater.* **361** 228
[16] Yoda R, Yokomaku T and Tsuji N 2010 *Mater. Character.* **61** 913
[17] Nye J F 1953 *Acta Metallur.* **1** 153-162
[18] Jiang J, Britton T B and Wilkinson A J 2013 *Ultramicroscopy* **125** 1-9
[19] Wilkinson A J and Randman D 2010 *Phil. Mag.* 90 1159-1177
[20] Ishido T, *et al.* 2007 *Ieice Electronics Express* **4** 775-781
[21] Tomita M, *et al.* 2011 *Jap. J. Appl. Phys.* **50** 010111
[22] Wilkinson A J 2006 *Appl. Phys. Lett.* **89** 241910
[23] Vaudin M D, *et al.* 2008 *Appl. Phys. Lett.* **93** 193116
[24] Villert S, *et al.* 2009 *J. Microsc.-Oxford* **233** 290-301
[25] Britton T B, *et al.* 2010 *Proc. Royal Society a-Mathem. Phys. Engng. Sci.* **466** 695-719
[26] Wilkinson A J and Britton T B 2012 *Materials Today* **15** 366-376
[27] Britton T B and Wilkinson A J 2012 *Acta Mater.* **60** 5773-5782
[28] Miyamoto G, *et al.* 2009 *Acta Mater.* **57** 1120-1131
[29] Wilkinson A J, *et al.* 2010 *J. Strain Anal. Engng. Design* **45** 365-376
[30] Ojima M, *et al.* 2011 *Acta Mater.* **59** 4177-4185
[31] Littlewood P D, Britton T B and Wilkinson A J 2011 *Acta Mater.* **59** 6489-6500
[32] Littlewood P D and Wilkinson A J 2012 *Acta Mater.* **60** 5516-5525
[33] Britton T and Wilkinson A 2012 *Ultramicrscopy* **114** 82-95
[34] Maurice C, Driver J H and Fortunier R 2012 *Ultramicroscopy* **113** 171-181
[35] Britton T B and Wilkinson A J 2011 *Ultramicroscopy* **111** 1395-1404
[36] Wilkinson A J, Dingley D J and Meaden G 2009 Strain mapping using electron back scatter diffraction. in: Electron backscatter diffraction in materials science. (Schwartz A J, *et al.*; Eds.); (Berlin-Heidelberg: Springer) 406
[37] Britton T B, *et al.* 2013 *Ultramicroscopy*, accepted for publication
[38] Dingley D J, *et al.* 2010 *J. Electron Microsc.* **59** S155-S163
[39] Dunne F P E, Kiwanuka R and Wilkinson A J 2012 *Proc. Royal Society a-Mathem. Phys. Engng. Sci.* **468** 2509-2531
[40] Karamched P S and Wilkinson A J 2011 *Acta Mater.* **59** 263-272
[41] Kartal M E, Dunne F P E and Wilkinson A J 2012 *Acta Mater.* **60** 5300-5310
[42] Winkelmann A, *et al.* 2007 *Ultramicroscopy* **107** 414-421
[43] Britton T B, *et al.* 2010 *Ultramicroscopy* **110** 1443-1453

Quantitative X-ray mapping, scatter diagrams and the generation of correction maps to obtain more information about your material

R Wuhrer[1] and K Moran[2]

[1] University of Western Sydney, Advanced Materials Characterisation Facility (AMCF), .O. Bag 1797, Au-2751 Penrith, New South Wales, Australia
[2] Moran Scientific Pty. Ltd., 4850 Oallen Ford Road, AU-2580 Bungonia, New South Wales, Australia

E-mail: richard.wuhrer@uws.edu.au

Abstract. Quantitative X-ray mapping with silicon drift detectors and multi-EDS detector systems have become an invaluable analysis technique and one of the most useful methods of X-ray microanalysis today. The time to perform an X-ray map has reduced considerably with the ability to map minor and trace elements very accurately due to the larger detector area and higher count rate detectors. Live X-ray imaging can now be performed with a significant amount of data collected in a matter of minutes.

A great deal of information can be obtained from X-ray maps. This includes; elemental relationship or scatter diagram creation, elemental ratio mapping, chemical phase mapping (CPM) and quantitative X-ray maps. In obtaining quantitative x-ray maps, we are able to easily generate atomic number (Z), absorption (A), fluorescence (F), theoretical back scatter coefficient (η), and quantitative total maps from each pixel in the image. This allows us to generate an image corresponding to each factor (for each element present). These images allow the user to predict and verify where they are likely to have problems in our images, and are especially helpful to look at possible interface artefacts.

The post-processing techniques to improve the quantitation of X-ray map data and the development of post processing techniques for improved characterisation are covered in this paper.

1. Introduction

X-ray mapping (XRM) was first developed by Duncumb and Cosslett in 1956 [1-2] and is the collection of characteristic X-rays as a function of the position of the scanning electron beam on the specimen [3]. This analytical technique provides an image related to the distribution and relative abundance of elements within a given specimen. This capability makes XRM particularly useful for; (i) identifying the location of individual elements, and (ii) mapping the spatial distribution of specific elements and chemical phases within a sample [4]. Figure 1 shows the backscattered electron (BSE) image and regions of interest X-ray maps (ROIM) from a hard-facing white cast iron with niobium carbide (NbC) particles incorporated. X-ray mapping has become one of the most popular and useful methods of X-ray microanalysis today used extensively in areas of materials science, chemistry, physics, geology, environmental sciences and biology.

Content from this work may be used under the terms of the Creative Commons Attribution 3.0 licence. Any further distribution of this work must maintain attribution to the author(s) and the title of the work, journal citation and DOI.
Published under licence by IOP Publishing Ltd

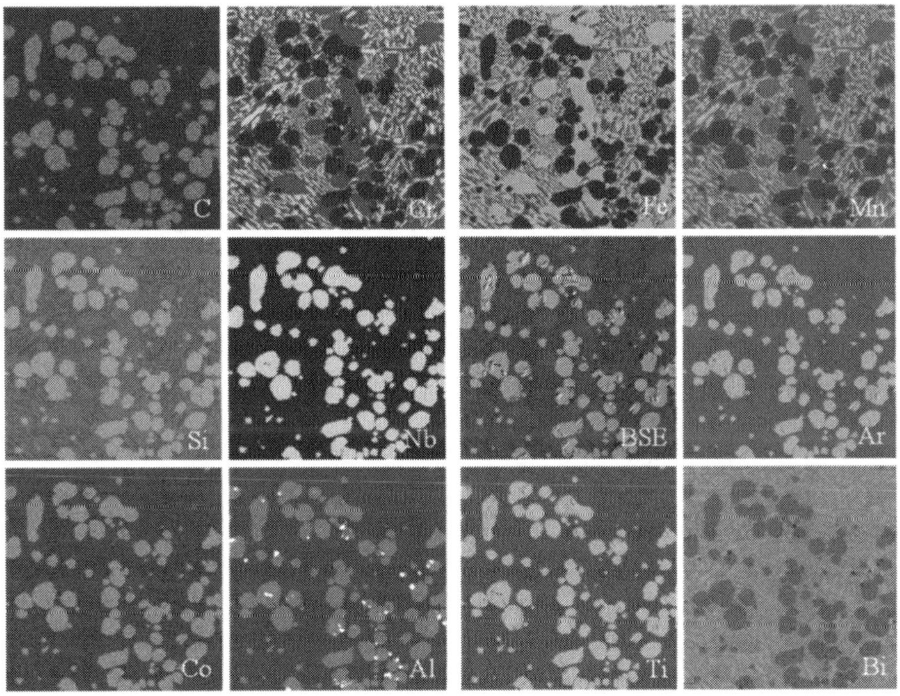

Figure 1. Regions of interest X-ray maps (ROIM) of a white cast iron hard-facing material with NbC particles. Map collected using a Be-window SDD at 20 keV, 512x512 pixels, 120 msec. X-ray maps for carbon, argon, cobalt, titanium and bismuth are artefacts. The detector cannot see carbon due to the Be-window, but does see Nb-Ln escape peak within the carbon ROI. Horizontal width of field (HWOF) = 300 μm.

X-ray mapping, in the past, was a slow process often being performed overnight. With the development of silicon drift detectors (SDD), it is possible to obtain live X-ray images. Currently there are very large area SDD's as well as multiple SDD detector systems capable of handling millions of counts per second. This has made XRM a more popular technique for characterising a material's microstructural properties by determining elemental distributions and phases in materials with subsequent chemical phase mapping (CPM) [3-5].

Silicon drift detectors allow the electron microprobes to operate at higher beam currents, which are more suited to wavelength-dispersive X-ray spectroscopy (WDS). Higher spatial resolution mapping is possible due to larger detection areas, improvements in the field emission guns (FEG) and lower keV mapping.

High pixel resolution X-ray maps of 512x512 pixels, with excellent counting statistics, can be acquired in a couple of hours. However, simple elemental X-ray maps can be misleading, so further post-processing (data manipulation) techniques may be required to obtain more information about a material's unique properties.

Our research aims at developing post processing techniques to improve the quantitation of X-ray map data and to develop further post-processing techniques for improved characterisation as an aid in assessing the practical properties of complex materials. These techniques include:

1. Atomic number, absorption and fluorescence (ZAF) corrected quantitative X-ray maps (QXRM), where the QXRM enables reliable quantitative results, that when grouped, can be an order of magnitude better than traditional analysis [6].
2. Scatter diagrams and rotational scatter diagrams for locating phases [7-8].
3. Pseudo colouring for locating phases. The colours of the elements can be rotated, which can often reveal further information (and other phases) in the material [4].
4. Elemental ratio maps. This is a powerful technique that helps to reveal subtle chemical variations in a sample. Ratio maps can also be created from different combinations of elements and from the different background regions near elements of the spectra [9].
5. Theoretical BSE maps. When generating quantitative X-ray maps, the computer can easily calculate the theoretical BSE coefficient for each point on the map. The comparison between the real and theoretical BSE map (image) gives us information about what is going on just below the surface of the sample [8].
6. ZAF maps, which give information on the correction that is required at specific points, or interfaces, of the different regions in the maps [9].
7. Totals mapping (wt%). The total at each point of the map can be calculated and mapped. These total maps give an idea of the quality of the analysis and also make it easy to determine problems such as missing elements and interaction volume effects [9-10].
8. Filtering to make phase selection easier, to produce cleaner images and aid in trace element location.
9. Identification of artefacts. Those effects in X-ray maps that should not be there.
10. X-ray map resolution determination through the use of Smart J package [11-12], can be used to do X-ray map resolutions for various elements compared to Monte Carlo simulations [13], aiding in determining interfacial boundary issues.

This paper will discuss several of the above mentioned techniques, all of which aid in obtaining a better understanding of a material's chemical and microstructural properties.

2. Experimental method

Current research work involves the use of a number of different SEM/EDS and microprobes. These include a multi-detector EDS and WDS X-ray mapping system installed on a JEOL 733 microprobe with two SDD-EDS X-ray detectors and four WDS detectors, a JEOL 8600 microprobe with three WDS detectors and an Si(Li) EDS X-ray detector and a dedicated X-ray mapping system, which is a JEOL 840 SEM with an UTW Si(Li) EDS detector all operated with a Moran Scientific microanalysis system.

3. Results and discussion

3.1. ZAF corrected quantitative X-ray maps (XRM versus QXRM)

Electron beam induced X-ray mapping (XRM), whether using EDS or WDS, has become a very useful characterisation tool for determining the elemental distribution in materials. There are many types of X-ray maps, which include 1) dot mapping, 2) regions of interest (ROI) mapping, 3) background subtracted and overlap corrected mapping, 4) quantitative mapping, and 5) full spectrum mapping (or spectrum mapping) [4, 6].

Regions of interest mapping, is where you set up an energy range (or number of channels) where an elemental line exists to see the distribution of the element of interest. Regions of interest are assigned as a single channel analyser (SCA) and a different colour is assigned for different elements [1, 6, 14]. X-ray maps are usually collected using raw counts from the elemental peaks of interest, however traditional ROIM does not always provide a completely accurate description of the material phases or elemental distribution as the integrated peak counts are only partially related to the true composition of the material and in fact may not be related at all [6]. The ROIM is very problematic when looking at moderate to low levels of elements and the accuracy of ROIM is dependent on what elements are present in the material. If the elements are present in reasonable amounts and have no overlaps with other elements then the ROIM may be near to correct. However, great care must be taken for the appearance of other artefacts [15], such as escape peaks and sum peaks that can occur on the EDS spectrum. Figure 1 shows several RIOM maps of the hard-facing white cast iron with NbC particles. However, some of the elements are not really present but represent a combination of density distributions, sum peaks, overlap errors and escape peaks. In this example; carbon is displaying the Nb-Ln escape peak generated in the carbon ROI; chromium, iron and niobium are substantially correct; manganese is predominantly the Cr-Kβ; silicon, argon, cobalt and titanium are predominantly a density distribution; bismuth is a Cr sum peak, and aluminium is a mix of real and density; titanium also contains the Nb-Lβ_1 sum peak.

For a true spatial elemental distribution we require background subtraction, overlap correction and inter-element correction. Quantitative X-ray mapping (QXRM) is where an X-ray intensity map for an element is corrected through standards based ZAF or another matrix correction procedure revealing the quantitative composition at each pixel location. Thus, the quantitative maps shown in figure 2 reveal the correct spatial element distribution. Generating ZAF corrected quantitative X-ray maps enables reliable quantitative results that can be an order of magnitude better than traditional analysis and is also far superior to ROIM where low levels of an element or elemental overlaps are present [6]. There are also extreme examples where the actual composition shows reverse correlation to the intensity such as with carbon in NbC compared with M_7C_3 high-chrome carbide due to absorption by niobium.

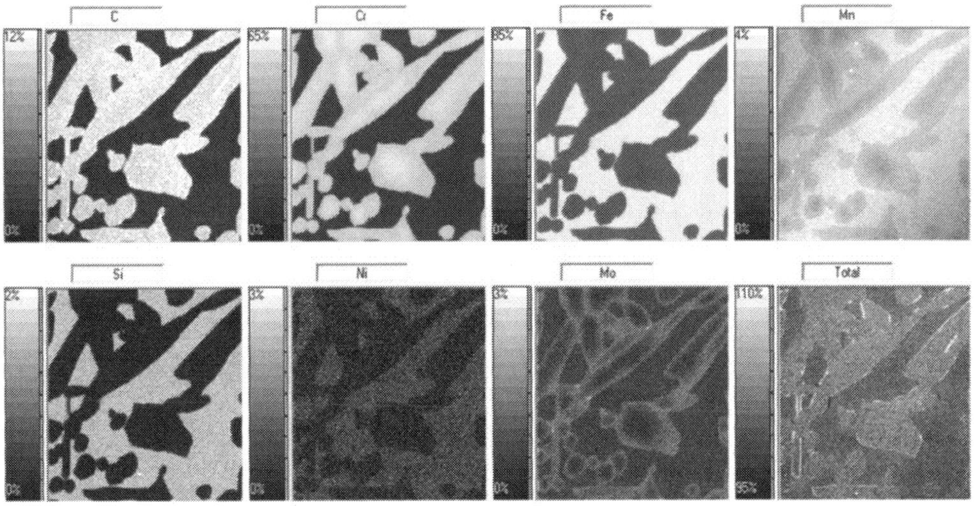

Figure 2. Quantitative X-ray maps of a white cast iron alloy, where the elemental maps have been corrected through standards based ZAF matrix correction procedures revealing the quantitative composition at each pixel location. Map collected using combined SDD-EDS (Be-window) and WDS at 20 keV, 512x512 pixels, 120 msec. Carbon, manganese and molybdenum were mapped using WDS. HWOF = 65 μm.

X-ray mapping and QXRM uses chemistry through use of either EDS and/or WDS analysis. For accurate QXRM, all elements need to be mapped, as some chemical phases are determined from very minor elemental variation and even elements that are difficult to analyse. Full spectrum X-ray mapping (FSXRM) which saves a spectrum for each pixel of an image makes the process of looking for other elements much easier at a later stage. The elements of interest do not need to be identified before starting the map, allowing the user to acquire data and recall all information at a later time away from the microscope. This form of mapping allows the user to produce new maps, construct spectra from any region of the image and quantitatively analyse these regions [16].

3.2. Scatter diagrams

The X-ray maps that are generated quite often have so much quantitative information that it is extremely complex to visually interpret them. Other techniques are required to correlate elements and phases in materials. One technique involves the use of scatter diagrams, which are pixel frequency versus element concentration profiles plotted against each other in two dimensions for selected elements within the sample [7]. Other terms used for scatter diagrams are concentration-concentration histograms (CHI) [17] or correlation diagrams. The scatter diagrams may be derived from ROIM or QXRM.

Scatter diagram generation is best understood by considering the information given in the linear intensity distribution histogram from each individual element X-ray (figure 3). An intensity histogram shows the frequency with which each value of the image intensity occurs against the full range of intensity values. From this information, a two-dimensional scatter diagram is produced, which is a plot of the frequency of occurrence of intensity values for one element against another for each pixel in the X-ray map (figures 3 and 4).

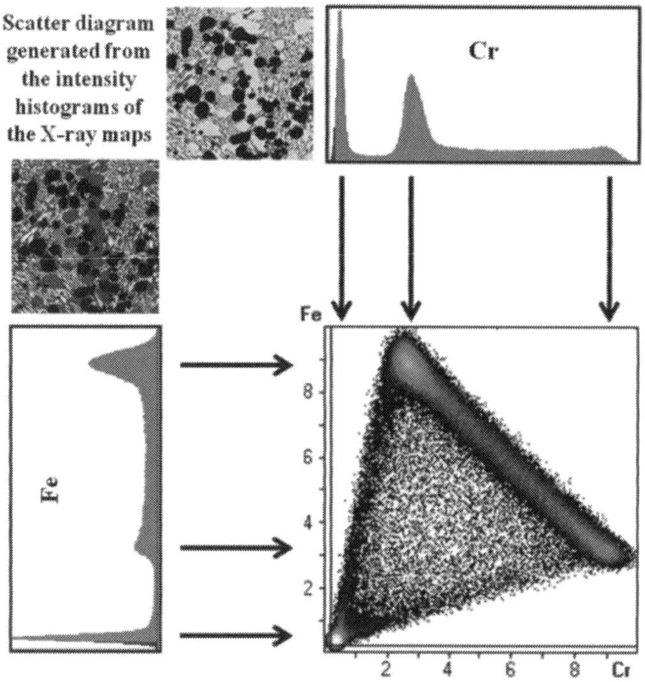

Figure 3. Scatter diagram generated from the chromium distribution plotted against the iron distribution. The vertical scale of the intensity histograms, represent the number of pixels and the horizontal scale represents the intensity of the pixel. The arrows indicate corresponding phases as clusters of points.

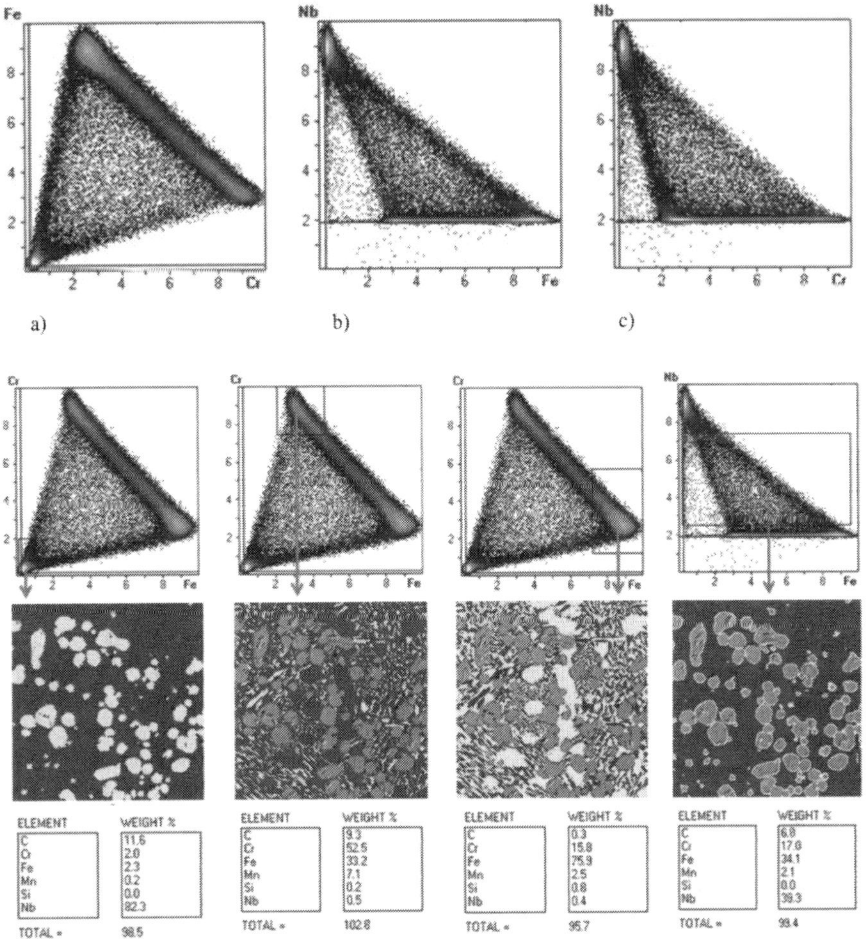

d)

Figure 4. A set of scatter diagrams showing the different clusters for the X-ray maps shown in figure 1. a) Iron versus chromium, b) niobium versus iron, and c) niobium versus chromium, d) scatter diagram cluster selection and the points then displayed and redrawn on the electron image with the quantification results from the selection of the areas. These clusters have a two sigma variation of ± 5 % relative. The quantification was calculated using combined EDS and WDS data.

These scatter diagrams provide a visual means of determining how elements are associated with one another in the specimen and make it easier to understand X-ray maps. There are many features observed when scatter diagrams are created. From these scatter diagrams we observe clusters, also referred to as nodes, which correspond to different chemical phases. The type of clustering observed in the scatter diagram, whether spherical, oval or linear, can give more information about the major and trace element distributions within phases as well as qualitative and quantitative phase information [7] through looking at the different node types, size and shape.

For any point on the scatter diagram of figure 3, the vertical axis of the scatter diagram shows the iron X-ray intensity at a pixel, whereas the horizontal axis shows the chromium X-ray intensity at the same pixel. The intense group of pixels at the bottom (co-ordinates 0, 0) corresponds to the low-Fe and low-Cr pixels, that is, the niobium carbides. Referring to figure 4, the reconstructed trace back (redrawn) image of these pixels shows the NbC (left image). This image shows the selected analysis points superimposed on the electron image which was collected while mapping, giving perfect correlation. The redrawn image from the high-Fe region of the scatter diagram reveals the matrix and the redrawn image from the high-Cr region of the scatter diagram reveals the M_7C_3-carbides.

Selecting areas on the scatter diagrams and observing where the points lie on the image is a very important part of the phase identification process. These selected analysis points may then be summed for a more accurate analysis in total or by selecting strategic areas on the image. Selecting areas on the electron or X-ray image, and showing where these points plot (retrace) on the scatter diagrams, is also important in locating "missing" clusters.

We can also observe linking (connection) between clusters indicating the boundaries between phases within a material as well as branching from clusters (some links may contain branches). Branching is defined as a link with a node at one end only, and is often referring to solid solutions (compositional variations) of the same phase or separate phases of similar chemistry [7]. Figure 4d shows that by selecting the links between the clusters, this linking is actually the boundary between the NbC particles and the matrix white cast iron and is most likely an artefact.

3.3. Colouring

Usually X-ray maps are displayed as grey scale images. The use of colour can enhance the visibility of intensity levels in a map because the human eye can distinguish many more shades of colour than grey levels [16].

Pseudo-colouring is often referred to as false colour and is used to display different elements. However, pseudo-colouring can be very helpful for locating phases, as shown in figure 5a. Pseudo-colouring is a simple method for determining elemental associations and is also called "primary colour imaging" [16]. This technique is where three elemental maps have been assigned the colours red, green and blue. Because the colours red, green and blue form specific colours when combined on a computer graphics monitor, elemental associations are clearly shown in a single pseudo-colour image. Figure 5 shows a pseudo-colour image of a polished section of a hard-facing white cast iron with NbC particles in which Fe has been assigned the colour red, Cr green, and Nb blue. From the combination of colours, it is clear that the orange/yellow region is a phase that contains both Fe and Cr, whereas if there were pink/purple regions this phase would contain both Fe and Nb.

The rotation of the colours between elements can often reveal further information (and other phases) in the material (elemental association), such as fine precipitates, trace elements, hairline cracks and small boundary interfaces that would be otherwise missed [7]. Indeed, the higher magnification pseudo-coloured X-ray map of the hard-facing white cast iron and rotating between different elements and colours reveals coring in some of the carbides in different areas of the hard-facing material (figure 5b).

3.4. Elemental ratio mapping

Elemental ratio mapping is a powerful technique that helps to reveal subtle chemical variations in a (figure 6). Ratio maps can also be created from different combinations of elements and from the different background regions near elements of the spectrum.

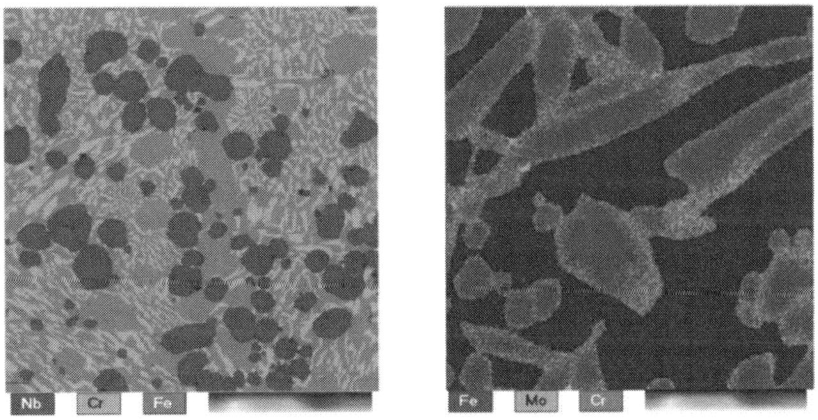

a) b)

Figure 5. a) Pseudo-colour map for the elements (Fe-red, Nb-blue and Cr-green). Maps collected at 20 keV, 512x512 pixel, 120 msec/pixel. HWOF = 300 μm. b) Higher magnification pseudo-colour map of another white cast iron for the elements (Cr-red, Fe-blue and Mo-green). HWOF = 65 μm.

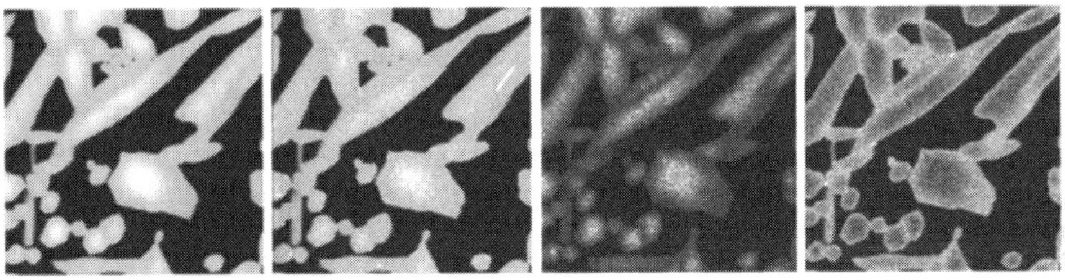

a) b) c) d)

Figure 6. Elemental ratio mapping of the quantitative X-ray maps shown in figure 2. a) Ratio map of chromium to iron (Cr/Fe), b) ratio map of carbon to iron (C/Fe) showing changes around the eutectic carbides, c) ratio map of chromium to molybdenum (Cr/Mo), and d) ratio map of molybdenum to iron (Mo/Fe). Map collected at 20 keV, 512x512 pixels, 120 msec, HWOF = 65 μm.

3.5. Theoretical BSE maps

When generating elemental X-ray maps, the computer can easily calculate the theoretical BSE coefficient (η), by looking at the quantitative data, and use this to generate the BSE coefficient for each point on the map (figure 7). The theoretical BSE map (or image) gives us information on what is going on just below the surface of the sample. The ratio of real BSE to the generated BSE can be used to identify problems such as interaction problems and ZAF correction problems.

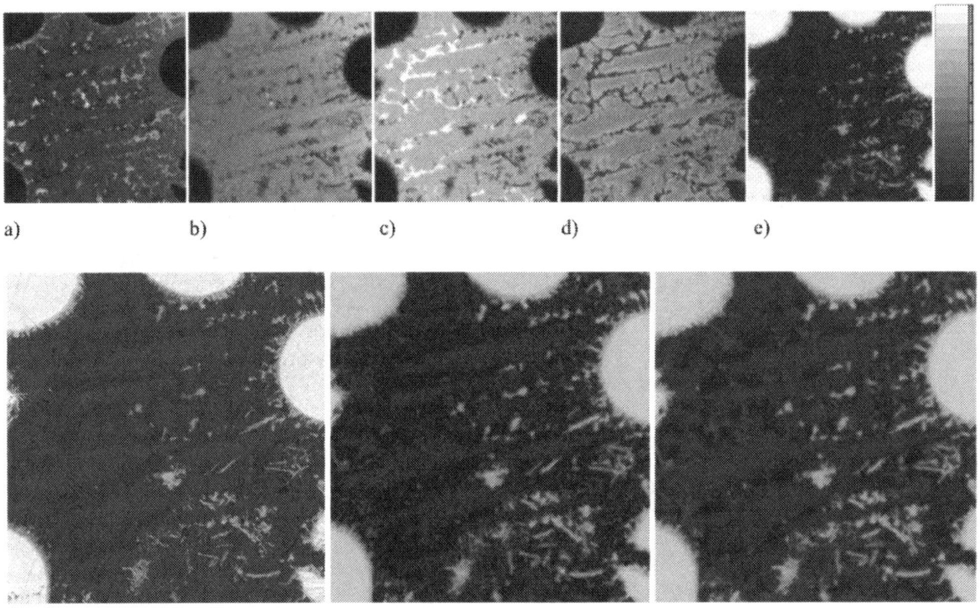

a) b) c) d) e)

f) g) h)

Figure 7. Quantitative (wt%) elemental X-ray maps produced from a hard facing bonded to a chrome steel. a) Cr 0 - 20 %, b) Fe 0 - 30 %, c) Ni 0 - 70 %, d) Cu 0 - 30 %, e) W 0 - 100 %, and BSE images using f) electrons, g) calculated BSE coefficient (η) using W-Lα, and h) calculated η using W-Mα. Map collected using a UTW-window Si(Li) EDS detector at 20 keV, 512x512 pixels, and 12 hours. HWOF = 100 μm.

Figures 7a-e show the quantitative elemental x-ray maps produced from a hard facing bonded to a chrome steel. Figures 7f-h shows three different BSE images, first from electrons, second from calculated η using W-Lα and third from calculated η using W-Mα respectively. The three BSE images show the effect of the depth distribution of the different signals that make up the images. The BSE image from high energy electrons shows the surface detail and atomic number differences to a depth of about 0.02 μm. The calculated η-image (from summing C_iZ_i) using W-Lα (8.396 keV), shows composition to a depth distribution of about 0.7 μm. The calculated η-map using W-Mα (1.777 keV), shows composition to a depth distribution of about 1 μm.

The electron image is sharp and shows surface scratches, whereas the theoretical image is diffuse from the subsurface effect of the excitation volume. The differences in the images can be quite dramatic. The calculated η- images are also affected by statistical variations, especially with the W-Lα line. This line is a low intensity line compared with the W-Mα line. There is almost a 3D effect (W-Mα line) in the calculated η-image, which can be seen when looking around the dendritic growths. The change in spatial resolution is also obvious between the three images.

Another technique to obtain a better understanding of the effect of low energy and high energy theoretical BSE mapping is to look at the ratio of the theoretical BSE to the Kα-line BSE (BSE/K BSE ratio map). Figure 8 shows the BSE image and figure 8b shows the BSE/K BSE ratio map indicating a reasonably flat grey level image with an even range, which means the correlation between the actual BSE with theoretical K BSE is very good. The grooves that are seen in these images are polishing grooves in the sample. The BSE/L BSE ratio map shows greater changes due to the L X-rays emanating from deeper within the sample.

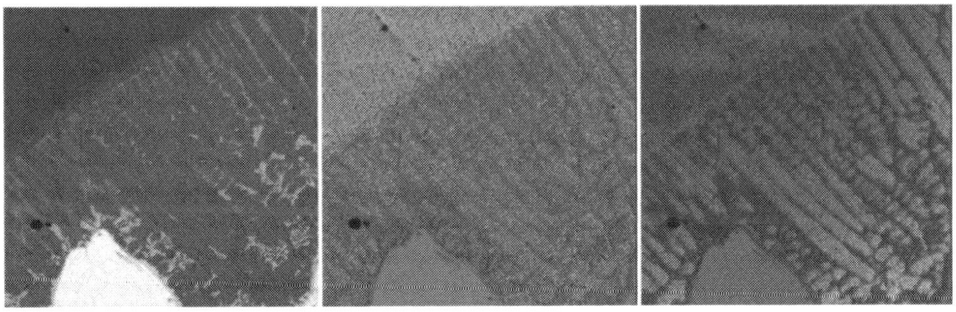

a) b) c)

Figure 8. Hard facing bonded to a chrome steel. a) BSE image collected with electrons, b) BSE/K BSE ratio map showing a reasonably flat grey level, and c) BSE/L BSE ratio map showing major changes (greater change in grey scale compared to BSE/K BSE). Map collected using a UTW-window Si(Li) EDS detector at 20 keV, 512x512 pixels over 12 hours. HWOF = 100 μm.

3.6. Atomic number, absorption and fluorescence (ZAF) maps

ZAF maps give information on the correction that is required at specific points, or interfaces, of the different regions in the maps. Figures 9a-c shows the *Z*-, *A*- and *F*-correction factor maps for chromium, respectively. Because of the elements present in this sample, there are large variations in the *Z*-, *A*- and *F*-correction factors. Chromium is a good example to show the fluorescence effect from iron, and atomic number effect from tungsten, with each element present having its own set of ZAF images.

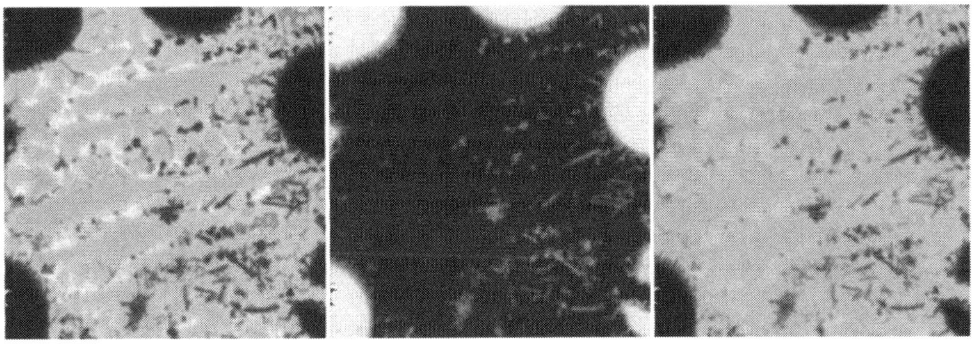

a) b) c)

Figure 9. Hard facing bonded to a chrome steel. These images are of the hard facing section only. a) Fluorescence factor image for chromium where the fluorescence factor varies considerably; in this case from 1 (black) to 1.1 (white), b) atomic number factor image for chromium where the atomic number factor varies considerably; in this case from 1 (black) to 1.37 (white), and c) absorption factor image for chromium where the absorption factor varies in this case from 0.9 (black) to 1 (white). Map collected using a Be-window Si(Li) EDS detector at 20 keV, 512x512 pixels and 12 hours. HWOF = 100 μm.

3.7. Totals mapping

When standards based analysis gives totals in the range 99 - 101 there is confidence in the accuracy of the results. This also applies to X-ray mapping. With X-ray mapping, the total at each pixel can be calculated and mapped (figure 10). Totals map then give us an idea of the quality of the analysis at each pixel and also makes it easy to determine problem areas within the map. As seen in figure 10, totals maps and total scatter diagrams can be generated. The total for each pixel is plotted against the concentration of an element for the same pixel. The totals image (figure 10c) shows the image for the range of 94 wt% to 103 wt% and reveals bright and dark edges which correspond to low and high absorption for carbon.

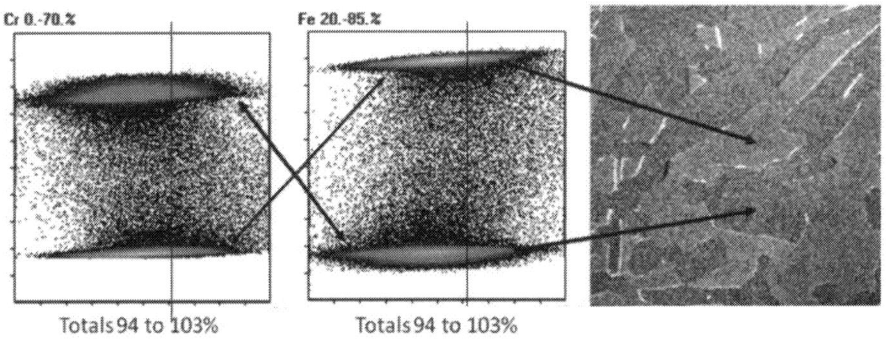

a) b) c)

Figure 10. a) Total (wt%) scatter diagram of Cr plotted (vertical) against Totals (Horizontal axis); b) Total scatter diagram of Fe plotted (vertical) against Totals (Horizontal axis); c) Totals image where the bright and dark edges correspond to low and high absorption for carbon. Arrows indicate the same phase for each element. The black vertical lines in figures 10a and b represent 100 wt%.

4. Conclusion

So much more information about a material can be obtained by further processing the X-ray maps through data manipulation. The collection of full spectrum X-ray maps (FSXRM) and the subsequent post processing of the data to obtain quantitative X-ray maps, elemental ratio maps, scatter diagrams, pseudo-coloured maps, theoretical BSE maps, totals maps as well as ZAF maps allows the user to 1) determine more information about their material through identifying the location of individual elements and phases, 2) map the spatial distribution of specific elements and chemical phases within a sample, 3) predict and verify where there is likely to be problems in the maps (images), and 4) help look at possible interface artefacts.

References

[1] Cosslett V E and Duncumb P 1956 Microanalysis by a flying spot X-ray method. *Nature* **177** 1172-1173

[2] Duncumb P and Cosslett V E 1657 A scanning microscope for X-ray emission pictures. in: *X-ray microscopy and microradiography*. (New Uork: Academic Press) 374-380

[3] Wuhrer R and Moran K 2006 Representation of microstructures through rapid X-ray mapping. *Mater. Australia* **38**(5) 12-14

[4] Wuhrer R, Moran K and Moran L 2006 Characterisation of materials through X-ray mapping. *Mater. Forum* **30** 63-70

[5] Wuhrer R, Moran K and Phillips M R 2006 X-ray mapping and post processing. *Microsc. Microanal.* **12** (Suppl 2) 1404CD-1405CD

[6] Moran K and Wuhrer R 2006 Quantitative bulk and trace element X-ray mapping using multiple detectors. *Mikrochim. Acta* **155** 59-66

[7] Moran K and Wuhrer R 2006 X-ray mapping and interpretation of scatter diagrams. *Mikrochim. Acta* **155** 209-217

[8] Wuhrer R, Moran K and Phillips M R 2008 Multi-detector X-ray mapping and generation of correction factor images for problem solving. *Microsc. Microanal.* **14** (Suppl 2) 1108CD-1109CD

[9] Moran K and Wuhrer R 2010 So, what other information can we obtain from X-ray maps. *Microsc. Microanal.* **16** (Suppl 2) 916-917

[10] Moran K and Wuhrer R 2010 Post processing of X-ray maps - Theoretical BSE mapping, totals mapping and ratio mapping. *Microsc. Microanal.* 16 (Suppl 2) 940-941

[11] IMAGE Java written by Wayne Rasband is available from http://rsb.ifo.nih.gov/ij/. The SMART-J plug-in is available from David Joy (djoy@utk.edu)

[12] Joy D C, Michael J and Griffin B 2010 Evaluating SEM performance from the contrast transfer function. in: *Proc. SPIE 7638, Metrology, Inspection, and Process Control for Microlithography XXIV* 76383J

[13] Gauvin R and Michaud P 2009 *Microsc. Microanal.* **15** 488-489

[14] Goldstein J I, Newbury D E, Echlin P, Joy D C, Romig Jr A D, Lyman C E, Fiori C and Lifshin E 2003 *Scanning electron microscopy and X-ray microanalysis, 3rd edition.* (New York: Kluwer, Academic Plenum)

[15] Wuhrer R and Moran K 2005 Artifacts in x-ray mapping. in: *AMAS VIII-The eight biennial symposium.* (Melbourne, Australia: University of Melbourne) 109

[16] Friel J J and Lyman C E 2006 X-ray mapping in electron-beam instruments. *Microsc. Microanal.* **12** 2-25

[17] Bright D S and Newbury D E 1991 Concentration histogram imaging: a scatter diagram technique for viewing two or three related images. *Anal Chem.* **63**, 243A-250A

[18] Bright D S and Newbury D E 2004 Maximum pixel spectrum: a new tool for detecting and recovering rare, unanticipated features from spectrum image data cubes. *J. Microsc.* **216** 186-193

AUTHOR INDEX